Embedded Systems

Series Editors

Nikil D. Dutt
Peter Marwedel
Grant Martin

For other titles published in this series, go to
http://www.springer.com/series/8563

Catherine H. Gebotys

Security in Embedded Devices

Springer

Catherine H. Gebotys
University of Waterloo
Department of Electrical & Computer Engineering
200 University Avenue W.
Waterloo, ON N2L 3G1
Canada
cgebotys@uwaterloo.ca

ISBN 978-1-4614-2519-9 e-ISBN 978-1-4419-1530-6
DOI 10.1007/978-1-4419-1530-6
Springer New York Dordrecht Heidelberg London

Softcover reprint of the hardcover 1st edition 2009

Printed on acid-free paper

Springer is part of Springer Science+Business Media (www.springer.com)

Preface

Security is increasingly widespread in many embedded systems. Although there are many books on cryptography and security, none that I am aware of focus on embedded systems, which have fascinating security issues with challenging design problems. From RFIDs to satellites, these embedded systems have important security requirements. There are excellent books on the mathematics behind cryptography with books specializing in areas such as elliptic curve cryptography. Many books examine software security including the design of secure algorithms focusing on buffer overflow attacks or viruses, wormholes, etc. There are some excellent books on security in systems; however, they also assume that readers have a background understanding in cryptography. Recently there is a book specifically focused on side channel attacks and analysis. However, there lacks a book that can help engineers, designers, computer scientists, and others in industry to appreciate the field of security in embedded systems. Mathematical books on cryptography are often too broad and difficult to understand due to their breadth and vast terminology. Software security books do not cover attacks or security design on embedded hardware. Books on side channel attack assume that the readers already have a thorough understanding of cryptography. This book's aim is to fill this gap by providing an overview of embedded system attacks, key properties, key uses, trusted platforms, public key cryptography, symmetric key cryptography, integrity, authentication, attacks, countermeasures, and reliable security. The ultimate goal is to provide the reader with an understanding of security issues so that they are ready to design security into their embedded system carefully, considering the security of the implementation.

This book is for the designers, engineers, computer scientists, and others in industry who need an understanding of security. This is not one-stop shopping, since we could not cram the entire field into one book. However, it is a first book to start with if the reader wishes to be introduced to the exciting field of embedded systems security. It could also be used for graduate course on introduction to Embedded Security or graduate students new to this area. This book may also be used for readers with a background in cryptography but who do not have an understanding of embedded systems and their security requirements or issues. Furthermore, readers interested in reading more about side channel analysis of real embedded systems would enjoy this book as well. Here, we examine the side channel impact of complex embedded systems, which have operating systems, cache misses, interrupts, etc. Other books covering side channels do not discuss complex embedded systems' problems such as alignment, interrupts, events, laboratory equipment limitations, and triggering.

Each chapter is focused on providing readers with an in-depth appreciation for the field. For example, the chapter on elliptic curve cryptography includes many examples illustrating how the mathematics work unlike most other texts. Readers are given the mathematics showing how and why protocols work, walked through point multiplication algorithms, and even at the lowest level the mathematics are illustrated to prove that points lie on a curve or to show how modular reduction and its algorithms work in binary fields. However, there are no mathematical proofs or theorems (that is left to the mathematicians who do it so elegantly). The chapter ends with a look at how (in)efficient elliptic curves are in hardware examining some energy, area, performance metrics to provide readers with a look at the impact of this type of security on an embedded system. Here, the focus is on understanding and appreciation. The book does not cover all algorithms or all protocols ever published; we leave that to other books. However, the aim is to give a more in-depth examination of the topic, as opposed to a broader coverage. Thus, many references and excellent texts are used in the references for further reading.

Security in embedded systems is a truly fascinating area with many great stories. I hope that the readers will enjoy reading this book as much as I have enjoyed writing it and researching in the field of embedded security.

Waterloo, ON Catherine H. Gebotys

Acknowledgments

I would like to thank my graduate students, postdoctoral fellow, and research associate for their suggestions on various chapters, specifically Reouven Elbaz, Solmaz Ghaznavi, Marcio Juliato, Patrick Longa, Edgar Santillan, and Brian White, and Amir Zadeh. Also I would like to thank my daughters Sabrina and Cassandra for readily creating several figures for this book. I would like to thank Charles B. Glaser for his encouragement in the writing of this book. Finally I have to thank my family, Robert, Kathleen, Sabrina, Cassandra, and Brennan, for their patience and encouragement during the writing of this book.

Contents

Chapter 1
Where Security Began

This chapter will briefly introduce important security concepts and terminology. It will also briefly look at the history of security along with the history of the side channel. The security concepts are discussed with respect to Alice and Bob to be consistent with the field of cryptography; however, throughout the remainder of the book we will assume that Alice and Bob can in fact be embedded devices.

The four main security concepts used today are as follows: *Confidentiality, integrity, authentication, and nonrepudiation*.

We will discuss these concepts using the communication of messages between point A and point B or specifically communications between Alice and Bob on the channel. This will maintain consistency with many other cryptographic texts that use Alice and Bob. The channel, shown in Fig. 1.1a, is a very general concept and could represent a wire (for communication over a wired network) or electromagnetic waves (for wireless communications using cell phones). Security is designed for this channel with Eve in mind. Eve is named after the eavesdropper. However, she is in general an attacker or adversary. As shown in Fig. 1.1b Eve can eavesdrop to see all data on the channel. In Fig. 1.1c Eve can intercept data on the channel, modify it, and send it on to the destination. Finally in Fig. 1.1d Eve can intercept messages and masquerade as Bob without Bob receiving any of his intended messages. Of course, depending upon specifics of the channel some or none of these attacks may be possible. Additionally there may be other attacks such as Eve initiating communication on the channel, or Eve masquerading as both Alice and Bob, in order to attack communications between Alice and Bob, etc.

If Alice wishes to send messages to Bob without Eve reading them, then she must employ the confidentiality principle. Typically an encryption of the messages is required. We generally assume that both Alice and Bob must possess a *key*. Typically they already share a symmetric key (shown at the top of Fig. 1.2a) or they can use a protocol to establish a symmetric key (or asymmetric keys). Each of them must also use an algorithm called a cipher, which will transform input data (input message) or *plaintext* into ciphertext. *Ciphertext* is used to refer to the encrypted plaintext that is generated by the cipher. In the simpler example of symmetric encryption if Alice is the sender, she will encrypt the plaintext and send this ciphertext to Bob. Bob will decrypt the ciphertext to obtain the plaintext that Alice sent. If some data is not encrypted, it is referred to as data sent in the *clear*. The attack shown in Fig. 1.1b

C.H. Gebotys, *Security in Embedded Devices*, Embedded Systems,
DOI 10.1007/978-1-4419-1530-6_1, © Springer Science+Business Media, LLC 2010

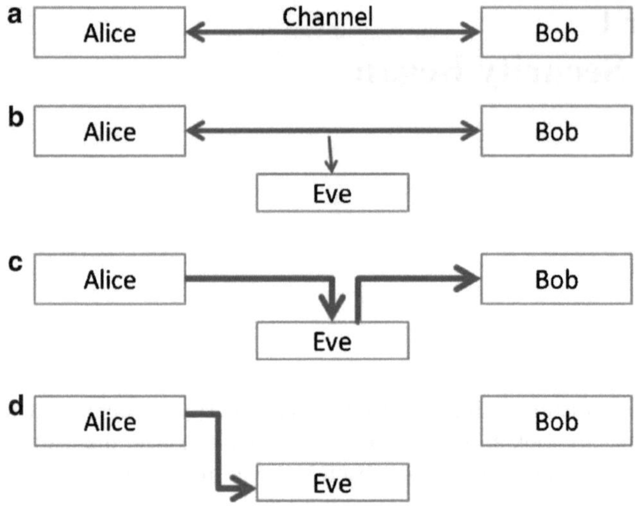

Fig. 1.1 (**a**) Alice, Bob and the channel, (**b**) Eve eavesdropping, (**c**) Eve modifying, and (**d**) Eve masquerading

is now thwarted since Eve does not possess a key that will decrypt the messages. Eve may, however, attempt an attack by performing *cryptanalysis*. Eve will collect all ciphertexts sent over the channel between Alice and Bob, as shown in Fig. 1.2b. She will use the ciphertexts in an attempt to determine the key. Eve may use a *brute force attack* that decrypts the ciphertexts using many different keys until the plaintext generated looks like a message. If the brute force attack is successful in determining the key Eve can decipher any message between Alice and Bob until they decide to change their keys.

Next consider the attack shown in Fig. 1.1c where Eve can modify messages over the channel. Assume Alice is ordering equipment from Bob. Alice does not only want confidentiality but she wants to make sure that Eve does not modify her order. Let us say she wants to purchase one PC. Eve may change this message in transit to 100 if there is no security mechanism available to protect Alice's message. Alice needs to use the integrity principle of security (integrity and authentication). Alice may use an integrity tag that can be verified by anyone (since there is no key involved). For example, Alice can create an integrity tag using a *one-way hash* of Alice's cheque for 100 PCs. This tag can be verified by numerous recipients such as the bank and Bob. However, Eve may be also able to insert a new integrity tag supporting a forged order for 100 PCs. Thus, Alice should use *authenticated encryption*. For example, Alice will generate and append a *data authentication tag* to her ciphertext. Bob uses the tag along with a computation with the ciphertext to verify that the ciphertext has not been tampered with. Alice can use a *message authentication code* (MAC) function to generate a data authentication tag or MAC tag. Unlike the one-way hash that has no secrecy, the MAC function uses a key. Use

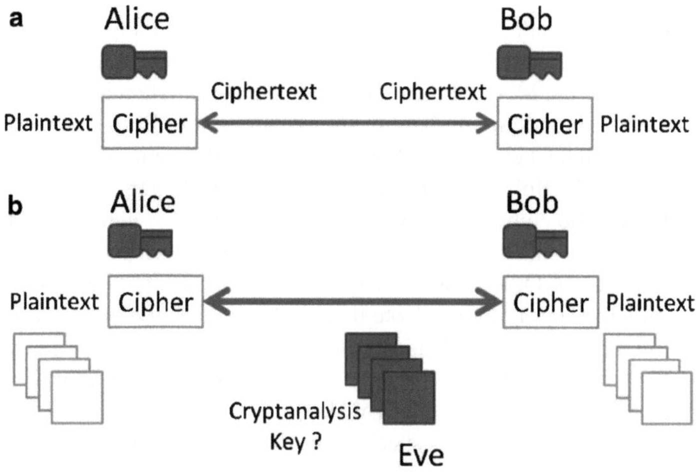

Fig. 1.2 Confidentiality in (**a**) and cryptanalysis in (**b**)

of the MAC ensures that only the intended recipient (who has the key) can verify the MAC and no one else. Thus if Eve does attempt to tamper with the authenticated message, Bob will not be able to authenticate it (since Eve does not have the correct key to perform the MAC) and Bob will discard the message. An example of the use of a MAC will be provided next. This example also introduces the concept of public key pairs and ephemeral keys. Further details on public keys will be provided in Chap. 5.

Consider a confidential and authenticated message transfer illustrated in Fig. 1.3 (which is a simplified form of ECIES encryption Protocol 5.3). In contrast to Fig. 1.2 where both Alice and Bob share the same key (symmetric cryptography), Fig. 1.3 illustrates an asymmetric key system (referred to as *public key cryptography or PKC*). In public key cryptography a user (such as Bob) has a pair of keys: specifically one is the *public key* (labeled with B) and the other is a *private key* (labeled with b). The public key does not require any secrecy and can be seen by anyone (e.g., Alice has a copy of it). In Fig. 1.3, Alice uses Bob's public key (shown on the left, B) along

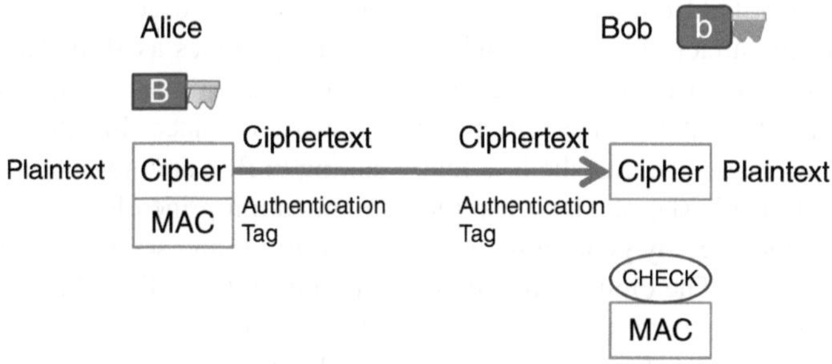

Fig. 1.3 Ciphertext and MAC tag sent over channel (simplified ECIES protocol 5.3)

with some random number (not shown). With these two values she computes one key to generate the ciphertext (k1) and a second key (k2) to generate an authentication tag (which is the MAC of the ciphertext). Both ciphertext and authentication tag (plus a modified random number not shown) are sent to Bob. Bob uses his own private key (b, from his key pair) and the modified random number to generate two new keys (k1 and k2). Bob then uses these two new keys to generate a MAC (using k2) and to decrypt the ciphertext (using k1). He compares his generated MAC with the MAC that Alice sent. If they are equal then Bob has verified that the data was not modified in transit. Otherwise if the two values are different he will discard the message completely and ignore it. Note that unlike the use of a one-way hash (providing only integrity not authentication), Bob is the only person who can verify the message. This is true since he is the only person who has possession of his secret key, b (which is used to generate the MAC key).

Note that Alice computed keys specifically for this message transfer in Fig. 1.3. She did this using not only Bob's public key but also a random number. The keys she computed are generally referred to as *ephemeral keys* since they are temporary keys used here only for this one message transfer. The next time she uses this protocol, she will generate a different random number in order to compute another set of ephemeral keys. Eve's job of cryptanalysis on the ciphertext is much more difficult than the case in Fig. 1.2 with symmetric keys, since every message transfer uses a different key. However, there are tradeoffs; this added security comes with a larger number of computations as will be detailed in Chaps. 4 and 5. It is of further interest to note that Alice used a one-way hash function in part of the computation of the ephemeral keys. Hence, there are many cryptographic functions typically involved in a single protocol.

Note also that Bob may want to be sure that the PC order received from Alice is not an older order that she made last year. For example, Eve may have saved this older PC order and has now transmitted it to Bob, or "replayed" it. This *replay* can be thwarted by using some timeliness factor in the computation of the data authentication. The timeliness parameter allows Bob to verify that the message is "fresh" and not last-years order. Use of a *nonce* (a data value that is used only once), a timestamp, sequence number, etc. can provide the required "freshness." This is referred to as *transaction authentication*, where the data and time are authenticated. Note that the ephemeral keys used in the example from Fig. 1.3 provide this freshness (since they were generated from a random number not previously used).

Consider the attack in Fig. 1.1d where Eve masquerades as Bob. To avoid this attack, Alice may wish to authenticate Bob. This is referred to as *data-origin authentication*, where the data origin may be a person or place. In other words she requires some assurance that she is actually talking to Bob and not Eve masquerading as Bob. For this she needs to use the authentication principle of security. Bob will sign his message by generating a *digital signature* (and sending it with the ciphertext to Alice). The ciphertext and signature are sent to Alice. Alice can check his digital signature to verify that Bob really did send his message. This digital signature is analogous to a handwritten signature. Of course, Alice should authenticate herself to Bob as well. Public key cryptography is used to support digital signatures. Further details will be provided in Chaps. 4 and 5.

Finally Bob may need assurance that Alice will stand by her word and pay for the ordered PC. To do this Alice signs her order and Bob can use this as proof that only Alice could have made that signature. This is referred to as *nonrepudiation*. It is the inability to deny the authenticity and integrity of a message or data. Typically nonrepudiation involves proving to a third party that a person did send a message.

The set of rules that two parties can use to communicate is referred to as a *protocol*. For example, consider the authentication principle where a user is to demonstrate that the user holds a secret without revealing the secret (over the channel). A *challenge-response* can achieve this. For example, Bob wishes to authenticate Alice. He sends her a challenge, c, and Alice sends back a response, r, which Bob uses to verify that Alice did hold the secret. The protocol below is an example of a challenge response authentication. Bob encrypts an integer, $E_k(m)$ [where $E_k()$ is an encryption algorithm or cipher that uses key k], and sends it to Alice. Alice decrypts the ciphertext, $D_k(c) = m$ [where $D_k()$ is a decryption algorithm], increments the number, encrypts it, $E_k(m + 1)$, and returns it to Bob. Bob can take this response and decrypt it to verify that Alice did increment the number.

Protocol 1.1. Challenge-Response

(1) *Bob* → *Alice*: $E_k(m) = c$
(2) *Alice*: $D_k(c) = m$
(3) *Alice* → *Bob*: $E_k(m + 1) = r$
(4) *Bob*: $D_k(r) =? = m + 1$

1.1 A Brief History of Cryptography

There are many reports that security started thousands of years ago back to the time of enciphered hieroglyphics in Egypt or enciphered words in the Hebrew scriptures. The most well known is perhaps the scytale (Britannica Web site), which involved writing letters on a piece of leather. When the leather was wound around a tapered baton a message would be revealed. This was perhaps the first example of a transposition cipher. It was said to have been used by the Spartans and ancient Greeks as early as 400 BC. Around 200 BC, a Polybius checker board or square was developed. This mapped letters into a pair of coordinates used to locate the letter on a grid. This coding of letters by coordinates was an early example of fractionation where plaintext symbols are translated into ciphertext symbols. This technique when combined with transposition is an important part of several ciphers used today.

The Caesar cipher was said to have been used by Julius Caesar. It was a simple substitution cipher or a monoalphabetic cipher meaning that one substitution and/or transposition is used. The cipher replaced each letter in the message with a letter a fixed distance away, k. For example, plaintext "hello" with $k = 2$ would become ciphertext "jgnnq." Using the English alphabet, where the letters are replaced with numbers from 1 to 26, we could generalize this cipher as $c_i = (p_i + k)\ mod\ 26$, for any given plaintext p (where p_i is the ith letter in the plaintext and similarly for ciphertext c and key k).

Leon Alberti devised the cipher wheel and invented perhaps one of the first attacks using a technique known as frequency analysis in the 1460s. The cipher wheel or cipher disk consisted of two disks each with the alphabet written around the edge of the circle. When one disk was rotated the alphabets from one disk to the other would line up differently, creating a monoalphabetic cipher. This cipher wheel made enciphering easier. However, it was also used for a polyalphabetic cipher. Leon Alberti suggested using two or more cipher alphabets and switching between them during encryption. His ideas were used to fully develop the Vigenere cipher later in the 1500s.

Frequency analysis, also developed by Leon Alberti, is based on the fact that letters in the English language are not used with the same probability. Some letters are more common than others. The most common letter in the English language is "e." For example, assume we are to crack a monoalphabetic enciphered message using the Caesar cipher. Assume also that the original message was written in English and the fixed distance between letters of the original message and letters of the enciphered message was k. The frequency of each letter in the enciphered message would be computed and the most common letter would be mapped to "e" in order to find the value of k. One would then decipher the message. If the message remained garbled then the next most common letter would be used to find k, etc. Other attacks could search for pairs or triplets of letters according to their frequency of use. This frequency analysis technique is much faster than a brute force attack where the enciphered message is deciphered for all possible values of k.

The Vigenere cipher was developed in 1585 and it was the first polyalphabetic substitution cipher. A polyalphabetic cipher is one that uses several substitutions and/or transpositions. In the Vigenere cipher, a key is represented as a string of "d" letters. Each letter, x, referred to a Caesar cipher whose ciphertext had $k = dist(\text{'}a\text{'}, x)$, where $dist(l_1, l_2)$ is the number of letters between l_1 and l_2 [e.g., $dist(\text{'}a\text{'}, \text{'}c\text{'})=2$]. For example, if the key was "a c e f" and the plaintext was "hello" the cipher text would be "h g p q o." The original key is repeated to generate key letters over the entire message. The mathematical form of this cipher is $c_i = (p_i + k_i) \; mod \; 26$. Attacks on this type of cipher are more difficult than that on the Caesar cipher. However, since the key sequence repeats, there is a chance that some frequently used words will repeat their ciphertext. If this occurs, one could deduce part of the repeating key and more importantly the key length. The key length is crucial, since once you have the key length you can slice up the ciphertext and use frequency analysis on a group of ciphertext letters related to each letter of the key. Note that if the key sequence is only used once and it is as long as the plaintext then this is referred to as a one-time pad. A one-time pad is an important concept in cryptography because it represents a theoretically unbreakable cipher.

It was most unfortunate for Mary Queen of Scots that her conspirators were not aware of the Vigenere cipher. Instead in 1586, with the help of a double agent who copied intercepted enciphered messages, Thomas Phelippes (at that time Europe's best cryptanalyst) used frequency analysis to crack Mary's cipher. The cipher was used to plan for Mary's release and Elizabeth's assassination (which never took place). The cipher was an extension of a monoalphabetic substitution cipher where 64 symbols were used to represent letters, words, and phrases. Although communications were decrypted, the arrests were not made until communications were deciphered, which clearly indicated that Mary endorsed Elizabeth's assassination. This communication arrived on 17 July and Mary was taken to trial and executed on February 8, 1587. This story illustrates that the danger of using a weak cipher can be worse than not using any cipher at all (Singh 1999).

The Jefferson cylinder in the 1790s was a cipher system using 36 wheels, each with a random arrangement of letters around the wheel. The discs were stacked on an axle. The order of the discs was the key. When the user wanted to generate ciphertext, the discs were each rotated until the plaintext was spelled out in one row of the system. The ciphertext was extracted by the user choosing any other row of the system.

The Wheatstone disk was developed in 1817 by Sir Charles Wheatstone (Mogollon 2007). This cipher system consisted of two concentric disks each with an alphabet on the periphery of each disk. The outer disc contained letters in alphabetical order with a blank space between the *z* and *a* letter, while the inner disk used a random ordering of letters. The clock-like hands were geared together in some fixed way. At any time they would each point to a different letter on each disk. However, the hands worked together such that when the big hand completed one revolution, the little hand would advance by one letter. The big hand would move clockwise advancing to each letter of the plaintext while the small hand would point to the ciphertext. If a double letter was found in the plaintext, the hands would not move, so some other letter such as *q* or *x* would be used for the ciphertext. This cipher system had the property that the ciphertext for a word was dependent upon the preceding plaintext word. This principle is referred to as chaining and is the basis for cipher modes today, which will be covered in Chap. 6.

The Vernam cipher was developed in 1917 and it was the very first stream cipher. It is based on the one-time pad, which is a series of random data. A pair of correspondents each has the same one-time pad that they use to encipher their messages. The plaintext is exclusive-ored with the random data in the one-time pad in order to generate the ciphertext. The data in the one-time pad can only be used "one time" as stated. Typically the one-time pad was a very long string of random letters, which was stored in a codebook. After the codebook had been completely used for encryption of many messages, the codebook was destroyed. Under the assumptions that the data are truly random in the one-time pad and that the random data are only used once, the one-time pad is theoretically unbreakable. Reports of compromised Vernam ciphers used by Russian KGBs who repeated the use of their one-time pads or codebooks are found in the VENONA files (NSA-Venona Web site). During a period from 1942 to 1944, they were reportedly able to decipher communications without any capture of the codebooks.

The Enigma (Greek word for riddle) rotor machine is perhaps one of the most well known cipher systems. It was used heavily during World War II. The Enigma was invented by Arthur Scherbius in Berlin in 1918 and broken in 1932 by a Polish mathematician Marian Rejewski. It performed a series of substitutions using mechanical and electrical connections. It had a series of rotor wheels with internal cross-connections providing substitution using a continuously changing alphabet. Initially it had three rotors and later the Germans added two more. At one instance in time, each rotor provided a substitution cipher. However, every time a letter was typed one or more rotors would rotate, thus changing the substitution. At certain times the rotation of the right rotor is carried to the middle rotor, etc. The Enigma in general provided a polyalphabetic cipher. There were other details including a ring setting that was equivalent to a session key. Interesting details of the Enigma can be found in Singh (1999). In 1949 Shannons communications theory was developed. The theory of entropy, definition of the perfect encryption (one-time pad), and principles of confusion and diffusion laid the foundations for modern cipher development.

In 1973 Horstel Feistel published his work on Feistel networks. Details are provided in Chap. 6; however, the Feistel network is still the basis of many ciphers used today.

The beginnings of public key cryptography were developed in 1975 by Whitfield Diffie and Martin Hellman. The Diffie–Hellman key exchange protocol was later published in 1976 (Diffie and Hellman 1976). This was reported to be developed in 1974 by Malcolm J. Williamson under classified work; however, no one knew about this work until 1997. In 1977, the DES cipher was standardized (see Chap. 6) and RSA was published.

In 1979 the first ATM machine was in use from IBM and later VISA. Earlier cards and ATM machines used DES to encrypt the data. They related the PIN to the account number in a secret way. For example, the concatenation of 11 digits (from the card issue date or the 11 beginning digits of account number) with the last five digits of the account number was encrypted with the PIN key. The first four digits of DES output were given to the customer as their "natural" PIN. Variations of this scheme were used. However, it is interesting to note that one bank's implementation was easily exploited by criminals who were able to withdraw large sums of money from other peoples' account (Virtual Exhibition in Informatics). This exploitation was possible because the bank had just encrypted the PIN on the card. Hence, attackers who knew the PIN number of a card were able to change the number on the strip (by using an account number from a discarded receipt) in order to use their PIN number to withdraw money from the other account number. Note that in this example, the encryption was secure but the implementation was not secure. Typically the implementation is weak and the focus of attacks.

In 1998, the most famous attack on smart cards was performed by researcher Paul Kocher, using the side channel. He is famous for cracking commercial smartcards, which were developed in the late 1990s (Kocher et al. 1999), and his work initiated the drive to develop sets of countermeasures to resist side channel analysis attacks.

More details will be provided in Chap. 8, however a brief look at the history of the side channel is provided next.

1.2 Brief History of the Side Channel

Side channel analysis hit the headlines in 1999 when a research named Paul Kocher took what was believed to be a tamper-resistant smartcard and broke it. What is interesting is how he broke it or obtained the secret key. He used the power side channel along with a technique known as differential analysis, which will be described in Chap. 8. This section will discuss the history of the side channel, which actually started many decades before the invention of differential analysis.

In mid-nineteenth century, Morse code was developed, followed later by telephone technology. Soon after radiotelegraphy was developed, it became apparent that eavesdropping may be possible. In 1915, the military discovered cross talk from their field wires, which were used to connect troops with their headquarters. Further investigations discovered that these interfering signals were actually enemy communications. They later discovered that even abandoned wires seemed to have leaked significant amounts of information to Germans. Clearing these wires was an important task during the war (Anderson 2001).

In 1918, Herbert Yardley discovered that classified information could leak from electric materials. Yardley and his staff, known as the Black Chamber, were engaged by the US Army to study combat telephones and covert radio transmitters in order to detect, exploit, and intercept signals (Siemon 2002) during World War I.

There were reports of data within a crypto device having modulated a signal on a tape of a nearby recording source. The EM emanations from a typewriter were reported to have been identified in the 1930s (VanTilborg 2005). In fact, it is rumored that they were worried about manufacturers of the typewriter putting bugs into the device in order to increase side channel signals for possible attackers. The first power drive typewriter was invented in the early 1900s. For example, the IBM Selectric typewriter was developed in 1961. Soon after the electric typewriter was developed it became apparent that an attacker in an adjacent room to where a person was typing a confidential letter could launch a side channel attack. This was performed by measuring the power dissipation from the power line in the wall while the typewriter was being used to type the letter. From examining the instantaneous power it could be determined which key was being pressed at what time. Thus, an attacker could recreate the entire confidential letter. For example, each time a key was hit on the typewriter, the power would produce a characteristic power spike. Each key of the typewriter would create an observably different spike. There were also reports of unusually high levels of EM signal emanations from an IBM selectric typewriter (McNamara 2004), which lead to the belief that emanations were being amplified for espionage purposes. During the 1970s, research into the leakage of data through EM emanations was classified and referred to as TEMPEST. Some of the TEMPEST findings were declassified in 1995 (Wolfe et al. 1970); however, there appear to be some recently declassified documents that provide interesting reading, such as Tempest-release (2007) and Tempest-release (2008).

In 1960, the British attempted to obtain communications from the French president, De Gaulle, who they thought might block the British from joining the European Economic Community (Wright 1987a, b; Kuhn and Anderson 1998). The British intelligence was eavesdropping on traffic in attempts to break the French diplomatic cipher. Though they were unsuccessful in breaking the cipher, Peter Wright, a MI5 scientist, and his assistant Toy Sale discovered a weak secondary signal associated with the enciphered signals. Using their own equipment to recover this signal, they determined that it was the plaintext that had leaked through the cipher machine. Years later, researchers at Bell Northern Research were testing the side channel of one of their high security phones (Simmons 2009). They had designed the phone with significant amounts of security and resistance to side channel emanations with special coatings on the phone casement, circuitry, etc. However, testing the phone in a chamber proved that there were still side channel emanations which could not yet be removed. They finally discovered that the phone cable was responsible for the emanations.

Though not truly a side channel by definition, there is a famous story related to EM emanations. In August 1945, Soviet children gave the US Ambassador Averell Harriman a wooden carving of the great seal of the USA (NSA Web site). This carving was hung in the Ambassador's office until 1952 when it was discovered that it had a bug in it. The bug was in fact a cavity that when activated by radio waves from attackers in a car outside the building would produce modulated waves, which when decoded by the attackers in the car would reveal conversations taking place in the office. There have also been numerous reports of microphones buried in buildings. Often these devices were hidden such as metal grids buried in concrete over ceilings of important rooms, as in the Department of State Communications area. Although these types of bugs are used to remotely listen to secret conversations, there was a much greater worry, specifically the threat that these bugs may be used to leak information from the side channels of the crypto-machines present in the offices (NSA Web site). In theory there was the potential that these emanations could radiate for considerable distances like radio waves. It is interesting to note that the Russians published a set of standards for suppression of radio frequency interference in 1954.

Side channel analysis is the focus of Chap. 8 and resisting side channel attacks is discussed in Chap. 9. The details of setting up a side channel analysis laboratory as well as examples of experiments are provided. EM acquisition as well as power acquisition are discussed and demonstrated through a number of experiments. The capture and analysis of real embedded systems are also studied including a PDA device and a contactless smart card.

1.3 Summary

The purpose of this book is to provide the fundamentals for understanding security in embedded devices. This is not to say that security for the more general case is well understood and implemented. For example, the quotation below indicates this is not the case.

> "...there's a lot of bad cryptography in the field due inexperienced programmers implementing systems which they do not understand" (Walton 2009).

The history of cryptography makes for fascinating reading, such as the Code Book (Singh 1999), Security Engineering (Anderson 2001), and recently unclassified documents at the NSA Web site (NSA 2009). The next chapter will start with some security concepts and give examples of the use of security in various embedded systems.

References

Anderson R (2001) Security engineering. Wiley, New York

Britannica Web site. History of cryptology – early cryptographic systems and applications. http://www.britannica.com/EBchecked/topic/145058/cryptology/25638/Early-cryptographic-systems-and-applications#ref=ref392544

Diffie W, Hellman ME (1976) New directions in cryptography. IEEE Trans Inform Theory IT-22(6):644–654

Kocher P, Jaffe J, Jun N B (1999) Differential power analysis. In: CRYPTO'99. Springer, New York, pp 388–397

Kuhn M, Anderson R (1998) Soft Tempest: hidden data transmission using eelctromagnetic emanations. In: Aucsmith D (ed) Information hiding, second international workshop, IH'98, Portland, OR, April 15–17, 1998, Proceedings, LNCS 1525, Springer, New York, pp 124–142

Mogollon M (2007) Cryptography and security services, Cybertech, Hershey, New York

McNamara (2004) The complete, unofficial tempest information page. http://www.eskimo.com/~joelm/tempestmisc.html

NSA (2009) National cryptologic museum – virtual tour. http://www.nsa.gov/about/cryptologic_heritage/museum/virtual_tour/museum_tour_text.shtml

NSA-Venona Web site. The Venona story, center for cryptologic history. http://www.nsa.gov/about/_files/cryptologic_heritage/publications/coldwar/venona_story.pdf

NSA Web site. The Center for cryptologic history. http://www.nsa.gov/about/cryptologic_heritage/center_crypt_history/index.shtml

Siemon (2002) Government levels of security enhanced with TERA cabling system. http://www.siemon.com/us/white_papers/06–03–02-tera-security-government.asp

Simmons D (2009) Discussions with Dave Simmons. July 2009

Singh S (1999) The code book. Anchor Books, Random House, New York

Tempest-release (2007) TEMPEST: a signal problem. http://www.nsa.gov/public_info/_files/cryptologic_spectrum/tempest.pdf

Tempest-release (2008) A history of U.S. communications security – the David G. Boak lectures. http://www.nsa.gov/public_info/_files/cryptologic_histories/history_comsec.pdf

VanTilborg HC (2005) Encyclopedia of cryptography and security. Springer, New York

Virtual Exhibition in Informatics. Making cash dispensers safer with DES. Virtual Exhibition in Informatics, Universitat Klagenfurt. http://cs-exhibitions.uni-klu.ac.at/index.php?id=264

Walton J (2009) Authenticated encryption. http://www.codeproject.com/KB/security/EncryptThenAuthenticate.aspx

Wolfe HM, Schmidt RP, Thompson TN (1970) Working against the tide, Part 1: COMSEC monitoring and analysis. http://www.nsa.gov/public_info/_files/cryptologic_quarterly/work_against_tide.pdf

Wright P (1987a) Spycatcher – The candid autobiography of a senior intelligence officer. William Heinemann, Australia

Wright P (1987b) Spycatcher, Penguin Viking, New York

Chapter 2
Introduction to Secure Embedded Systems

Security is increasingly widespread in many embedded systems. Embedded systems requiring security range from the smallest RFID tag to satellites orbiting the earth. This widespread need for security is expected to continue for many more decades. Future growth services include identify control (e-passport, e-VISA), public services (e-administration, e-commerce, e-banking, transportation), communication (SIM card, PDAs), retail business (inventory systems), health care (patient monitoring, RFID, pharmaceuticals supply chain), and entertainment (games, movie industry). For example, the global shipment of smart cards exceeded five billion units in 2008. This is expected to increase by 11% through 2012 (RNCOS 2009). In 2008, 70% of the shipment was attributable to mobile subscribers. The use of contactless smart cards is expected to grow by 30% through 2012 (RNCOS 2009). This section will briefly introduce some types of security attacks on embedded systems and then overview some interesting embedded systems describing their security requirements.

Embedding security into devices is not a straightforward process. First the type of security functionality to embed into the device must be determined. This is often a challenge since specifying security requirements largely depends upon attack or threat models, which may not be fully known at the time. Designers must also ensure that their implementations are secure, since this is typically the focus of attacks. Unlike other embedded constraints such as energy, performance, and cost, which can be verified and quantified, the verification of security is often not possible (apart from functionality). In general, the security cannot be quantified nor can it be readily verified due to the possibility of unforeseen future attacks. From a security point of view, a complete understanding of the device from the process level and up is necessary in order to verify that the security and its implementation are sound. This section will discuss attacks and the need for security in some interesting embedded systems.

Alice and Bob now represent two connected embedded devices. For example, they may represent a processor connected to memory or a FPGA connected to nonvolatile memory or two components networked on a chip or a RFID tag communicating with a tag reader, etc. In these situations the attack may be focused directly on one of the devices or the entire system.

C.H. Gebotys, *Security in Embedded Devices*, Embedded Systems,
DOI 10.1007/978-1-4419-1530-6_2, © Springer Science+Business Media, LLC 2010

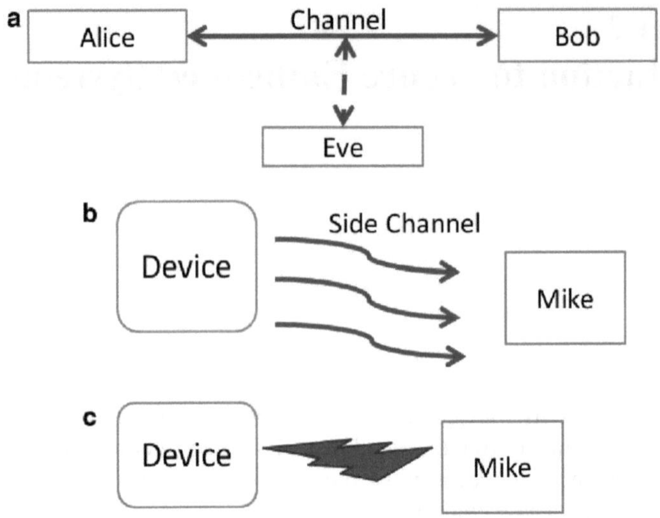

Fig. 2.1 Channels in embedded cryptography in (**a**) and (**b**) and attacking device in (**c**) without a channel

Eve has a wider range of attacks, which she may launch on a wide variety of embedded systems. For example, attacks on embedded devices can include the general unauthorized access to an asset, use of an unauthorized device, making the device execute unauthorized code, and cloning of a device (Jun 2008).

There is also Malicious Mike who does not depend upon a channel like Eve. Mike can directly tamper with a device, as shown in Fig. 2.1c, and he may also be able to run a side channel attack. Figure 2.1b illustrates the side channel where unintentional information is leaked from the device through power, EM, etc. Mike's objective is to obtain sufficient information leaked from the side channel in order to determine the key. In a general sense, Mike may modify the temperature, supply voltage, etc, in order to obtain better measurements from the side channel or he may explicitly attack the device through dumping memory, tampering, decapsulating chips, or other means.

Embedded systems are very different from those systems generally thought of as general-purpose CPUs (Marwedel 2006). General-purpose CPUs are generally designed for high performance supporting full programmability and supporting a wide range of workloads. This is unlike embedded systems which generally utilize processors and/or devices embedded in specific systems that are highly constrained. Embedded devices may be constrained by cost, memory, energy, clock frequency, word size, mass, volume, etc. Typically these embedded devices are a part of our everyday life (automobiles, entertainment systems, etc). The networking of these embedded devices has lead to a new emerging topic of great interest known as pervasive or ubiquitous computing. Security is an important requirement for these embedded devices.

There are many reports of attacks on various systems including networks, PCs, RFID tags, cell phones, automobiles, satellites, etc. A few are listed below.

> "Millions of dollars lost in identity theft… eleven people from five different countries have been charged with the biggest identity theft and computer hacking case in American history. … accessible wireless computer networks… to remotely capture sensitive information such as the card numbers, passwords and account information. The account information was then stored on encrypted computer databases. … then sold some of the credit and debit card numbers" (Landers 2008)

> "… to inject audio to, and record audio from bypassing cars that have an unconnected Bluetooth handsfree unit running… " (Schneier 2005)

> "… for mobile phone manufacturers to make a greater effort and fix the Bluetooth security problems in their handsets… " (Kotadia 2004)

> "James Van Bokkelen is about to be robbed.… plans to use a cheap, homemade USB device to swipe the office key out of Van Bokkelen's back pocket… 'I just need to bump into James and get my hand within a few inches of him'… " (Newitz 2006)

> "Hacking a commercial satellite that's been up there more than 10 years is very easy for some people, if you have the right equipment "(Kent 2006)

It is interesting to note that many of these reports of attacks are on embedded systems. Typically the cryptography is strong but the implementation is weak and often the focus of attack. Often embedded devices are complex and contain proprietary IP. Attackers have to gain as much information as possible concerning the system under attack. This includes sources of information such as company patents, errata, white papers, blogs, etc. Equipment supporting attacks include probes, ROM reader, logic analyzer, scopes, disassemblers, etc. There are events that provide workshops on hacking hardware such as the Blackhat conference where one can even find presentations describing attacks, such as Grand (2005) and Tarnovsky (2008). Alternatively there are companies that offer white hat services for "ethical hacking." It is generally a good rule to assume that the attacker will know almost everything about the device except the secret keys, etc. This will ensure that implemented security is thorough and robust as possible. As they say there is no security in obscurity.

Although there is no universal taxonomy of attacks for embedded systems, there have been some classifications for attacks (VanTilborg 2005). However, the definition of attack terminology does vary. For example, some define an invasive attack to mean physically tampering with the device or decapsulating chips while others refer to opening up the device or injecting faults into the device without damaging the device, etc. Some general attack definitions are provided below.

Probing attack or penetration attack – these attacks are generally active and invasive. The attacker typically breaks into the boundary of the device, intercepts communication lines, or dumps memory contents from the device. The attack focuses on the bus structures of the embedded system. Typically accessible busses are probed so that data can be made observable as well as possibly controllable. Most often these are busses on the PCB, which are visible or which pass through pins that are accessible. An example of this is the attack of the Xbox (Huang 2002), also known as a boot attack, where bus probing on the PCB allowed the attacker to obtain the secret key used to authenticate programs.

Designers sometimes consider on-chip buses as well. For example, decapsulating the chip and placing tiny probes on the silicon is a concern in high-security devices such as credit cards based on smart card or chip card technology. The concern here is that an attacker can obtain sufficient information from probing the internals of chips and then clone many smartcards with this information.

Monitoring attack. These are passive and noninvasive attacks. The side channel is an example of this type of attack. This attack is often referred to as a "lunchtime attack." For example, attackers can temporarily take a credit card in order to charge the lunch meal to it, but additionally run the attack on the card and return the card to the owner who is unaware that their secrets have been stolen. If an attacker is well versed in attacking cards, it would only take minutes to acquire the traces for later analysis. The attacker can now clone credit cards and thus steal money from the owner. This is not only a concern for these chip cards but many other embedded systems including PDAs, cellphones, and even automobiles. For example an attacker may be able to rent a car, extract the secret key from the remote keyless entry device and later return to steal the car.

Manipulation attack. This attack is also noninvasive, and the device boundary is left intact. However, the attacker can change the temperature, supply voltage, clock, etc. For example, spiking the supply voltage may inject faults into the device. These faults may be an injection of zeros, which would cause the cryptographic algorithm to reveal the key. Alternatively this type of attack may also be used in conjunction with the monitoring attack, for example, providing important boosting of signals so that a side channel attack is easier to perform.

Substitution attack. In this attack the module is removed and replaced with an emulating device. For example, a module that communicates with a trusted platform module (TPM) may be replaced by a malicious device, which succeeds in obtaining a sufficient number of plaintext, ciphertext pairs in order to mount an attack to obtain the secret key. Another example here is replacing chips in automobiles with modchips, which provide higher levels of performance typically only available to more expensive car models. There are also modchips for gamestations, which reportedly bypass the authentication allowing users to play unauthenticated programs on their stations.

Additional attack terminology is outlined below.

Replay attack is an attack where data sent to and from the device is recorded and used later in time. This replay attack can be done to impersonate another authentic user or authentic device. For example, an attacker may wish to impersonate Alice and exchange confidential messages intended for Alice with Bob.

Modification attack is an active invasive attack supporting the modification of internal connections or memory. The software version of a modchip is an example of this.

Spoofing attack replaces part of the message with their own data. For example, an order on the internet for one PC becomes attacked and ends up being an order for 100 PCs.

Splicing attack allows the attacker to permute data from the original message, so the device receives the permuted data.

The next section will examine some different embedded systems outlining some possible attacks and security requirements.

2.1 Contact Chip Card or Smart Card

There are three varieties of smart cards: contact, contactless, and hybrid (which can operate in either mode). Contactless smartcards will be discussed together with RFID tags in Sect. 2.2. Applications of the smartcard are numerous. For example, smart cards are used in the financial industry (credit cards, debit cards, etc.), the telecommunications industry (SIM, Phone cards), and many other applications such as authorized entry cards, e-passports, satellite/pay-TV consoles, etc. Today smartcards consist of a single die, which contains a general processor in addition to ROM, EEPROM, and (nonvolatile and volatile) memory units. Some embedded dies also contain one or more crypto-processors. In most of these commercial applications important constraints include low cost and high security.

There have been reports of phone card cloning, pay-TV cloning, and malicious ATM machines. For example, the hacking of Pay-TV cards was achieved through the use of voltage supply spikes. These voltage spikes cause some values to be zeroed in the device. When the data values were zeroed the key used in the cryptographic algorithm was easily obtained and cards were cloned with the key (Kuhn and Anderson 1996; Anderson 2001).

The side channel attack on smartcards was discovered by researcher Paul Kocher (Kocher et al. 1999). He was able to extract the secret keys out of many smartcards. His work initiated the drive to develop sets of countermeasures to resist side channel analysis attacks. Side channel analysis and more details on these types of attacks on real embedded systems will be discussed in Chap. 8 with countermeasures for resisting side channel analysis in Chap. 9. Some possible attack points for smartcard-based ATM systems are shown in Fig. 2.2 including the ATM points of attack and side channel in Fig. 2.2a, b, respectively.

As with most high-security embedded systems, reliability is a crucial aspect. Ross Anderson (Anderson 2001) describes an incident where due to a fault in the card reader (possibly misaligned swiping of the card through the reader), an individual's card was wrongly determined to be forged. The individual was beaten up and put into jail. However, it was determined later that the card was in fact not forged but the reader made errors in reading it. It is likely that the errors were masked and therefore went undetected in the normal checksum. Even though the probability of masking errors was likely very small, a more reliable approach using several checksums should have been employed or alternatively the reader should have been able to run its own self-tests to verify its correct operation and alignment. It is interesting to note that the cryptographic checksum did function correctly and detected the errors.

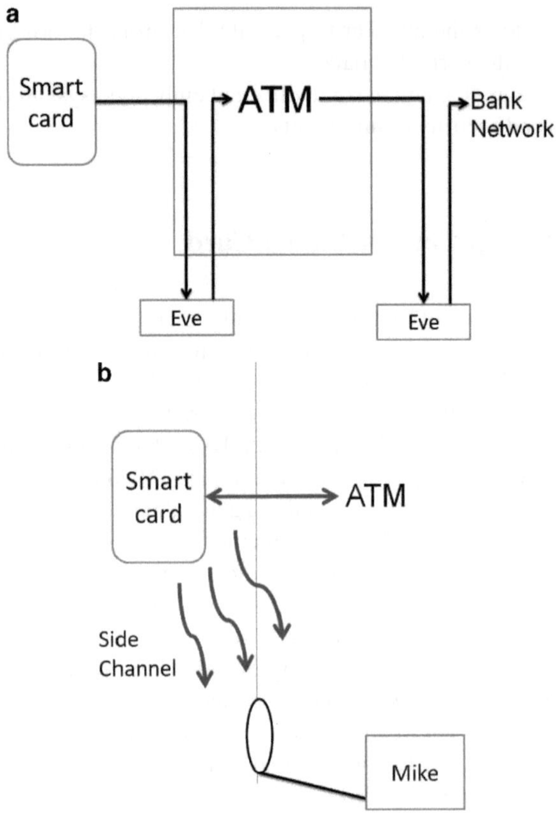

Fig. 2.2 Illustration of possible attacks points in contact card-based ATM in (**a**) and side channel probe on boundary of contactless card-based ATM in (**b**)

Clearly high levels of security are required for smartcard-based financial applications including authentication of the reader and card in addition to confidentiality, integrity, and nonrepudiation. The next section will discuss the newest type of smartcard, the contactless smartcard, in addition to RFID tags.

2.2 Contactless SmartCards and RFID Tags

Both near field contactless smartcards, or proximity cards, and the HF (high frequency) tags have an operating frequency of 13.56 MHz and use the standard ISO/IEC 14443 RF power and signal interface (ISO14443 1999). Types of applications that may use RFID/contactless-card technology include passports, toys, dishwashers, pet identification, credit/debit cards, real-time asset management or supply chain management, etc. RFID tags have found many other uses, such as wireless sensor networks and generally offer low-cost ubiquitous computing.

Contactless cards and RFID tags include an antenna (which generally spans the periphery of the smart card or tag). This antenna provides the necessary power from the EM field provided by the reader. The reader generates a sinusoidal field with a 13.56-MHz carrier, which supplies the card/tag with operating energy. In general tags/cards can read up to 30 cm, 1 m, and 7 m away for (125–135 kHz) low, high (13.56 MHz), and ultrahigh (2.45 GHz) frequency RFIDs, respectively. Alternatively some active RFID tags have batteries and can read over 100 m away (Rieback et al. 2006). Both card/tag and reader modulate this field in order to communicate. The reader and card/tag are inductively coupled and the reader's field can be treated as purely magnetic. In type A card/tags, 100% amplitude shift keying and a modified miller code are used for reader to card communications. The card/tag to reader communication on the other hand uses load modulation and employs on/off keying with a Manchester code. The card/tag switches an additional load into the field in order to generate a subcarrier (of approximately 847 kHz). They use tiny peaks of radio energy above and below the carrier signal to transmit tag/card data. The reader can sense when the card/tag switches into this load modulation. This appears as a reduction in amplitude. Other standards for smart cards include ISO 7810 (ISO7810 1999) and 7816 (ISO7816 1987–2005) in addition to ISO 14443 (ISO14443 1999).

Figure 2.3a illustrates a passive eavesdropping attack by Eve who can see all data communicated between the card/tag and the reader. The experimental setup for launching this attack is described in Chap. 8. Additionally Eve may be able to modify the data during these communications [see tag spoofing in Rieback et al. (2006)]. Since these devices are typically near field devices, a relay type of system could be used to activate and communicate with a reader as shown in Fig. 2.3b.

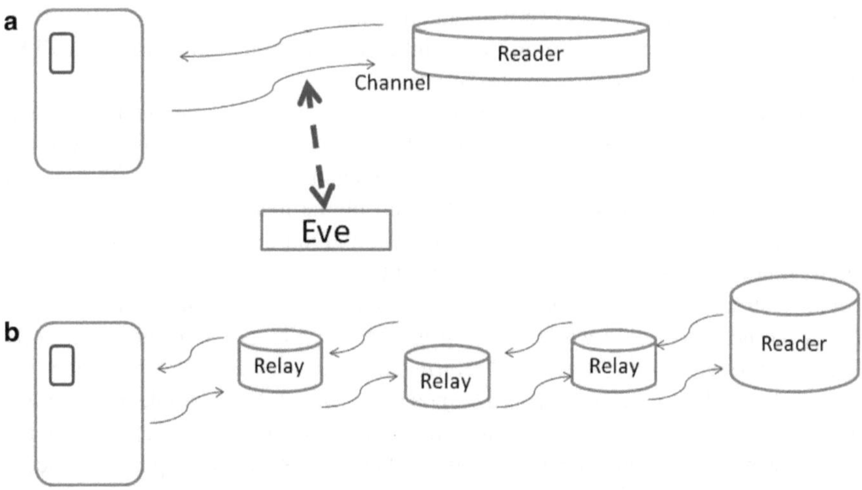

Fig. 2.3 Eavesdropping on near field contactless device in (**a**) and far field relay for challenge response attack in (**b**)

For example, consider an attacker who wishes to unlock a door but does not have the required authentic contactless smartcard. The door control typically sends out a challenge to any card within the vicinity. If the card returns the correct response the door control will unlock and open the door. Assume that someone with the authentic card is nearby but outside of the EM field of the door control. The attacker can use several special cards or tags to relay the challenge to the authentic card whose response can be relayed back to the door control. This complete challenge response would allow the attacker to gain access through the door. This attack has been demonstrated in Hancke (2005). There are other demonstrated attacks such as one which "kills" RFID tags (Oren and Shamir 2006). Examples of other possible attacks include performing unauthorized tag reading, denial of service attacks, covert tracking, clandestine location tracking of people or objects, snooping, etc.

Clearly there is a great need for security in card/RFID systems. However, RFID tags can generally only accommodate 8,000–15,000 gates, and this is typically only sufficient for some built-in logic with EEPROM. In fact, the majority of RFID tags cannot even authenticate readers due to their strict low-cost and low-power requirements. Thus, extremely efficient or ultralightweight security is a crucial requirement for RFID technology. Both card/tag and reader (mutual) authentication along with confidentiality and checks on timing are important to thwart malicious reader attacks.

2.3 Cell Phones and PDAs

It is not surprising that cell phones and PDAs may contain security largely for confidential communications, secure e-mail and internet communications, etc. However, there are other uses for security in these embedded systems. Some of these will be discussed below along with examples of attacks.

A reported use of security in some cellphones is battery authentication. It was rumored that if the battery authentication failed (e.g., the battery brand was not the same as the cell phone brand) power management was turned off. This use of security encouraged users to purchase the cell phone manufactured batteries that lasted longer due to the use of power management. Examples of this authentication can be found in presentations such as Paar (2008).

Consider a communications device such as a PDA or cell phone, which the user leaves on the lunch table while washing up. Meanwhile, the attacker temporarily takes the device, runs the attack, and returns the device to the lunch table before the owner returns. This is referred to as a "lunch time" attack. For example, if the attack was successful in obtaining the key, all communications can now be decrypted or the attacker can even masquerade as the owner. These attacks are an important concern because in these cases the owners are not aware that their device has been attacked. Hence, more dangerous attacks are possible in this type of situation because the attacker has a longer period of time before the owners suspect that their secrets have been stolen. The specific attack may involve dumping memory contents (hoping that the key is found somewhere), interrupting the device during a cryptographic operation and then dumping cache (hoping that the key was temporarily in cache), or running a side channel attack to extract the key (since the key may not be stored in accessible memory in the first place and in fact the key may not be accessible even to the owner of the device). It is also interesting to note that the attackers may be the authorities wishing to monitor communications of a suspected criminal.

Attacks on the GSM in cell phones and ciphers used for cell phone standards have been performed (Rao et al. 2002). Others have even reverse engineered PDAs to expose designs and proprietary methods. Attacks that dump memory in order to obtain data may be performed. Side channel attacks are also relevant for these types of devices since often their keys are not even known to the user of the device (Gebotys et al. 2005). PDA forensics is also an emerging area of interest (Jansen and Ayers 2004). For example, there are a large number of cellphones lost each year, and there is the potential that some data on these devices may be highly sensitive or dangerous in the wrong hands. There are likely many other types of attacks on these types of devices that people use every day. Details of an attack and countermeasures for a PDA will be described in Chaps. 8 and 9.

2.4 Automobiles

> "Embedded security will be an enabling technology for the majority of car IT sytems such as telematics, infotainment, secure software download, and ad hoc networks. escar is the premier international workshop which provides a forum for a systematic treatment of this emerging field" (ESCAR).

Today upper end automobiles include a sophisticated network of embedded processors and components, compared with their counterparts 50 years ago, which were completely mechanical based. The electronics in upper-scale automobiles include drive-by-wire designs with over 60 embedded processors (Marwedel and Gebotys 2004). For example, to illustrate the implications of the drive-by-wire system, consider a driver pressing his foot on the brake pedal. In older car models the pedal would be mechanically connected to the braking mechanism at the wheel. In a drive-by-wire system, the foot on the pedal generally causes a brake command to be placed on the bus. It is transferred to a control circuit near the brake, which reads the command and controls the braking. This sophisticated network includes control traffic (such as the brake command) as well as sensor traffic. The future automobile will have internet access ports (for automatic software updates, road condition monitoring, real-time diagnosis, etc.), satellite transmission capabilities, biometric ignition, etc.

In general technological advances have made the automobile more difficult to "attack." For example, auto theft used to involve duplicating metal keys. Later an auto thief would use a transmitter at a parking lot, which would enumerate and transmit codes until one of the codes unlocked a car (whose horn would sound and whose lights would conveniently flash for further identification). With biometrics, auto theft has become even more difficult. However, by examining the input/output ports or access points of the automobile network, attacks may be possible. For example, tire sensors as well as side view mirrors and other locations have network access ports, which provide possible attack entry points. Researchers have indicated that it is possible to launch attacks through the automobile bus and place commands onto the bus (Hoppe and Dittman 2007). For example, at an access point such as the side view mirror it may be possible to access the bus and place commands onto the network such as "unlock" doors. However, there are other attacks that may possibly be launched. An example is the use of Bluetooth to eavesdrop on automobiles within range or to inject audio into a neighboring car (Schneier 2005). Other attacks involve replacing chips in cars in order to attain higher performance than available for the existing car model. This attack is referred to as a modchip attack and is in fact now legal in some countries.

The automobile is a truly wireless mobile device today. With Bluetooth, GPS, and Internet, it has seen a dramatic increase in functionality. However, at the same time there are increased security needs in this area. Consider possible implications of a virus obtained from the Internet, which infects the drive-by-wire network within the car. Or alternatively consider an attack through any of the wireless interfaces, which causes a denial of service attack on the car network. This denial of service attack on the car network may lead to a brake failure. Although most current automobiles have a mechanical system in place for braking back up, there are plans to move completely to digital control in the near future.

However, there are other needs for embedded security in the automobile, which are not that obvious. One of these is litigation. Because of the high complexity of the automobile system, electronic components, chips, and platforms are typically multivendor. Thus, in the case of an accident or other type of failure, litigation is a huge concern. Security is required to provide proof that chips did or did not have a role to play in an accident. Here nonrepudiation is of great legal concern and can be supported by security. For example, an important event within the electronics of the automobile is signed by the chip which made such a decision. These events and digital signatures are logged so that in case of an accident the cause and liable party (such as the chip vendor) can be determined. Since accidents may also be caused by erroneous information received wirelessly from other sources (such as in a vehicular ad hoc network or VANET), typically these received messages are also accompanied by digital signatures. For example in the case of the VANET (Paar 2009), an automobile may receive GPS, speed, and other information along with digital signatures, from nearby cars. This information is used to compute trajectories which are used to indicate any potential car collisions. The automobile may not directly act upon a computed potential collision, other than to gently 'wake up' the driver (perhaps by vibrating the steering wheel). The signatures are likely there for liability purposes. For example assume that data is incorrectly computed by chip A within car X and then sent to car Y such that car Y does not predict the oncoming collision. The digital signature signed by chip A verifies that chip A made a computation error and hence the vendor of chip A is liable in the collision of car X and Y. It is also interesting to note that 256-bit ECDSA signatures (see Chap. 5) are recommended to accompany the wireless messages sent within the VANET. Much smaller signatures (128-bit or less) would have been sufficient, however it is likely a decision made by the legal/political profession (without a sufficient understanding of cryptography) who insist that the 'highest' standardized levels of security be used.

Driver safety has always been a focus in the design of automobiles; hence, it is no surprise that security is a major concern in this area. Security is such a concern that a new conference, ESCAR Security in Automobiles (http://www.escar.info/), was established to encourage researchers to advance this area of study. Clearly there are many needs for security in automobiles. The safety critical aspect of automobiles makes security implementations very challenging. Additionally the security implementations must consider not only resistance to attacks, but also reactions to possible attacks in order to support design for safety and security.

2.5 Game Stations

Game stations also have security requirements to prevent unauthorized games from being played. However, there are other reasons for security in game stations. Some systems, such as the Xbox, are built upon general-purpose workstations, which have extremely high performance. Thus, the makers do not want the game station to be purchased at low cost and used as a high-performance general-purpose computer (otherwise no one would purchase the same hardware selling as a high priced workstation). Many game stations use authenticode or they authenticate their programs (Shamir and VanSomeren 1998). Other embedded systems using "authenticode" approaches include mission critical systems, systems needing to limit third party add-ons, and security programs themselves.

Consider the Xbox game station that was attacked by Andrew Huang, a PhD student from MIT (Huang 2002). It contains a key responsible for decrypting and verifying a bootloader that decrypts and verifies a kernel image. The kernel image in turn authenticates and checks the integrity of applications loaded on the Xbox. This chaining is referred to as the chain of trust. The bootloader and kernel are stored in a flash ROM, whereas the secret key and crypto algorithm are stored in a secure boot block (hard-coded in an ASIC) and the decryption is executed by the Pentium CPU. The Pentium CPU is connected to the secure boot block using a high-speed bus that is not encrypted. The bus was assumed to be secure due to its high speed. The flash ROM also contained a decoy boot block. By examining and analyzing data on the high-speed bus the key was obtained (Huang 2002). Once the secret boot procedure was determined, any program could be run on the Xbox. Furthermore, the value of the key was the same used in all Xbox stations. This illustrates the importance of the notion that any cryptosystem is only as secure as the secrecy of its key.

2.6 Satellites

There are estimated to be over 8,000 satellites orbiting the earth (NASA Web site), and many are interested in launching more satellites such as commercial telecommunications companies, military, government, research (ESA), etc. Satellites may typically remain in orbit for a period of time often beyond the mission time. There have been numerous (largely unsubstantiated) reports of attacks on satellites, including hackers taking over control of satellites (Xnet 2007). There are also many threat models for satellite systems (GAO 2002), which have been developed. Hence, there is a need for security in satellites, despite the fact that few satellites have security. The top constraints of a satellite system include volume, mass, and power. Since this embedded system is exposed to harsh radiation, reliability and remote reprogrammability are a necessity.

Confidentiality is a concern in high-security satellite systems. Authentication of the base station should be considered since communication with a malicious base station may be a concern. Integrity may not be a concern if hackers are technically unable to modify communications to or from the satellite. On the other hand, sometimes satellites will require revised programs to be sent up to the satellite. Integrity of these programs is important and may be the focus of attacks especially in mission-critical cases. Further details on security in satellites will be discussed in Chap. 10.

2.7 FPGA, Networks on a Chip

FPGA systems are used in many systems providing reprogrammability through reconfiguration. Security is an important concern in many FPGA systems. In particular, FPGAs contain many IP cores. The IP core design information is stored in external nonvolatile memory and loaded into the FPGA during power up of the circuit. This channel between the memory and FPGA is the focus of several attacks. There are many possible attacks (Drimer 2007) on this type of embedded system largely due to the well-defined channel and power-up configuration.

An attacker could acquire the data sent to the FPGA during power up in order to clone the FPGA design. This is illustrated in Fig. 2.4. These cloning attacks can be thwarted by using ciphertext, which is then decrypted within the FPGA system. In addition to confidentiality, which is important for secure FPGAs, authentication may also be important. For example, attacks may involve modifying the configuration bit stream or ciphertext; hence, authenticated bitstreams are an important issue in security for FPGAs.

In FPGA scenarios, a replay attack may also be possible. For example, the attacker records a version of the encrypted bitstream. Later after hardware updates have been performed in the FPGA, the attacker resends the older bitstream to the device, causing the device to be reprogrammed with this older version of hardware.

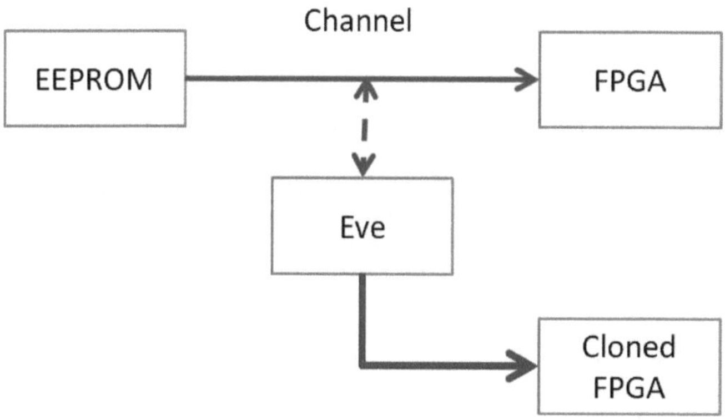

Fig. 2.4 Attack on FPGA systems

Thus, authenticated bitstreams should include a timeliness property, which is typically achieved with nonces, timestamps, etc. Section 10.6 will further discuss FPGA security and Sect. 3.8 will examine NoC security issues.

2.8 Summary

Efforts to incorporate security into many of these embedded systems have only recently been initiated. There are many more embedded devices that will likely also have to incorporate some form of security. Some of these implementations will be safety critical systems such as the automobile. Hence, the secure and reliable implementation of cryptographic functions will be a necessity. Security standards and guidelines extended for embedded use will likely be an important step so that widespread secure implementations will be achieved. The remainder of this book will cover the following:

1. Understanding of security principles with appreciation for the mathematics and underlying computations (Chaps. 3–7)
2. Understanding of attacks and countermeasures for real embedded devices (Chaps. 8 and 9)
3. Understanding of design for reliability in secure embedded systems (Chap. 10)
4. Interactions of security with other design concepts and standards (Chap. 11)

This focus is important for embedded designers to gain an understanding of security principles, security constructs, security attacks, and other issues. The next chapter will start with one of the most important aspects of a security system, the key. It will introduce some properties of the key and various functions associated with the key over its lifetime in an embedded system.

References

Anderson R (2001) Security engineering. Wiley, New York

Drimer S (2007) Volatile FPGA design security – a survey. Computer Laboratory, University of Cambridge, Cambridge, UK. http://www.cl.cam.ac.uk/~sd410

Gao (2002) Critical infrastructure protection: commercial satellite security should be more fully addressed. USGAO, GAO-02–781. http://www.gao.gov/new.items/d02781.pdf

Gebotys C, Ho S, Tiu CC (2005) EM analysis of Rijndael an ECC on a wireless java-based PDA, CHES 2005, LNCS 3659 GmbH 250–265

Grand J (2005) Advanced hardware hacking techniques. Defcon 12. http://grandideastudio.com/wp-content/uploads/advanced_hardware_hacking_slides.pdf

Hancke GP (2005) A practical relay attack on ISO 14443 proximity cards. http://www.cl.cam.ac.uk/~gh275/distance.pdf

Hoppe T, Dittman J (2007) Sniffing/replay attacks on CAN buses: a simulated attack on the electronic window lift classified using an adapted CERT taxonomy. Workshop on Embedded Security in Systems (WESS)

Huang A (2002) Keeping secrets in hardware: the Microsoft Xbox case study. MIT AI lab, AI Memo 2002–08. http://www.ai.mit.edu

ISO 7816 (1987–2005) Identification cards – integrated circuit cards, 1st edn. International Organization for Standardization, ISO/IEC, Geneva, Switzerland

ISO 14443 (1999) Identification cards – contactless integrated circuit(s) cards – proximity cards. Final committee draft. International Organization for Standardization, ISO/IEC, Geneva, Switzerland

ISO 7810 (2003) Identification cards – physical characteristics. Final draft. International Organization for Standardization, ISO/IEC, Geneva, Switzerland

Jansen W, Ayers R (2004) Guidelines on PDA forensics SP800–72. http://csrc.nist.gov/publications/nistpubs/800--72/sp800--72.pdf

Jun B (2008) Protecting consumer electronics, HT1–108, RSA 2008 presentation

Kent J (2006) Security fears raised at conference. http://news.bbc.co.uk/2/hi/technology/5399050.stm

Kocher P, Jaffe J, Jun NB (1999) Differential power analysis. In: CRYPTO'99. Springer, New York, pp 388–397

Kotadia M (2004) Bluetooth phone hacking tools 'spreading quickly'. http://networks.silicon.com/mobile/0,39024665,39118440,00.htmG

Kuhn M, Anderson R (1996) Tamper resistance – a cautionary note. Second USENIXworkshop on electronic commerce, Oakland, CA, pp 1–11

Landers K (2008) Millions of dollars lost in identity theft. Transcript from AM. http://www.abc.net.au/am/content/2008/s2325433.htm

Marwedel P (2006) Embedded system design, 2nd edn. Birkhauser, Springer, New York

Marwedel P, Gebotys C (2004) Panel on secure and safety-critical vs. insecure, non safety-critical embedded systems: do they require completely different design approaches? In: ACM Proc of CODES+ISSS'04, 8–10 Sept 2004, Stockholm, Sweden, pp 72–73

NASA Web site. http://techtran.msfc.nasa.gov/SBIR/tether.html

Newitz A (2006) The RFID hacking underground. WIRED, issue 14.05. http://www.wired.com/wired/archive/14.05/rfid.html

Oren Y, Shamir A (2006) Power analysis of RFID tags. http://www.wisdom.weizmann.ac.il/~yossio/rfid.

Paar C (2008) New directions in lightweight cryptographic primitives for RFID applications. In: RFID CUSP workshop, John Hopkins University, presentation, Baltimore, MD

Paar C (2009) Crypto Engineering: Some History and Some Case studies, CHES 2009 Presentation http://www.iacr.org/workshops/ches/ches2009/presentations/07_Invited_Talk_II/CHES2009_paar.pdf

Rao JR, Rohatgi P, Scherzer H, Tinguely S (2002) Partitioning attacks: or how to rapidly clone some GSM cards. IEEE Symp Security Privacy 31–41

Rieback M et al (2006) A platform for RFID security and privacy administration. In: Proceedings of the 20th conference on large installation system administration, Washington, DC

RNCOS (2009) Smart card market forecast to 2012. RNCOS. http://www.reportbuyer.com/banking_finance/debit_credit_cards/smart_card_market_forecast_2012.html

Schneier (2005) Eavesdropping on bluetooth automobiles, Schneier on security. http://www.schneier.com/blog/archives/2005/08/eavesdropping_o.html

Shamir A, Van Someren N (1998) Playing hide and seek with stored keys. In: Financial cryptography 1998, Springer, Berlin, 10.1007/3–540–48390-X 1999. LNCS 1648:118–124

Tarnovsky C (2008) Security failures in secure devices. Black Hat briefings and training. http://www.blackhat.com/presentations/bh-europe-08/Tarnovsky/Presentation/bh-eu-08-tarnovsky.pdf

VanTilborg HCA (2005) Encyclopedia of cryptography and security. Springer, New York

Xnet (2007) Hackers control a British military communications satellite, xnet solutions, http://www.890 xnet.com.pk/news/2007/05/hackers-control-british-military.html

Chapter 3
The Key

This chapter will introduce the key, which is perhaps one of the most important aspects of any cryptographic system. A system is only as secure as the key. This was noted by Auguste Kerckhoffs in the nineteenth century and has since become known as Kerchoffs' principle. Random values, secrets, keys, physically unclonable functions (PUFs), and trusted key storage will be covered in this chapter along with functions used to create, update, or freshen keys. Other important characteristics to be discussed include key length, key lifetime, key backup, key destruction, key recovery, and key revocation. An introduction to security equivalence is also provided since this relies heavily on key length. Various methods for storage of keys will be outlined along with a discussion of trusted platform modules (TPMs). In addition, a networks on chip (NoC) view of keys and their storage and use are also provided.

The specific value of keys may also be constrained by the cryptographic application in which they are used. For example, typically there are fewer constraints on symmetric keys as compared with public–private key pairs where special mathematical properties are required. Various techniques for generating keys and tests for randomness will be discussed in this chapter. In cryptographic theory, there are two general types of keys, symmetric keys and public (or asymmetric) keys (see Chaps. 4 and 5). There are different approaches to implementing public key schemes such as those based upon integer factorization (such as RSA), discrete logarithms (DSA, ElGamal), both discussed in Chap. 4 and Elliptical curve discrete logarithms (ECC), discussed in Chap. 5. Key distribution center, or KDC, key translation center, or KTC, and certificate authorities (CA) are third parties or trusted sources and are discussed further in Chap. 4. A more in-depth discussion of other key systems, such as identity key systems, can be found in Menezes et al. (1996). This chapter will discuss key generation and characteristics including key length, an important parameter dictating security, as well as how keys are restricted to be used with different sets of applications.

C.H. Gebotys, *Security in Embedded Devices*, Embedded Systems,
DOI 10.1007/978-1-4419-1530-6_3, © Springer Science+Business Media, LLC 2010

3.1 Key Randomness

Keys can be generally obtained from sources of randomness within the embedded system itself or through computations performed on these sources of randomness. Alternatively keys can be obtained from cryptographically secure pseudorandom number generators executed by circuitry or software running on the chip. System information, such as status registers, system clock, interrupt vectors, date, time, etc., is often used as a seed for the cryptographic pseudorandom number generators.

A measure of the amount of information in data is often referred to as entropy. Entropy was developed by Shannon in 1948 using concepts from information theory. It is defined using a mathematical measure of information or uncertainty. As an example if we consider a coin toss, the two possible outcomes, heads or tails, can be encoded with 1-bit (each outcome has a probability of $1/2$). Thus the entropy of the coin toss experiment is 1-bit. The random variable of this experiment, X, would take on one of the two variables, each representing a probability. In this example, we have the average number of bits in an encoding of X is $\Pr_{\text{heads}}$ (No. of bits representing heads) $+ \Pr_{\text{tails}}$ (No. of bits representing tails) $= 1/2(1) + 1/2(1) = 1$. If we have n coins, the average length of bits in an encoding of x would be n-bits. More formally entropy is defined as shown in (3.1).

$$H(X) = -\sum_{x \in X} \Pr(x) \log_2 \Pr(x). \tag{3.1}$$

For the experiment with a single coin toss, the entropy would be calculated as $H(X) = -(2)(1/2)\log_2(1/2) = -\log_2(1/2) = 1$. For an experiment with n equally likely outcomes, the entropy would be calculated as shown in (3.2)

$$\begin{aligned} H(X) &= -1/n(\log_2(1/n)) - \cdots - (1/n)\log_2(1/n) \\ &= -1/n(\log_2 1 - \log_2 n) - \cdots - 1/n(\log_2 1 - \log_2 n) \\ &= -1/n(0 - \log_2 n) - \cdots - 1/n(0 - \log_2 n) \\ &= \log_2 n. \end{aligned} \tag{3.2}$$

In this case, the number of bits required to encode this experiment (Stinson 2006) is equal to the entropy which is $\log_2 n$. Typically the average length encoding is close to the entropy. Note that one cannot increase the entropy by increasing the number of bits in the encoding, unless the number of outcomes actually increases. In cryptography, the entropy of a cryptosystem is a measure of the size of the key space. Thus a cipher using a 128-bit key has an entropy of 128-bits. In practice although we might use a pseudorandom number generator to produce our keys, the seed should have sufficient entropy in order to generate a strong enough key. For example, one might have a user scribble on a notepad and based upon the coordinates which were used in the scribble a seed may be generated. Alternatively, the seed may be the key and the pseudorandom number generator output may be the key stream used in a stream cipher.

In practice it is very difficult to have a k-bit truly random number generator (RNG), but if we did its entropy would be k-bits. As discussed in Chap. 6, linear feedback shift registers and other constructs generate pseudorandom numbers. They are not truly random since they are deterministically produced by the state and feedback arrangement of the shift registers. In these cases, the entropy of the generator can never exceed the entropy of the seed. Hence getting sufficient entropy in the seed is important.

Ideally a nondeterministic reliable source for the seed would be advantageous. Examples suggested by researchers include thermal noise, atmospheric noise, or nuclear decay (Jun and Kocher 1999). However, the difficulty with using a nondeterministic source is that it is difficult to measure the randomness and the reliability. An example of a nondeterministic noise source is a random noise generator which measures the thermal noise. This circuit measures the amplified voltage across two undriven resistors. The noise is correlated to local environmental characteristics such as electromagnetic (EM) radiation, power supply fluctuations, temperature, etc. The signals are subtracted from two adjacent resistors to remove any possible coupling effects (Jun and Kocher 1999). Other sources include the drift between two free-running oscillators (Jun and Kocher 1999). There are other proposed architectures for truly RNGs. These include free running ring oscillators, such as Fibonacci ring oscillators or Galois ring oscillators. In these two cases, a LFSR-like structure is used except the D-flipflops are replaced by inverters. Figure 3.1 illustrates an example of the Galois and Fibonacci ring oscillators.

Some analysis has been performed to show that the Fibonacci and Galois ring oscillators perform better than classical ring oscillators (Dichtl and Golic 2007), producing higher orders of magnitude of entropy. More recently a ring oscillator which uses the metastability property to produce randomness (Vasyltsov et al. 2008) was shown to produce higher entropy as well as higher throughput than the classical ring oscillators based on jitter. Further discussion on these types of circuits is provided in Sect. 3.2.

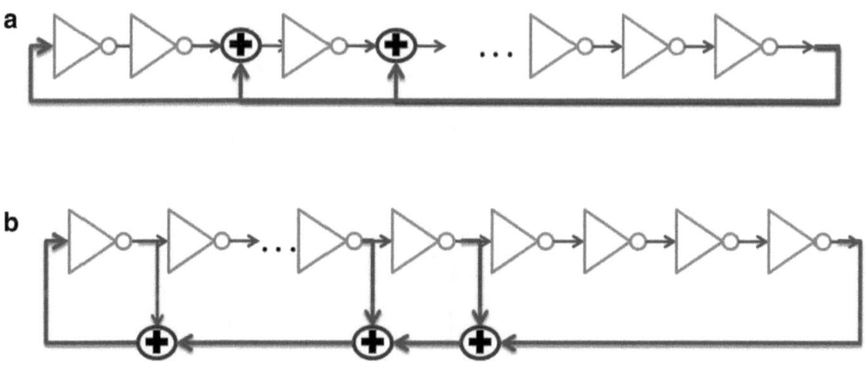

Fig. 3.1 Galois in (**a**) and Fibonacci in (**b**) ring oscillators

3.2 Physically Unclonable Functions

Physically unclonable functions, or PUFs, can be used to derive secrets from physical characteristics of integrated circuits. When the physical characteristic is timing and delay, it is analogous to nondeterministic means for extracting random values from noise which was briefly discussed in Sect. 3.1. However, in general unlike these ring oscillators, PUFs may extract one random value or secret as opposed to a key stream. In both cases, the circuit must be powered-on in order to obtain the random value(s). Process variation has been used to identify circuits, for example by examining the drain current (Lofstrom et al. 2000). However, a secure form of identification for the IC is required in cryptographic applications. Other research on PUFs examined physical one-way functions (Pappu 2001) using optical medium with laser light. Acoustic and coating PUFs measure the capacitance of a medium covering the IC or acoustic reflections (Skoric et al. 2005). Silicon PUFs on the other hand are those which are based on the specific process timing and delays within the devices on silicon (Suh and Devadas 2007). It is well known that timing will vary among die on the same wafer due to processing variations, wafer curvature, etc. In fact when chips are fabricated, a report on process corners is typically performed using several test die within each wafer to identify the variation in timing across the wafer.

An arbiter PUF is shown in Fig. 3.2. In this circuit, the user input, $x(0), \ldots x(n)$, controls the multiplexors and the circuit is activated with one rising edge of a signal on the left-hand side input. The circuit outputs one-bit of random data at the right-hand side which is latched into the D-flipflop. When a multiplexor control input is one, the top and bottom data input from the left are passed through the multiplexor. Whereas when the multiplexor control is a zero, the top and bottom data are switched. Consider a rising edge on the input data at the left feeding the left pair of multiplexors. At the end of the circuit the data from the top is fed into the data of a clocked flipflop, whereas the bottom data is fed into the clock. If the top data

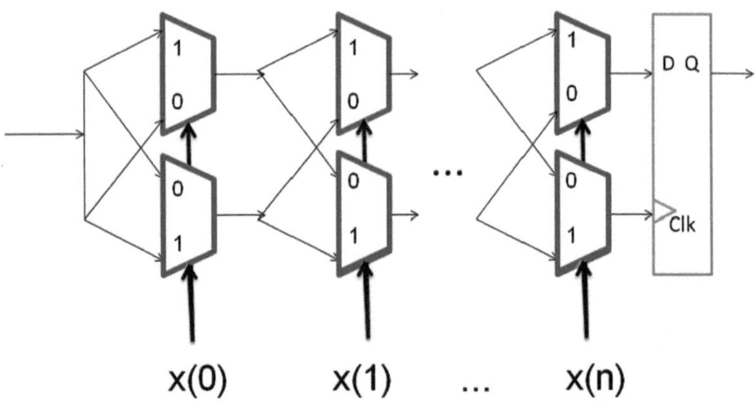

Fig. 3.2 Arbiter PUF example

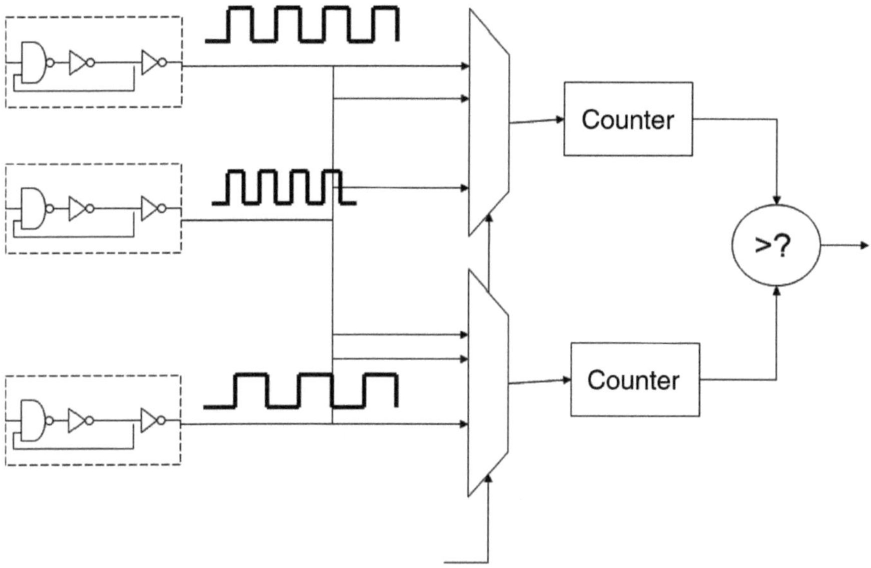

Fig. 3.3 Ring oscillator PUF example

arrives before the bottom data it will be latched into the flipflop otherwise a zero is latched into the data. Further security can be added by including a PUF output as a multiplexor control or exclusive-oring multiple outputs together. This arbiter PUF generates a 1-bit random data value and uses a n-bit input, $x(i)$, $i = 1, \ldots, n$. In order to generate m-bit output, m arbiter PUFs could be used. Alternatively, a seed could be used with a LFSR to generate k different x's which can be used as input to the arbiter PUF which can then generate k bits.

A different approach to designing a PUF is shown in Fig. 3.3. It is analogous to the previously discussed ring oscillator approach (in Sect. 3.1) and is called a ring oscillator PUF. A delay loop is replicated a number of times on the left-hand side. Each ring oscillator is connected to both multiplexors. Although the oscillators are identical, their delays and thus frequency of oscillation will vary due to process variation. The controls select a pair of different oscillators which are output through the pair of multiplexors. The multiplexor output is connected to counters whose difference creates the single-bit random data output. Through counting many cycles, the difference in frequencies will dominate over routing skews (Suh and Devadas 2007). However, since temperature variation and supply variations will modify the circuit delays error correction is also needed. Alternatively each oscillator could be used independently to generate random bits as described in Sect. 3.1.

The ring oscillator PUF is easier to analyze with respect to entropy, has higher reliability, and is easier to implement than the arbiter PUF. The arbiter PUF consumes less energy and is faster and smaller than the ring oscillator PUF (Suh and Devadas 2007).

These PUFs, however, are not sufficient for generating keys, since the random values for a given chip may vary over time due to noise, temperature, etc. Thus to

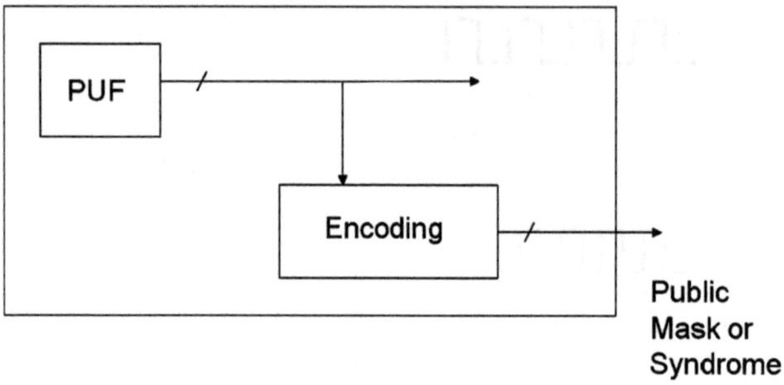

Fig. 3.4 Error correction codes or syndromes generated for PUF output

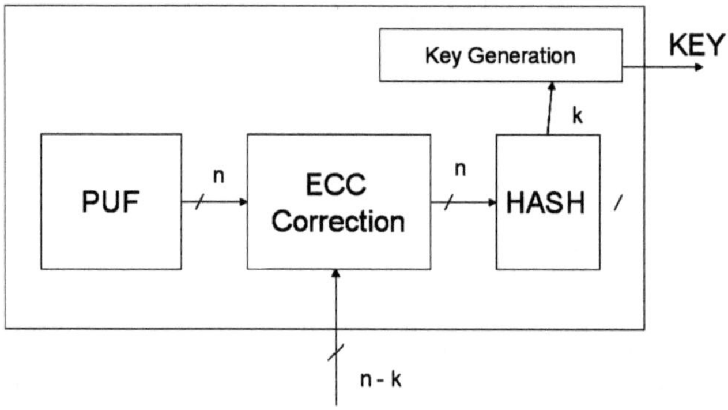

Fig. 3.5 Generation of key from PUF

add consistency to the error-prone random number generation by PUFs, there is an initialization phase which includes error correction and then a key generation phase (Suh and Devadas 2007). The PUF outputs a specific value for which an error correcting code is generated (see Chap. 10 for discussion of hamming codes) as illustrated in Fig. 3.4. Other information may also be stored with the error correcting code, such as the value of x for an arbiter PUF. These values can be stored in the clear. Later when the PUF is used to regenerate the value, the codes are accessed and used to correct any errors in the PUF output. Additional bits are generated from the PUF circuit so that attackers cannot obtain all bits of information that the PUF outputs from the syndrome. For example, BCH (n, k, d) coding can be used to correct $(d-1)/2$ errors at most out of n-bits using a $(n-k)$-bit syndrome. To generate a key, after error correction of the PUF output, the secret value can be hashed to generate the correct key size. Alternatively the hashed PUF output can be used as a seed for a key generation process as shown in Fig. 3.5. Studies with PUFs indicated that on one piece of silicon a PUF design had less than 1% variation on a FPGA due to temperature (20–120°C) and supply voltage (±10%) variations (Suh and Devadas 2007). The AEGIS processor uses a security module similar to TPM but located on chip (Suh and Devadas 2007). It uses a PUF to generate volatile keys and also checks the integrity of off-chip memory, in particular programs, using hash trees, which are discussed in Chap. 7.

3.3 Key Lifetime, Freshness, Updating

The lifetime of a key is often referred to as a cryptoperiod. In theory, the longer the lifetime of the key, the greater is the chance that the key may be compromised. Furthermore there is a risk for greater damage as well due to keys with longer lifetimes. These keys also may provide larger amounts of data to the attacker thus making the attacker's job easier in general. For example, more plaintext–ciphertext pairs may be collected for attacks. If keys are compromised, the user should immediately contact the Certificate Authority, KDC, credit card authority, etc. However the user may not be aware of the compromise. In this case, damage may have significant implications.

The cryptoperiod could be defined relative to time or the amount of data. The system is considered to be more secure if cryptoperiods are shorter. However, the shorter the cryptoperiod the more computations are required. For example, there are computations required for key updating or defining new keys, as well as the updating of stored encrypted data. For this reason, keys used for communication typically have shorter cryptoperiods than those used for encrypting large amounts of stored data. Details and recommendations on the cryptoperiod for various types of keys can be found in Barker et al. (2007).

For the case of asymmetric keys, the lifetime of the certificate may be different from the cryptoperiod of each key from the key pair. The cryptoperiod of the private signature key may be shorter than the lifetime of the public signature verification key. The longer lifetime of the public signature verification key is not a major concern. Symmetric key wrapping keys which encrypt a large number of keys should have a relatively shorter cryptoperiod. Whereas a symmetric master key can have a much longer cryptoperiod since it is used to derive keys for single use or a small number of sessions.

It is possible to update keys using an update function as opposed to performing a new key establishment/agreement protocol or sending encrypted keys to parties as discussed in Chap. 4. For example, both parties can take their old keys and using a one-way hash function, H_o, and possibly also a key derivation function, *KDF*(), create new keys using (3.3). Key derivation functions generally map a secret into a key.

$$k_{\text{new}} = KDF(H_o(k_{\text{old}})). \tag{3.3}$$

Unfortunately the security of this new key in equation (3.3) is generally equivalent to the security of the old key. For example, an attacker who has learned the value of the old key can also apply the same hash to obtain the new key. Thus alternatively a MAC, discussed in Chap. 7, will be used as a key derivation function since an attacker will not have access to the secret key used to compute the MAC.

Key control has also been addressed in the past using control vectors. Assume we have a control vector, V_c, which specifies the uses and restrictions for that key. The computation for generating a key encrypting key, k, may involve hashing the control vector and exclusive-oring this value with the master key, K_m, as shown in (3.4).

$$\begin{aligned} H(V_c) \oplus K_m &= k \\ E_k(k_s). \end{aligned} \tag{3.4}$$

The session key, k_s, is then encrypted with this key and stored with the control vector. The values stored are V_c, $E_k(k_s)$. When the session key needs to be used again the computations in (3.5) are performed (Schneier 1996).

$$\begin{aligned} H(V_c) \oplus K_m &= k, \\ D_k(E_k(k_s)) &= k_s. \end{aligned} \tag{3.5}$$

There are many other key updating functions possible including the concept of a forward cryptographic transformation, or $F_f()$, as shown in (3.6). Any security compromises of a key, k_t (key created at time t), which was updated using a forward cryptographic transformation on a key k_{t-1}, does not threaten past ciphertext created with past keys k_{t-1} or earlier. This could be an important consideration since often attackers are interested in decrypting past data which may have been encrypted with older keys. The forward cryptographic transformation is in definition equivalent to a one-way function, described in Chap. 7.

$$k_t = F_f(k_{t-1}). \tag{3.6}$$

An interesting example of this concept is presented in Kocher (1999) incorporated with a tree structure providing the additional resistance from side channel analysis, which will be discussed in Chap. 7.

When the lifetime of a key ends, it is important to destroy the key. Often applications will leave data in cache or memory locations unless a user specifically takes steps to erase the data. Deleting it from memory is often not sufficient due to memory reminance. For example, data previously stored at a memory location can often be observed even after zeros are stored to that location in an attempt to clear the previous data. Hence for secure memory, the memory locations should be written over many times with zeros or all ones, etc. so that an attacker having access to the device cannot observe any data such as keys. Steps may be taken to erase even past or older keys, since they may be used to decrypt past confidential data, etc.

3.4 Key Length

The security of a cryptosystem largely depends upon the length of the key and the strength of the algorithm. The later is more important. However, both should be suited to the embedded system of interest. This section will focus on the key length and it will assume that a strong enough cryptographic algorithm is used in conjunction with this key. Strong cryptographic algorithms are discussed in Chaps. 4–7.

In general, a measure of the security of an algorithm is the amount of time it takes to obtain the key using the best-known attack algorithm. If there are no attack algorithms known for the cryptographic algorithm, then a brute force attack is assumed. In a brute force attack, one assumes that all possible keys (or on average $1/2$ of all possible keys) must be tried in order to determine what the secret key is.

Table 3.1 Equivalent security of brute force attack on cipher to best known attack on PKC (Menezes et al. 1996)

No. of bits for Cipher BF attack	80	112	128	192	256
No. of bits for PKC					
ECC parameter n	160	224	256	384	512
RSA parameter n	1,024	2,048	3,072	8,192	15,360

Table 3.2 Approximate time for brute force attack on cipher by 200M encryptions per second device

Bits	8	16	32	64	80	112	128	192
Time	1×10^{-6} s	0.0003 s	21 s	3×10^{3} years	2×10^{8} years	8×10^{17} years	5×10^{22} years	9×10^{41} years

This process known as cryptanalysis will be discussed further in Chap. 8. Table 3.1 illustrates equivalent security using the best-known attack algorithm on public key cryptographic (PKC) algorithms (Menezes et al. 1996). The time for the attack is shown as equivalent to the time for a brute force attack (BF attack) on some ciphers. For example, the brute force attack on an 80-bit cipher is equivalent to the time to run the best-known attack of a 160-bit discrete logarithm or elliptic curve cryptographic algorithm. This is also shown as equivalent to the time required to perform the best-known attack on a 1,024-bit RSA cryptographic algorithm. Thus a 160-bit ECC system has equivalent security to a 1,024-bit RSA system.

Cryptanalysis involves determining the secret key so that ciphertext can be converted to plaintext. For example, assume that the attacker has a block of plaintext and the corresponding block of ciphertext. Furthermore assuming the cryptographic algorithm is known, all possible keys are attempted until the correct ciphertext (or plaintext) corresponding to the input plaintext (or ciphertext) is obtained. In the brute force attack, for a m-bit key there would be 2^m possible keys. Other more efficient attack algorithms are often developed and can be used to attack security algorithms, such as the quadratic sieve algorithm for attacking the product of two large prime numbers (Schneier 1996). The viability of the brute force attack relies largely on the key length assuming that the cryptographic algorithm is strong.

Determining what size of key to use in any system depends upon what amount of cryptographic security is required. This depends upon the application, implications of attack, etc. However, key length is not the only factor affecting system security. Key lifetime and other parameters also have a significant role in the security of the system. Often one must assume that highly parallel attacks on systems could be launched in order to obtain secret keys. But as key lengths increase, attacks generally grow in difficulty. Consider a 64-bit key; the size of the key space is $2^{64} = 18,446,744,073,709,551,616$. Now assume you have a device which performs 200 million encryptions per second (200Meps). Searching the entire key space would take approximately 3,000 years ($2^{64}/[(200\times10^6)(60\times60\times24\times365)]$). Other times are shown in Table 3.2 for this same device.

Table 3.3 Approximate distance and communication delay to satellite

Satellite orbit	Distance from earth (km)	Communication delay from earth to satellite
LEO	80	2.4 ms
GEO	35,000	1.07 years
Mars	78,000,000	2.3×10^3

Table 3.4 Approximate time for brute force attack of cipher on a satellite

Orbit	8-Bits	16-Bits	32-Bits	64-Bits	80-Bits	112-Bits
LEO	1 s	157 s	119 days	1.4×10^9 yrs	9×10^{13} years	4×10^{23} years
GEO	274 s	1 day	146 years	6×10^{11} years	4×10^{16} years	2×10^{26} years
Mars	7 days	5 years	3×10^5 years	1×10^{15} years	8×10^{19} years	4×10^{29} years

Table 3.5 Approximate security equivalence of cipher and satellite

	No. bits of cipher				
Satellite	16-bits	32-bits	64-bits	80-bits	112-bits
Orbit	LEO	Mars	LEO	GEO	LEO
	(157 s)	(3×10^5 years)	(1.4×10^9 years)	(4×10^{16} years)	(4×10^{23} years)
Computer	32-bits	64-bits	80-bits	112-bits	128-bits
	(215 s)	(3×10^3 years)	(1.9×10^8 years)	(8×10^{17} years)	(5×10^{22} years)

Consider a security system embedded in a satellite. This example will be further detailed in Sect. 10.5.1.2. Assume that an attack on this embedded satellite security involves guessing and sending a secret to the satellite under attack (which is analogous to guessing a password on a computer login). Clearly unlike the attack on a ciphers secret key, the satellite attack time is dominated by the time it takes for the secret guess to be transmitted to the satellite. Table 3.3 illustrates the distance from the satellite to earth and the communication delay for transmitting data from earth to a satellite in different orbits, LEO, GEO, or Mars. The speed of light in a vacuum is used to calculate the delays. For example, the low earth orbit is approximately 80 km from earth and requires 2.4 ms to transmit a message to the satellite from earth. Assume that the only known attack on the secure satellite is a brute force one and the attacker can send as many guesses as possible to the satellite. Table 3.4 lists the attack time for exhaustively sending all possible secret guesses to the satellite in different orbits. Table 3.5 presents a security comparison of the satellite security strength to brute force attacks on ciphers. It is interesting to note that based upon results in this table, the security required in the satellite is far less than that normally required in a terrestrial system, where a 200Meps device is available. For example, a 32-bit cipher attack is roughly equivalent to 16-bit security in the LEO orbit. This is obtained by calculating a brute force attack using a 200Meps device on the 32-bit key (shown in brackets in the last row of Table 3.5) and finding the equivalent attack time for the satellite system (shown in brackets in second row of Table 3.5). This is a rough comparison just to give an indication of similar attack times and therefore similar security strengths. As a further example, an 80-bit satellite security system in GEO orbit requiring 4×10^{16} years has similar security strength to a 112-bit brute force attack on a 112-bit cipher (which requires 8×10^{17} years) with a 200Meps device. Hence in all cases fewer bits are required in an embedded satellite security system than that in a terrestrial embedded system.

For many ciphers, there are more powerful attacks such as differential cryptanalysis. Differential cryptanalysis is a chosen plaintext attack. Here specific plaintext–ciphertext pairs are used and the differences between plaintexts are used to correlate with differences in the corresponding ciphertexts in order to deduce characteristics of the key. This type of attack broke a 56-bit key DES cipher (refer to Chap. 6) in 1990 using only 200 plaintext–ciphertext pairs (Biham and Shamir 1990). Many other attacks have also been announced. Although DES was standardized in 1977, in FIPS PUB 46, the more secure variant supporting longer keys, triple DES (3DES or TDEA) was not officially standardized in ANS X9.52 until 1998.

Attack time is typically measured using the fastest attack algorithm with current technologies. With this approach key size can be determined according to specifics of the embedded system. However, the risk with this technique is that future algorithms may be developed to crack the security more efficiently. Thus often designers will use higher levels of security than necessary to ensure adequate security for the lifetime of the key. However, in some embedded systems such as a microsatellite where mass, cost, and energy dissipation are crucial, it may not make sense to use more than 112-bits of security since it is equivalent and slightly stronger than the current 128-bit standards. For example, using more than 112-bits of security will cost more area and energy dissipation. Since embedded systems are heavily constrained, it becomes necessary to analyze whether security strength is adequate or sufficient for the current and future technologies, since one can rarely afford to overdesign the system.

3.5 Key Storage and Authentication Issues

In general keys should be stored in encrypted form and/or in secure memory. This is especially important in embedded systems where attackers may be able to perform memory dumps. For example, researchers in Shamir and Van Someren (1998) found that even within gigabytes of memory, keys stored in the clear could be found. This was largely due to the randomness property of keys which differentiated them from other data such as program data and code. Alternatives for key storage include detachable devices, spreading the key out over more of the program, or using the key as a constant in a program and encrypting the program. Keys may also be split into parts. Each part may be stored on a separate device or different system location. For example, half of a key was stored on a chip card and the other half in a terminal. In this way if the chip card was lost the entire key would not be compromised (Schneier 1996).

Assurance of key integrity and authenticity is also important. This includes ensuring that the key came from an authentic source and is in an appropriate format. The integrity of keys can be authenticated using MACs or digital signatures. Other general checking includes that asymmetric keys have the correct mathematical properties and do not represent known weak keys, etc.

Key escrow also known as key recovery allows third parties gain access to keys, which normally are owned by just one person. For example, it may be necessary to decrypt data due to a court order. This is quite a controversial area involving many privacy and trust issues. Readers can refer to Abelson et al. (1998) for more discussion. However, consider the storey of a colleague who was well versed in cryptography. He had unexpectedly passed away and the family was left with his laptop where he had stored all his important information including lawyers, contacts, accounts, etc. Unfortunately all his laptop data was securely encrypted and no one knew the password. Hence key escrow may be an important consideration. An alternate approach for key escrow is analogous to house keys. Alice can store her keys on a card or device. Then Alice can give the device to Bob until she returns. Even if Bob attempts to use the key on her computer, it will be audited or a log of this activity will be apparent meanwhile Bob cannot learn the key assuming it is not output from the device.

3.6 Key Types

Typically there are many different types of keys used in any system. For example, there are signing keys, signature verification keys, authentication keys (such as MAC keys), encryption keys, key wrapping keys, key transport keys (or key encrypting keys), RNG keys, symmetric master keys, key agreement keys, ephemeral key agreement keys, and authorization keys (Barker et al. 2007). There are many different types of keys; however, managing all the keys is often quite complex. On the other hand, using the same key for more than one cryptographic process may weaken the security provided by the protocol.

For example, often there are two master keys for a security coprocessor: one for encrypting symmetric keys and one for encrypting asymmetric keys. These master keys are typically stored in a tamper-resistant storage module. No other key can encrypt the master keys, thus they are often referred to as clear keys (since they are stored in the clear in secure memory).

In addition to the value of the key, there is other information which needs to be stored with the key. For example, the owner of the key, the lifetime of the key, the key identifier, the specified uses of the key, authorized users of the key, integrity checksums on the key, and other information can be stored. Often this information is referred to as a control vector which identifies the key type, what other keys this key can encrypt, and whether the coprocessor can export this key or not and allowed uses for this key. In general to minimize risk a key should not be used for several purposes, since if it is compromised the implications would be larger.

One of the major problems with passwords, which are a type of key, is the lack of randomness. For example, in order to secure password storage on a computer, typically the hash of the password is stored, not the password. When a user enters their password the computer hashes it and compares it with the stored value and if it matches allows the user to log in. However, an attacker who obtains a list of hashed passwords can use a dictionary attack to determine the passwords. This attack is viable since users normally do not choose random types of passwords. Hence in order to build some resistance to attacks into the system, the notion of "salt" was developed. Salt is a random string, concatenated with the password and input to the hash function. The output of the hash is then stored along with the salt. An attacker with access only to the hashed values now has to guess all possible random salt values for each password guess in order to attack the system, thus making the attack more difficult. Also this allows two system users with the same password to be supported. The concept of using "salt" is analogous to an initialization vector, which is discussed in Chap. 6, which allows two identical messages to encrypt to two different ciphertexts. The salt, like other values such as a nonce, initialization value (IV), challenge, etc., is often used in different protocols yet do not have to be protected. These random values can often be transmitted or stored in the clear unlike keys and secrets.

3.7 Trusted Platform Module (TPM)

In general, a trusted platform is one that is secure from attack such as software attacks, or physical attacks, etc. Ideally even if the platform is stolen it should be secure. Ideally key storage should therefore reside in a trusted platform. Additionally computations which access these keys or confidential data should also be trusted. Examining a typical simple computing system today, each processor has on-chip cache as well as off-chip memory. Data typically resides in these memory modules and protecting their integrity is of utmost concern in security applications. There are several approaches to protecting the data. For example, encrypted data may be stored in memory; however, key management becomes important in this case. Alternatively storing data in a trusted platform may be useful. Alternatively ensuring data integrity, specifically ensuring that data has not been changed intentionally may be of interest in memory outside of a trusted platform (see discussion of hash trees in Sect. 7.3).

TPMs are chips that have been designed to securely store keys, certificates, and passwords. An example of a TPM and its content is illustrated in Fig. 3.6. The TPM has the functionality to generate keys (symmetric and public keys) (de)encrypt, create digital signatures, etc. It can be used to securely store platform configuration information, perform password or authorization checks, generate session keys, and encrypt data to be stored on other devices such as flash memory and disk drives. The TPM specification addresses authentication and attestation (proving a platform is trustworthy and has not been breached). Remote attestation is also possible allowing

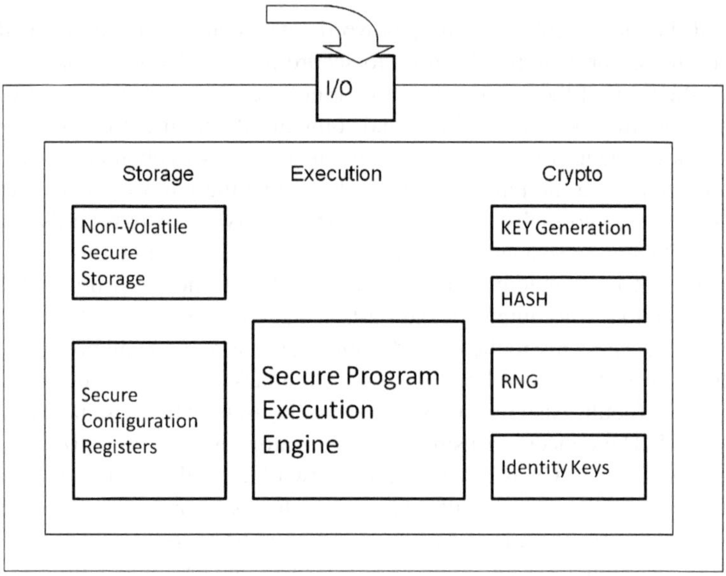

Fig. 3.6 Example of TPM contents

the TPM to trust information from other TPMs. It is believed that since the TPM is implemented in hardware, it therefore has higher security (or better protection from software attacks) than by providing this functionality in software. For example, even if the device is stolen it is believed that the security in the TPM makes access to keys inside the device harder to attain.

The TPM chip typically contains protected storage and cryptographic engines. Volatile and nonvolatile memory is provided typically for key storage, certificate storage, and password storage. As well it contains a RNG (typically a FIPS 140-2 certified generator), hash modules (such as SHA-1), as well as private and public key functionality (such as a RSA engine and key generation module). The incorporation of elliptic curve cryptography as well as more secure hashes is planned. For example, the AT97SC3203S (Atmel 2008) contains ROM, EEPROM (for RSA keys), 8-bit RISC CPU, SRAM, hardware asymmetric cryptographic engine, RNG, GPIO interface, and physical security circuitry.

There is a platform configuration register (PCR) which holds hash values of the platform configuration information. The TPM will sign the PCR state with a TPM key to verify the state of the platform at that time. For example, the TPM may hash the state information before the shutdown process for endpoint integrity.

The TPM includes a key hierarchy where the storage root key (SRK) is protected by the TPM along with an endorsement key (EK). The private key from the EK pair can never leave the TPM. It is referred to as a nonmigratable key and represents a unique identifier for the TPM chip. For example, when the TPM is manufactured by Company *A*, the RSA key pair or endorsement key may be created by EK generation service with Company *A* (Kinney 2006). Once this EK key pair is created for the TPM chip it cannot be modified or changed. There is a X509 certificate used to authenticate the public key of the EK. The TPM vendor can produce the certificate or the TPM vendor gives the OEM the information sufficient to produce the certificate. The SRK is the crypto key of the owner. Both the EK and SRK are the most important keys within the TPM. These can be classified as a migratable key (can be used outside of TPM or migrate to other TPM) or a certifiable migratable key (CMK). The CMK can be migrated but with control and the TPM can certify properties of this key (Hardjono and Kazmierczak 2008). There are other keys also which may be used within the TPM or stored externally. There are different types of keys supported such as signing keys, storage keys (encrypts keys for storage), and other keys.

The TPM specification was developed as an open specification by the trusted computing group, a nonprofit corporation with international membership and industry participation from semiconductor, system software, networking, and service provider companies (TCG Web site). One of the objectives was to provide global interoperability, made more difficult with proprietary approaches. These chips are manufactured by various companies including ATMEL and STMicroelectronics and others. Application software and application software development environments are also available for the TPM.

3.8 Network on Chip Security

Although the TPM was generally developed for PC platforms, the idea of having trusted hardware is a universal theme which is applicable to embedded systems. A alternate approach to a TPM chip, is on-chip key storage and on-chip secure computations. Specifically, place the trusted components on the same silicon as other processing components in the embedded system. This approach makes tampering much more difficult since one cannot probe on-chip busses as easily as off-chip busses. Even side channel analysis is more difficult when off-chip transfer of critical data is not possible, as discussed in Chap. 8. This section will discuss one proposal for supporting on-chip trusted hardware (Gebotys and Gebotys 2003). Other on-chip approaches such as (Suh and Devadas 2007) have been developed, however they generally allocate a single core which can only be used to execute security functions unlike the more flexible approach to be described next.

On-chip keys must be protected otherwise malicious software downloaded to the chip may be able to read keys and send their values to remote users on the larger external network. In general, it will be of primary importance to protect the secure IP cores from un-trusted software and prevent exposure of keys outside of the system on chip (SoC) and even within the SoC communication network. A secure IP core is a core which computes with these keys, typically performing some cryptographic algorithm. Unlike TPM, where there is one secure chip which performs all cryptographic functions, this proposal suggests several modules or cores which may perform security functions. This avoids any security bottleneck in the system.

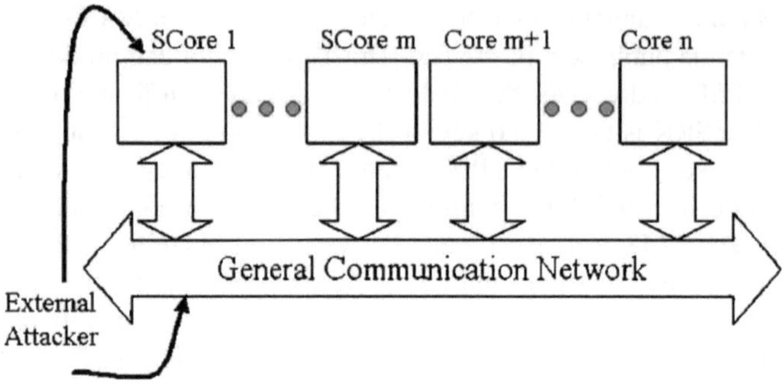

Fig. 3.7 Network on chip with security cores, *SCore i*

Additionally it may be possible that the secure cores could also perform normal computations thus supporting a wider range of functionality.

In a NoC there may be ordinary cores, *Core i*, as well as security cores, *SCore j* as shown in Fig. 3.7. At the network level, a security wrapper supporting symmetric key cryptography is implemented for communication between IP cores on the NoC which are involved in security applications. A security wrapper for each core and a central key-keeper core are proposed to ensure that unencrypted keys do not leave any core and the NoC itself. Security for NoCs has not previously been studied; however, this approach has advantages for IP core vendors providing further protection of not only their hardware IP but also their software which will run on their IP core.

The NoC will be required to support security for several reasons such as (1) wireless communication or IP-enabled applications requiring user authentication, encryption; (2) product authentication (preventing counterfeit products); (3) NoC chip authentication; and (4) even IP-core authentication on the NoC itself. We will assume that NoCs have m secure cores (or *SCores* in Fig. 3.1) and another $n - m$ other cores, where $m \geq 1$. A secure core is defined as a hardware IP core which can execute one or more security applications (such as encryption, authentication, key exchange, etc.) and it may also be able to execute other general nonsecurity application as well. This NoC scenario suits current secure IP cores on the market today as well where authentication and encryption requires several secure cores such as an AES core, a SHA-1 core, etc. The NoC should in either case be resistant to external attacks which can obtain data from the communication network (through external I/O pins attached to the communication network or EM radiation from it), shown as bottom arrow in Fig. 3.7, or through power analysis attacks on individual core power pins (see top arrow in Fig. 3.3). Key distribution is important, since keys are updated from time to time. Thus the NoC will require a mechanism to allow downloading of the key onto the chip and into the memory of the secure IP core. Furthermore malicious software running on a regular core (*Core*) could attempt to acquire keys or send false keys to security cores. Protection against malicious software and untrusted cores must be supported in the NoC. The proposed security methodology at the network level will be described next. It supports key protection and secure transmission on the communication network.

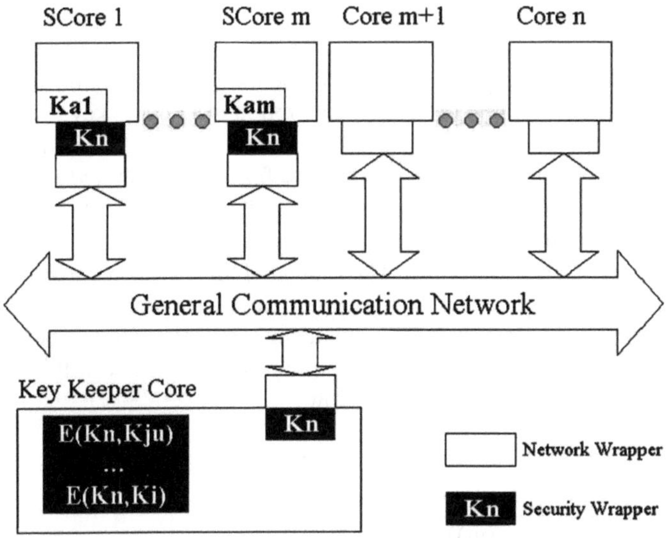

Fig. 3.8 Key-keeper core in silicon with NoC

The network level methodology for security is based on symmetric key cryptography. A secure core can be an AES core, a SHA-1 core, a prime field operation elliptic curve core, etc… Each secure core (i) has its own security wrapper, see shaded (*Kn*) boxes in Fig. 3.4. The security wrapper is located between the secure core and the network wrapper (see Fig. 3.8. The keys stored in the security wrapper are (1) the network master key, *Kn*, and (2) the working key *Kn'* (generated from the master key *Kn*), both stored in nonvolatile memory within the security wrapper and (3) message authentication code (MAC) keys, *Kmac i* and *Kmac k* (key keepers MAC key), are also stored. Since *SCores* may often receive keys from the key keeper (i.e., user keys for encryption, or authentication), the key keeper's MAC key (*Kmac k*) is stored for convenience in the security wrapper for quick authentication. If sufficient memory is available and security requirements are high, it is possible that MAC keys of other *SCores* could be also stored in the security wrapper. Inside the secure core, an authentication key (*Kai* in Fig. 3.8) is stored in nonvolatile memory, for use in core authentication and core software authentication. The security wrapper performs a number of functions including (1) encryption of messages from the secure core to the network $E()$, (2) decryption of messages from the network to the secure core $D()$, (3) create a hash of a message $H()$, and (4) create a MAC *MAC*(). A key-keeper core in the NoC (see Fig. 3.8) is responsible for secure key distribution on the NoC. This core is also a secure core (and therefore has its own security wrapper). It stores encrypted keys, such as encrypted keys for individual applications, encrypted user private keys (*Kju*) (i.e., authentication or encryption keys to be used for decryption of audio, video, email, ebanking, or signature generation or authentication of outgoing messages, etc.), other users public keys (*Ki*) (i.e., for encryption of outgoing audio, video, email, signature verification, etc), and *SCore*

MAC keys (*Kmac i*). The key keeper is also responsible for updating the master network key (*Kn*) at random times using a symmetric key exchange protocol. The following terminology will be used to describe the functionality of the security in the security wrapper: $E_a(m)$ represents encryption of m using key a; $D_b(c)$ is decryption of c using key b; $H(m)$ is a x-bit hash of a message m; and $MAC_k(m)$ is a n-bit MAC of message m using key k. Also for *SCore i*, let *tcnt j* represents the count of how many messages were sent to *SCore j*, *rcnt j* represents how many message were received from *SCore j*, and ',' represents concatenation. Consider sending a message from *SCore i* to *SCore j*. The following three steps are involved (1) generate a working key (since it is well known that one should not use the same key to transmit different messages), (2) provide authentication (step 2 below) and (3) encrypt message and signature. These three steps are performed within the security wrapper of *SCore i* and are illustrated below:

Protocol 3.1. Message transfer *SCore i* to *SCore j*

1. *SCore i*: $H(Kn, tcnt\,j) = Kn', tcnt\,j{+}{+}$
2. *SCore i*: $MAC_{Kmac\,i}(H(m)) = s$
3. *SCore i*: $E_{Kn'}(m, s) = c$
4. *SCore i* $\rightarrow$ *SCore j*: c
5. *SCore j*: $Kn' = H(Kn, rcnt\,i), rcnt\,i{+}{+}$
6. *SCore j*: $D_{Kn'}(c) = m, s$
7. *SCore j*: $MAC_{Kmac\,i}(H(m)) =?= s$

Only *SCore*'s have access to the MAC keys. Hence an untrusted core cannot obtain the MAC keys. It is assumed that the *SCores* only run trusted software (and do not have any malicious codes). The security methodology tries to minimize the chance that the untrusted cores may be able to obtain the network master key, through changing it at random times. However, even if the master network key is obtained, the working keys and MAC keys would be difficult to obtain. This basic protocol, for exchanging messages, supports key transfer from the key keeper to *SCores*, the updating of new authenticated keys which have been sent to the portable device, and others. For example, using this basic protocol new authenticated user keys can be authenticated using authentication SCores, decrypted using AES SCores, and reencrypted with working network keys and sent back to the key keeper for storage. Each transmission on the network ensures that unencrypted keys or messages are never sent between *SCores*. Hash tags (*tcnt j* on line 1 and *rcnt i* on line 5 of Protocol 3.1) ensure that the data has not changed during transmission. MACs attached to encrypted message are used to verify that the encrypted message came from a trusted source.

The internal SCore authentication key (key located in *SCore*, not in security wrapper) which has not been used so far is important for a number of security purposes. For example, secure cores could receive new software upgrades (wirelessly) from their IP core vendors. Since only trusted software is assumed to be running on security cores, this new software would have to be authenticated. This is performed by sending to the NoC executables authenticated with the internal private authentication key of the secure core, *Kai* in Fig. 3.8, which only the vendor of the IP core can do since they are the only ones who know the secure core's private key, *Kai*. Again this upgrade can be performed remotely or wirelessly. Authentication (for example with crypto checksums to ensure that all security code executables are authenticated) could occur within the *SCore* (not the security wrapper) to safeguard the private *SCore* authentication key. Additionally the vendor of the IP core can use the core's private authentication key to prevent illegal use of the core in unauthorized NoCs (which have not paid license fees, etc. for use of the core). The core vendors would create an activation key *(K_activation)* for their core to be used in the NoC from the cores *Kai* and the IDs of the other cores in the network. Legal uses of the core would receive this activation key from the core vendor and store it encrypted in the key-keeper core. For example, on reset of the NoC the secure core would receive an activation key (*K_activation*) from the key keeper and verify this key by calculating its own security function ($F()$, which would be encryption for an encryption core, etc.) with its private key (*Kai*) and the concatenation of all other core IDs (obtained from polling all other cores) on the NoC. For example, if $K\_activation = F(Kai,\ core\ ID\ 1||\ core\ ID\ 2||\ldots core\ ID\ n)$, then the secure core will function properly else it will not operate and shut down permanently. Standards would have to exist so that each core would have a core ID number.

Ensuring confidentiality of communications on the on-chip network and authenticity of executables which utilize private keys, a secure SoC environment for security applications is possible. Design for security involves secure design of hardware cores, secure algorithm design, and secure network protocols, all important for NoCs of the future. More details on this proposal can be found in (Gebotys and Gebotys 2003) and (Gebotys and Zhang 2003).

3.9 Summary

There are many interesting ways of generating randomness for use in keys. In addition to strength of the key, the key lifetime, key updates, and key management are important considerations for any system. It is typically not sufficient to assume that one random key will remain secure for the lifetime of the embedded system. The attack on the Xbox (Huang 2002), discussed in Chaps. 2 and 8, is a good example. Here only one fixed key was used in all stations, with no consideration of key updating or any key management. It was likely assumed that the key was so secure that it would never have to be changed, since the same key value was used in all Xbox systems. On the other hand, key management does not come for free, hence key properties and computations should be optimized for security objectives of the specific embedded system.

In addition to standard security problems which include finding truly random numbers or seeds of sufficient entropy and secure storage for keys, embedded systems often require extra security such as the protection of IP within a design. This is especially the case for FPGAs where design IP can be loaded on and off the device easily through JTAG ports in order for configuration data (bitstream) to be stored or loaded. The protection of this IP as well as ensuring its integrity may be crucial to the embedded system. Embedded systems vary greatly and so do their security requirements. Establishing and maintaining secure strong keys and determining safe storage for these keys are challenges for embedded systems. Functions and protocols which utilize these keys will be described in Chap. 4.

References

Abelson H et al (1998) The risks of key recovery, key escrow, trusted third party and encryption. Ad Hoc Group of Cryptographers and Computer Scientists. http://www.cdt.org/crypto/risks98/

Atmel (2008) AT97SC3203S trusted platform module, Vers 1.2, SMBus Two-wire interface summary document. http://www.atmel.com/dyn/resources/prod_documents/5132s.pdf

Barker E, Barker W, Burr W, Polk W, Smid M (2007) Recommendation for key management – Part 1: General (Revised). NIST Special publication 800–57

Biham E, Shamir A (1990) Differential cryptanalysis of DES-like cryptosystems. In: Advances in cryptology, CRYPTO '90. doi = 10.1.1.56.1367. Springer, Berlin, pp 2–21

Dichtl M, Golic JDj (2007) High-Speed true random number generation with logic gates only. In: Paillier P, Verbauwhede I (eds) CHES 2007, LNCS, vol. 4727. Springer, Berlin

Gebotys C, Gebotys R (2003) A framework for security on NoC technologies. In: IEEE proceedings of annual symposium on VLSI

Gebotys C, Zhang Y (2003) Security wrappers and power analysis for SoC technologies. In: Proceedings of ACM/IEEE/IFIP international conference on hardware/software codesign and system synthesis, pp 162–167

Hardjono T, Kazmierczak G (2008) Overview of the TPM key management standard. TCG presentation. http://www.trustedcomputinggroup.org/

Huang A (2002) Keeping secrets in hardware: the Microsoft xbox study. MIT AI laboratory, AI Memo 2002–2008. http://www.ai.mit.edu

Jun B, Kocher P (1999) The Intel random number generator. Cryptography research, Inc. White paper prepared for Intel corporation

Kinney S (2006) Trusted platform module basics: using TPM in embedded systems. Newnes, Oxford.

Kocher PC (1999) Leak-resistant cryptographic indexed key update. US Patent 6,539,092, 2003

Lofstrom K, Daasch WR, Taylor D (2000) Identification circuit using device mismatch. In: Proceedings of ISSCC 2000

Menezes A, van Oorschot P, Vanstone S (1996) Handbook of applied cryptography. CRC, Boca Raton, FL

Pappu R (2001) Physical one-way functions. PhD thesis, Massachusetts Institute of Technology

Schneier B (1996) Applied cryptography. Wiley, New York

Shamir A, Van Someren N (1998) Playing hide and seek with stored keys. In: Financial cryptography 1998, LNCS, vol. 1648. doi: 10.1007/3–540–48390-X 1999. Springer, Berlin, pp 118–124

Skoric B, Tuyls P, Ophey W (2005) Robust key extraction from physical unclonable functions. In: Proceedings of the applied cryptography and network security conference 2005, LNCS, vol. 3531

Stinson DR (2006) Cryptography, 3rd edn. Chapman and Hall/CRC, London/Boca Raton, FL

Suh GE, Devadas S (2007) Physical unclonable functions for device authentication and secret key generation

TCG Web site. http://www.trustedcomputinggroup.org/

Vasyltsov I et al (2008) Fast digital TRNG based on metastable ring oscillator, CHES 2008, LNCS, vol. 5154, IACR. Springer, Berlin, pp 146–163

Chapter 4
Using Keys

Diffie imagined two strangers meeting via the Internet, and wondered how they could send each other an encrypted message.

(Singh 1999)

As discussed in Chap. 3, keys must remain protected from would be attackers in order to provide security. However, they must be accessed by trusted users or devices in order for security computations to be performed. These security computations include protocols not only for communicating or establishing temporary keys but also for communicating confidential messages, signatures, etc. This chapter will introduce the basic protocols which use keys to transport confidential messages, establish keys, to authenticate messages, or to create a digital signature. General protocols are introduced first for both shared key or symmetric keys and public key security. Detailed mathematics are also introduced for public key cryptography, specifically integer factorization, and discrete logarithms. Elliptic curve cryptography (ECC) is covered in a separate chapter. Implications for embedded systems will be discussed at the end of Chap. 5 so that the techniques in this chapter are compared to that of ECC.

In cryptographic theory, there are two general types of keys: symmetric keys and asymmetric keys. The symmetric key schemes utilize the same key value at both ends of the communication channel. Asymmetric key schemes are referred to as public key schemes, where a public key is paired with a private key and both are used in the application. In this case, the owner of the public and private key pair must always maintain confidentiality of the private key, but the public key can be made available to anyone. For example, in order to communicate over a channel with another party only the public key of both parties is required. There are different approaches to implementing public key schemes such as those based upon integer factorization (such as RSA), discrete logarithms (DSA, ElGamal), and elliptical curve discrete logarithms (ECC). The later will be covered in Chap. 5 since it is viewed as most efficient and thus important for embedded systems. It is assumed in all protocols covered in this chapter that an attacker is capable of seeing data transferred from one user to another user in the communication channel (such as an eavesdropper). Also an attacker is able to substitute the data being transmitted

C.H. Gebotys, *Security in Embedded Devices*, Embedded Systems,
DOI 10.1007/978-1-4419-1530-6_4,

or stop a transmission or create a transmission to any user. A box with latches or a lock which can hold messages (Singh 1999) will be used in this chapter to illustrate several concepts.

Key transport refers to the secure transfer of a key from one party to another party. Key establishment or agreement refers to two or more parties separately deriving the same secret key without other parties being able to derive or obtain this key. In addition to Alice and Bob who wish to communicate, there is typically a third party involved which is known as a key distribution center, or KDC, for symmetric key operations. In general, there may also be a key translation center, or KTC, which for example receives an encrypted session key from Alice, proceeds to decrypt the session key with Alice's key, reencrypts the session key with Bob's key, and finally sends the reencrypted session key to Bob. There may also be "key access" servers which operate similar to a KTC by sending encrypted keys to authenticated users. In asymmetric key operations, the third party is referred to as a certificate authority (CA). This third party must be a trusted source, in other words it should never be compromised and any user should not be able to masquerade as a KDC or CA.

Mathematical definitions and terminologies used in this chapter are given below:

$x \bmod n$	It is equal to the remainder, r, from dividing n into x ($x = qn + r$, q is quotient, r is remainder).
Z_n	The set of integers between 0 and $(n - 1)$, or the integers modulo n (e.g., Z_2 is the binary field), where addition, subtraction, and multiplication operations are performed modulo n.
Z_n^*	When n is prime, Z_n^* is known as the multiplicative group of Z_n and is equal to the set of integers between 1 and $(n - 1)$.
u^{-1}	It is the (multiplicative) inverse of u, $u \in Z_n$ where $(u\,u^{-1}) = 1 \bmod n$ (if u^{-1} exists then u is said to be invertible, u is invertible if $gcd(u, n) = 1$).
Order	The order of u ($u \in Z_n^*$) is the smallest t such that $u^t = 1 \bmod n$. The order of Z_n^* is the number of elements in Z_n^* specifically $\lvert Z_n^* \rvert$.
$w \mid a$	$w \mid a$ or w *divides* a means there exists an integer c, where $a = wc$ (*or* w is a factor of a *or* w is a divisor of a).
$gcd(a, n)$	*Greatest common divisor* (gcd) is the largest positive integer that divides both a and n.
Relatively prime	a is relatively prime (or coprime) to n means $gcd(a, n) = 1$.
$\phi(n)$	Euler's (totient) function is the number of integers in $[1, n]$ which are relatively prime to n.

Generator	If $u \in Z_n^*$ and order of u is $n-1$, then u is called a *generator* (or primitive element of Z_n^*). Alternatively if $u \in Z_n^*$ and has order $\phi(n)$, then it is a generator. For example if the powers of u create all elements in the set Z_n^*, then u is a generator. Consider $u = 3$, $u \in Z_5^*$. u is a generator since the first four powers of u are: $3^0 = 1$, $3^1 = 3$, $3^2 \; mod\, 5 = 4$, $3^3 \; mod\, 5 = 2$). Note that if Z_n^* has a generator, it is said to be cyclic
Order ($q \; mod\, p$)	g has order ($q \; mod\, p$), then g is a generator of the unique cyclic group of order q in Z_p^*, where p, q are both prime.

Terminology:

$L_k()/U_k()$ function to lock (or close the lock associated with key k)/unlock with a key k

$E_k()/D_k()$ function to en/decrypt using key k

Pi/*Si* Public/secret key for user i

Alice→*Bob*:D Alice sends data D (possibly defined by computations) to Bob. If D is an operation, such as $E_k(m)$, then it is assumed that Alice performed the operation and sent the result to Bob.

Bob: A *opn* $B = C$ This defines an operation, *opn*, performed by Bob on two operands (A and B) whose output will be called C. All operations are performed on the left-hand side of the equals sign. Any operations shown on the right-hand side of the equals sign is used just for definition and no operation is performed (unless it is a simple operation such as a Hash that is on the right-hand side of the $=?=$ operator).

KDC →*Bob:* $T = D$ *opn* E This indicates that T is sent to Bob, but this operation, *opn*, is not performed in the protocol by Bob. (It was likely previously performed at some other time by KDC in this example.)

w, z Data w and z are concatenated

$H(m)$ is a one-way hash function operating on a data m.

$=?=$ refers to checking that the left-hand side is equal to the right-hand side.

$S_k(m)$ is a signature on message m made with key k.

4.1 No Shared Keys

To start a discussion of using keys, we can first begin by assuming that two users have randomly selected their keys and now they wish to confidentially send messages to each other. Although they have no shared keys, they can still use a protocol which provides confidentiality. This protocol will first be described using a double latched box to illustrate the basic principles. Finally the general protocol will be presented.

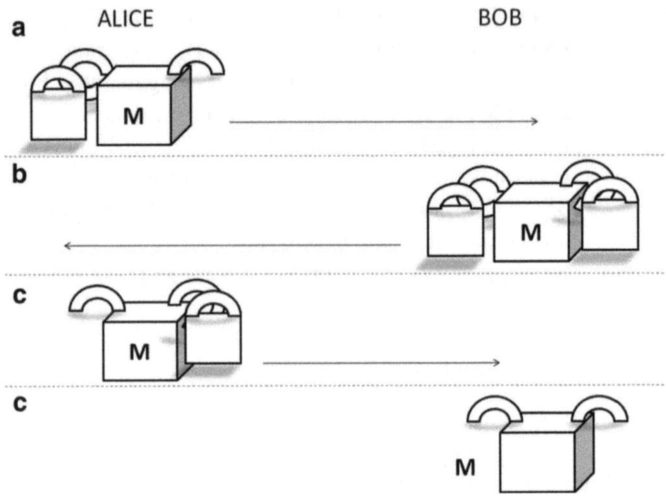

Fig. 4.1 Illustration of Shamir's no-key protocol using a double latched box

Perhaps the simplest method of visualizing the transfer of confidential information between two parties is using a box which has a double-latched lid. The message to be sent (***M*** in Fig. 4.1a) will be placed in the box. For illustration purposes only, one can assume there are two latches, one on the right and one on the left of the box, as shown in Fig. 4.1a. Assume that each user has her/his own unique secret key and lock. Also assume that each user does not know or have a copy of the other user's secret key. Assume that the box is locked if there is at least one lock on either latch. For example, in Fig. 4.1a, the box is locked by a lock on the left and the latch on the right is open. In Fig. 4.1b, the box is shown with a lock on both the left and a lock on the right. In Fig. 4.1c, only the right latch has a lock. Also assume that the box is unlocked only if both the latches do not have locks on them (as in Fig. 4.1d). Now we can start to discuss the no-key protocol for confidentially transferring a message from Alice to Bob, developed by Adi Shamir, using Fig. 4.1. Alice places a message, ***M***, for Bob into the box. She then attaches her lock on the left-hand side latch, thus locking the box closed, as shown in Fig. 4.1a. She then sends this locked box to Bob, indicated by the first arrow to the right in Fig. 4.1a. When Bob receives this locked box, he places his lock on the right-hand side of the box and sends the box back to Alice, as shown in Fig. 4.1b. Next Alice removes her lock and returns the still locked box back to Bob in Fig. 4.1c. Finally Bob can open the box by unlocking his lock using his secret key in Fig. 4.1d. Bob can then finally access the message found in the box from Alice. This scheme illustrates how two secret unshared keys together with respective locks (one owned by Alice and one owned by Bob) can be used to transfer a message confidentially over a channel between two parties.

In an electronic implementation of this scheme, the process of locking/unlocking the box with a secret key can be represented by encryption/decryption with a secret key. For example, assume that the process of locking a lock (paired with a key k) on the latch of the box which contains message m can be represented as $E_k(m)$. Similarly assume that unlocking the lock with key k can be represented as decryption or $D_k(m)$. The protocol is shown below:

Protocol 4.1. No-shared key message transfer

(1) *Alice* $\rightarrow$ *Bob*: $E_{Ka}(m)$
(2) *Bob* $\rightarrow$ *Alice*: $E_{Kb}(E_{Ka}(m))$
(3) *Alice* $\rightarrow$ *Bob*: $D_{Ka}(E_{Kb}(E_{Ka}(m)))$
Bob: $D_{kb}(D_{Ka}(E_{Kb}(E_{Ka}(m)))) = m$

The no-shared key message transfer protocol starts with Alice sending Bob her message encrypted with Alice's own secret key in step (1) of Protocol 4.1. Then Bob encrypts the data with his own secret key and send this doubly encrypted data back to Alice in step (2). Alice receives this data and decrypts with her secret key (which she originally used to encrypt the original message). She then sends the resulting data to Bob who decrypts using his private key which finally reveals the original message Alice wished for Bob to receive in step (3).

Under this approach the encryption must provide for the fact that placing a "lock" on the latch can be removed later after another lock has been placed. This implementation requires that the following property must hold:

$$E_{k1}(E_{k2}(m)) = E_{k2}(E_{k1}(m)).$$

If the property holds then the resulting operation performed by Bob in step (3) becomes $D_{kb}(D_{ka}(E_{kb}(E_{ka}(m)))) = D_{kb}(D_{ka}(E_{ka}(E_{kb}(m)))) = D_{kb}(E_{kb}(m)) = m$. In general, this condition is not satisfied by most encryption algorithms (such as 3DES, AES, etc.). However, this property is satisfied by modular exponentiation where encryption with a key Ka, takes the message, m (an integer) and raises it to the power Ka, modulo p (where p is a large prime, and $m < p$, $Ka < (p-1)$, and Ka is coprime to $(p-1)$). Decryption involves raising the value to the inverse of Ka modulo $(p-1)$. When the data being transferred is a key (which is m in the above protocol), it is called Shamir's no-key protocol (see Protocol 12.22 in (Menezes et al. 1996)). It is interesting to note that no-key establishment or communication is required, so Alice or Bob at any time can use whatever secret key (s)he likes. Three passes are required in this protocol, thus it is typically intended to be used to establish a session key between Alice and Bob. Once the session key is established more efficient single pass communication can then be used to transfer data between parties. For example, Sect. 4.2 will discuss protocols for transferring data when there is already a shared key among two parties.

4.2 Using a Preexisting Shared Key

The assumption of users or devices already having established shared keys may be appropriate in many embedded systems (unlike the case of typical networked systems). For example, it is often possible to establish keys in hardware devices or chips before they are shipped out or put into their embedded systems. Keys may be predetermined and embedded into the die during fabrication (if it is appropriate for the product to share the same fixed key) or programmed later after the chip is fabricated by storing the value into the chip. Alternatively keys may be defined later when the system is in the field. For example, satellites may have their keys loaded into the system immediately before launching. This is preferred since satellites are often launched in foreign countries where high security is required (e.g., military or government satellites). Loading keys earlier into the satellite may be a cause for concern since it allows for possible prelaunch attacks on the satellite. Attacks may consist of dumping the memory from the satellite device in order to extract keys well before launch time. Alternatively measurements of the side channel during attempts to authenticate access into the satellite might also reveal embedded keys. Thus key establishment algorithms may not always be necessary in many embedded systems. Of course in many cases, it may still be important to use temporary session keys to further protect the master keys embedded in the device and these temporary keys may be automatically derived from the master key. This section will assume that keys are already shared between relevant parties. Protocols for transferring temporary session keys between two parties will be described.

Protocol 4.2. Shared key message transfer

(1) *Alice*: $E_k(m) = c$
(2) *Alice* $\rightarrow$ *Bob*: c
Bob: $D_k(c) = m$

In the shared key message transfer protocol, Alice encrypts her message with the shared (or session) key in step (1) of Protocol 4.2. She then transfers the encrypted message to Bob, who decrypts it with the same shared key to obtain the message. If the message is actually a key then this is referred to as a key wrapping protocol. A key may be transferred this way to Bob to be used in all future communications with Alice. This provides a method for updating session keys. Of course Bob needs to know that the encrypted key came from Alice. Authentication will be further discussed in Sect. 4.3, where better procedures for updating keys or sending encrypted messages are presented.

4.3 Using Keys in Conventional Crypto

In the 1980s, the banks were having difficulties managing keys for transferring data in between the bank offices (Smith 2002). This prompted the need for a key distribution center. A variation of the original proposal, developed by Branstad (1987), eventually lead to standardization of the key distribution center in ANSI X9.17. This standard was later updated in FIPS 171 and now both are retired (see http://www.itl.nist.gov/fipspubs/withdraw.htm) though used in routers and modems (e.g., http://www.msen.com/fievel/mmill/X9.17.html). KDCs were used to generate temporary keys for communication between a pair of branch offices. In the examples which follow, we use two individuals or devices who wish to communicate, again named Alice and Bob. It is important to note that in general Bob may be a server such as a mail server and the KDC may be an authenticator server.

In conventional cryptography, there is always a trusted third party. For example Alice shares a secret key, *Ka*, with a key distribution center, KDC. Bob also shares a secret key, *Kb*, with the KDC. When Alice wishes to communicate with Bob, she requests a session key from the KDC for this communication and the KDC then sends encrypted session keys to both Alice and Bob. In the KDC session key transfer protocol, we illustrate the request to the KDC using *'from Alice'*, *'to Bob'* to indicate the request from Alice to talk to Bob (e.g., the two users involved in the communications) in step (1).

Protocol 4.3. KDC session key transfer

(1) *Alice* $\rightarrow$ *KDC*: *'from Alice'*, *'to Bob'*
(2) *KDC* $\rightarrow$ *Alice*: $E_{Ka}(k)$
Alice: $D_{Ka}(E_{Ka}(k)) = k$
(3) *KDC* $\rightarrow$ *Bob*: $E_{Kb}(k)$
Bob: $D_{Kb}(E_{Kb}(k)) = k$

The encrypted session keys, for example $E_{Ka}(k)$ and $E_{Kb}(k)$, are sent to Alice and Bob by the KDC in step (2) and (3), respectively. Each user decrypts them using their own keys *Ka* and *Kb*, to obtain the session key, k.

There are many problems with the simple KDC session key transfer protocol. We will examine one problem as follows: an attacker, Eve, could modify Alice's request in line (1) to the KDC by substituting Eve for Bob, specifically as *'from Alice'*, *'to Eve'*. Thus Eve, pretending to be Bob, could then read all of Alice's messages. In order to prevent this attack, Alice needs to verify that the data from KDC is what she had asked for. Hence we can modify the scheme by having the KDC send the users name back to verify the names were not changed as follows:

Protocol 4.4. Improved KDC session key transfer

(1) *Alice* $\rightarrow$ *KDC*: *'from Alice'*, *'request to Bob'*
(2) *KDC* $\rightarrow$ *Alice*: $E_{Ka}(k,$ *'to Bob'*$)$
Alice: $D_{Ka}(E_{Ka}(k,$ *'to Bob'*$))=k$, *'to Bob'*
Alice verifies 'to Bob'
(3) *KDC* $\rightarrow$ *Bob*: $E_{Kb}(k,$ *'for Alice'*$)$
Bob: $D_{Kb}(E_{Kb}(k,$ *'for Alice'*$))=k$, *'for Alice'*

In the above modified protocol, Alice can decrypt the message from *KDC* in line (2) to obtain the session key and in addition to obtain the second part of the decrypted message, specifically 'to Bob,' in order to verify that the session key is intended for Bob (not an attacker such as Eve).

However, there are other issues with this protocol such as how does Bob know that the message from the KDC is not being replayed from an attacker who stored this from an earlier point in time. For example, the attacker may have just determined the session keys value somehow and can use it to decipher messages from Bob. The same applies to Alice, specifically is she sure the message in line (2) came from the KDC and is not being replayed? In general, more information in addition to the destination is incorporated into the KDC response. Section 4.3.1 will continue to address these types of forgery problems of KDC.

4.3.1 Needham–Schroeder

The Needham–Schroeder protocol supports communication with a KDC in order to facilitate a session key for Alice and Bob and also to prevent forgery. The protocol incorporates nonces and a challenge–response to prevent replay attacks and forgery. It also differs from the previous protocols presented earlier in that the KDC sends sufficient information only to Alice and does not send any messages to Bob. The simplified Needham–Schroeder protocol is referred to as the Needham–Schroeder shared key Protocol 12.26 in Menezes et al. (1996). The data for Bob which is sent to Alice will be referred to as $Ticket_B$ in the protocol. This also reduces the number of outgoing messages from the KDC.

Protocol 4.5. Simplified Needham–Schroeder Key Establishment

(1) *Alice* $\rightarrow$ *KDC*: *'from Alice'*, *'to Bob'*, *nonce*
(2) *KDC* $\rightarrow$ *Alice*: $E_{Ka}(k, nonce, to\ Bob, Ticket_B = E_{Kb}(k, from\ Alice))$
    Alice: $D_{Ka}(E_{Ka}(k, nonce, to\ Bob, Ticket_B)) = k, nonce, to\ Bob, Ticket_B$.
    Alice verifies nonce.
(3) *Alice* $\rightarrow$ *Bob*: $Ticket_B$
    Bob: $D_{Kb}(Ticket_B) = k, from\ Alice$
    Bob: $E_k(Bobs\_nonce) = Check_A$
(4) *Bob* $\rightarrow$ *Alice*: $Check_A$
    Alice: $D_k(Check_A) = Bobs\_nonce$
    Alice: $E_k(Bobs\_nonce\text{-}1) = Check_B$
(5) *Alice* $\rightarrow$ *Bob*: $Check_B$
    Bob: $D_k(Check_B) =?= Bobs\_nonce\text{-}1$

In the simplified Needham–Schroeder protocol (Menezes et al. 1996), Alice creates a nonce and sends it to the KDC in line (1) in addition to the usual request. In line (2) Alice verifies (e.g., "*Alice verifies nonce*") that the KDC sent her the same nonce that Alice sent the KDC in line (1) so that she can be assured that the message received is not being replayed. In line (3) Alice forwards the encrypted message or $Ticket_B$, to Bob. In line (4) Bob wishes to verify that Alice has some session key and hence sends her a nonce. Bob can also be assured in line (5) that the message sent has not been replayed by completing a challenge–response with Alice and to ensure that Alice does have the other shared session key. Bob still has no way of knowing whether line (3) is replayed and the attacker has learnt the session key. Hence the concept of using timestamps and validity period will be introduced next to resist this type of forgery.

4.3.2 Kerberos

Unlike the past described protocols, the Kerberos authentication protocol (RFC 1510) uses time stamps for freshness and a validity period to indicate the period of time the session key is valid. Freshness is important to ensure that older messages are not being replayed. This helps to prevent further damage from attacks where the session key might have been compromised. The timestamps were proposed by other researchers (Denning and Sacco 1981). The simplified Kerberos protocol [called the basic Kerberos authenticated Protocol 12.24 in Menezes et al. (1996)] also uses the term tickets to describe encrypted copies of a temporary secret. In the Kerberos protocol, the KDC is referred to as the Kerberized server (KS).

Protocol 4.6. Simplified Kerberos

(1) *Alice* $\rightarrow$ *KS*: *'from Alice'*, *'to Bob'*, *nonce*
(2) *KS* $\rightarrow$ *Alice*: $Ticket_B = E_{Kb}(k$, *'from Alice'*, *validity_period*),
$E_{Ka}(k$, *nonce*, *validity_period*, *'to Bob'*)
Alice: $D_{Ka}(E_{Ka}(k$, *nonce*, *validity_period*, *'to Bob'*))=k,
nonce, *validity_period*, *'to Bob'*
Alice verifies nonce and destination.
Alice: E_k(*'from Alice'*, *Alices_Time_stamp*)=*Authenticator*
(3) *Alice* $\rightarrow$ *Bob*: $Ticket_B$, *Authenticator*
Bob: $D_{Kb}(Ticket_B) = k$, *'from Alice'*, *validity_period*
Bob: D_k(*Authenticator*)=*'from Alice'*, *Alices_Time_stamp*
Bob: E_k(*Alices_Time_stamp*)=$Check_A$
(4) *Bob* $\rightarrow$ *Alice*: $Check_A$
Alice: $D_k(Check_A) =? =$ *Alices_Time_stamp*

In line (3) Bob checks that Alice's time stamp is reasonable (according to the standard), and if so, sends back the time-stamp to Alice for verification. The protocol below authenticates Alice to Bob in line (4) and establishes a session key (Menezes et al. 1996).

The Kerberos system was proposed for networks of computers and it makes several assumptions including the clocks are synchronized among all computers. The keys used for individuals were based upon the user's passwords. Thus the compromise of a user's account or computer may be possible. The Kerberos was designed to minimize the network implications of a compromised computer or user account. Loss of a session key leads to short-term damage usually confined to the parties of concern. However, the main weakness of the system is having a compromised KDC. In this case, the entire database of master keys would be compromised and security loss would be inevitable since the attacker could now masquerade as the KDC.

There are other problems with the shared key concept. A solution to some of these problems is provided with public key approaches to be described in Chap. 5. As will be seen, the key distribution center will be called a certificate authority or CA. It sends authenticated and encrypted public keys to each user. However, each user also has a secret private key which the CA does not know (unlike the KDC approach). As will be discussed next, this key pair (public and private) supports more security concepts such as digital signatures which are essential for ecommerce and other applications.

4.4 Public Key Approaches

The discovery of public key schemes has made a tremendous impact on the security field with respect to invention of digital signatures and key establishment. The development of public key systems started as early as 1952 when the challenge–response approach was suggested by Horst Feistel as a solution to the ground-to-aircraft identification problem studied at the Air Force Cambridge Research Center (Diffie 1988). This together with the suggestion by M.V. Wilkes in 1972 of using one-way functions for computer password storage provided the two building blocks necessary for the development of public key systems (Diffie 1988). Specifically the one-way function which is easy to compute but very difficult to invert provided the mechanisms for the public key with a trapdoor function provided by the private key. For example the user can easily encrypt their message with the recipient's public key to provide the one-way function. Then the trapdoor function is provided by recipient who uses their private key to decrypt the message. Whereas the challenge–response approach lead the way to the digital signature by posing a question, the challenger is unable to answer, but whose answer he could easily check for correctness. For example, the user (or challenger) could request a message be signed. This user is not able to produce the digital signature but upon receipt could use the public key to confirm the signature was in fact correct. In 1974 a paper written by Ralph Merkle described the key distribution problem, introducing the concept of sending data to a recipient where an eavesdropper who also obtains the data has "1 million times 1 million" attempts to find the key compared with the intended recipient who has an easier time obtaining the key with only "1 million" computations (Diffie 1988). The Diffie–Hellman scheme was published in 1976 and it was suggested even earlier by a British Scientist (Diffie 1988). In 1977 the RSA approach had been developed and was discussed in a Scientific American article by Martin Gandener. It was subsequently published a year later in Rivest et al. (1978).

Public key schemes are well known for their use in providing digital signatures. As well they are known for their ease in establishing and communicating temporary keys among a party. Public key schemes involve a key pair, also known as a public key and a private key. These schemes have two important properties.

The first property of the public/private key pair is:

Anything encrypted with one key can be decrypted with the other key.

Or stated mathematically for a given public/private key pair, K_p/K_s (for public, secret keys respectively), then

$$D_{K_s}\left(E_{K_p}(m)\right) = m \text{ or } D_{K_p}\left(E_{K_s}(m)\right) = m.$$

The first property is very useful for authentication. For example Alice can create her own signature, by typically encrypting her message or hash of her message (or ciphertext, etc.) with her private key. This signature is concatenated to the data or ciphertext to be transmitted. Bob, who receives the transmission, can now verify Alice's signature by decrypting it using Alice's public key and checking that it is equal to the message. Since Alice is the only one to hold the private key associated with Alice's public key which many users may hold, she is the only one to have signed the message assuming her signature has been verified. Note that Alice must hold Bob's authenticated public key, and often this may be transmitted in the clear, hence the second property of not being able to derive the private key from the public key must hold.

The second property of the public/private key pair is:

Given the public key, no one can use it to derive the private key.

The second property is crucial in that since many users will have access to public keys, it is important that this information cannot be used to extract any information about the secret key. Next we will illustrate some protocols based on public key cryptography (PKC).

4.4.1 Protocols

This section will describe public key approaches to encryption, digital signatures, and key establishment. The detailed mathematics is postponed until the following section, Sect. 4.4.2.

4.4.1.1 Basic Encrypt

The encrypt function using asymmetric keys will be discussed briefly in this section. A lock is used again to illustrate these concepts.

Consider Fig. 4.2 where an unlocked lock is shown in (a) on the left-hand side under Alice. In this illustration, Bob wishes to send Alice a message but wants to be sure no one else will see it. Bob requests an unlocked lock from Alice (or from the certificate authority). Specifically Alice's unlocked lock represents her "Public Key." Anyone upon receipt of Alice's unlocked lock can use it to lock a message and send the locked message to Alice. Only Alice has access to the secret private key for her particular lock. Thus although anyone can lock a message in a box using Alice's unlocked lock (i.e., use Alice's public key to encrypt a message), no one other than Alice can unlock the box (i.e., decrypt the message). The locking and unlocking is used to represent encryption and decryption, respectively. For illustration purposes, the message, $\boldsymbol{M}$, will appear "in the lock" to represent the fact that the message is inside a box locked with that lock.

So examining Fig. 4.2, first Bob requests and receives an unlocked lock from Alice in (a). He then locks his message, $\boldsymbol{M}$, with Alice's lock and sends it to Alice

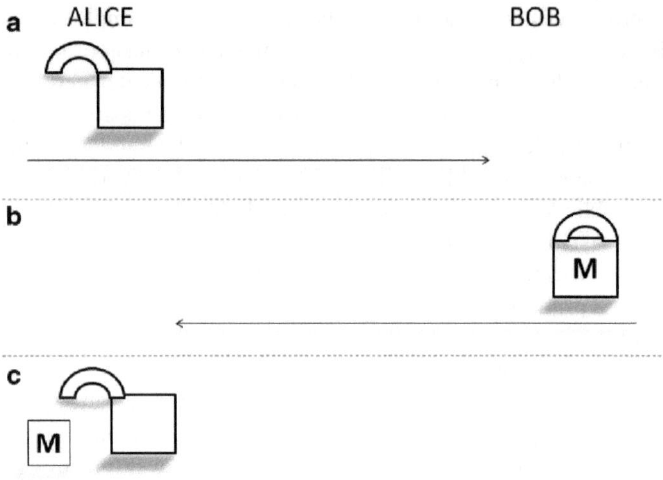

Fig. 4.2 Illustration of public and private key pairs using a lock

in Fig. 4.2b. Alice unlocks her lock to obtain the message in Fig. 4.2c. This protocol is clearly more efficient than the previous protocol of Fig. 4.1, since it only has 2 passes or 1 pass if Bob already has access to Alice's public key (or unlocked lock). Thus the process of locking has been used to represent the encryption of the message or creation of the ciphertext, c, $c = E_{Pa}(m)$ [i.e., Bob uses Alice's public key, Pa, with the encryption algorithm $E_{Pa}(m)$]. When Alice performs the unlocking, she is decrypting the ciphertext to obtain the message, m [i.e., Alice uses her secret key, Sa, with the decryption algorithm $D_{Sa}(c)$, where $D_{Sa}(c) = D_{Sa}(E_{Pa}(m)) = m$]. This is a simplified realization of how general public key systems work. Figure 4.2 is further detailed in the protocol below where steps (1)–(3) refer to Fig. 4.2a, b.

Protocol 4.7. Simplified PKC data transfer

(1) $Alice \rightarrow Bob$: Pa
(2) $Bob \rightarrow Alice$: $E_{Pa}(m)$
(3) $Alice$: $D_{Sa}(E_{Pa}(m)) = m$

4.4.1.2 Digital Signature

A digital signature is one of the most crucial functions for which public key cryptography is famous. Without public key or asymmetric systems, the secure use of a digital signature would not be possible. For example with only one secret shared key, only trusted sources who have authority to have a copy of your key can verify your signature. To have anyone other than the "bank" verify your check is very risky since you must "trust" them with your secret key. Even trusting banks may be risky (Smith 2002). In contrast, the public key systems allow one person to create a digital signature that can be safely verified by anyone without having to trust them. In fact

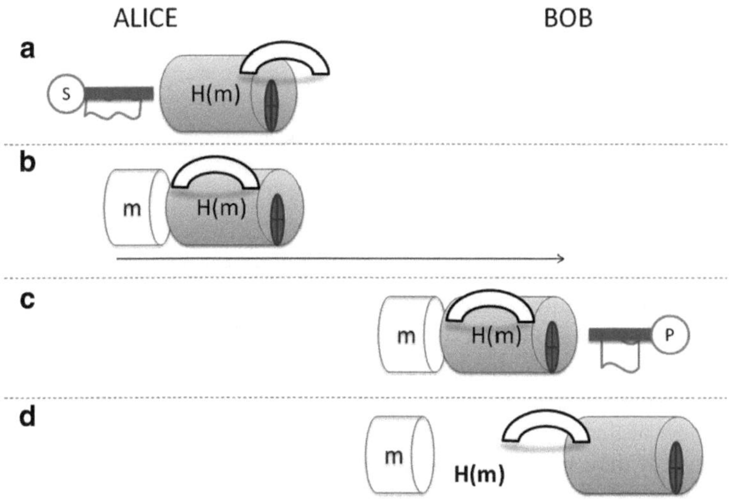

Fig. 4.3 Illustration of digital signature using public key approach

the digital signature can be verified without any knowledge of the secret key used to sign the data. All that is needed is access to the individual's public key in order to verify the signature. The user's secret key remains safe since deriving the secret key from the corresponding public key is assumed to be a hard problem. The digital signature will be illustrated with the same public/secret key pairs as discussed earlier; however, it is often the case that a separate public/secret key pair would be used solely for signing and another public/secret key pair would be used for encryption.

The digital signature function can be explained using a similar lock illustration to that used earlier in Chap. 4. Consider Fig. 4.3a where Alice has taken the hash of a message, $\boldsymbol{H(m)}$, and locked it in a special cylinder with her secret key on the left. In this example, we choose to use the hash of the message in order to create smaller sized data for signing as opposed to signing the larger entire message. This special cylinder is different from the box used to illustrate cryptographic protocols in previous sections of this book. This cylinder has a special key for locking and a second different key for unlocking. The locking/unlocking key pair for this special cylinder is Alice's secret/public key pair. So if Alice wishes to send a message to Bob, $\boldsymbol{m}$, (perhaps even in the clear illustrated in Fig. 4.3) and she wants Bob to be sure that it was Alice who sent this message, then she sends the special cylinder locked with her secret key in addition to the message, as in Fig. 4.3b. For example inside the cylinder she would put the hash of her message, $\boldsymbol{H(m)}$. When Bob receives the message, $\boldsymbol{m}$, and special cylinder, he would use Alice's public key to open the special cylinder and obtain the hash, $\boldsymbol{H(m)}$, as shown in Fig. 4.3c. Now to verify that Alice really did send this message, he would proceed to take the hash of the message $\boldsymbol{m}$ and compare it to the hash obtained from the special cylinder, $\boldsymbol{H(m)}$. If the hashes were equivalent, Bob would be assured that the message, $\boldsymbol{m}$, did indeed come from Alice, since she is the only person in possession of the secret key which locked that special cylinder. The simplified authenticated data transfer protocol is given in Protocol 4.8 where the digital signature concept provides authentication. It assumes that Bob has an authenticated copy of Alice's public key.

Protocol 4.8. Simplified authenticated data transfer

(1) *Alice*: $E_{Sa}(H(m)) = s$
(2) *Alice* $\rightarrow$ *Bob*: m, s
Bob: $D_{Pa}(s) =?= H(m)$

In step (1) of the simplified authenticated data transfer protocol step, Alice computes the hash of message m that she is about to send to Bob. She then signs the hash using a digital signature s. In step (2) Alice sends the message in the clear along with the signature to Bob. Bob uses Alice's public key to transform the signature into the hash of the message and he compares this value to his computed hash of the message. If they are equal then he is assured that the message was from Alice, since she was the only one to have signed the message since she is the only one to have the secret key, *Sa*. Thus Bob authenticates Alice. Of course there may be confidentiality requirements on the message as well which is shown in Protocol 4.9, named simplified authenticated and encrypted data transfer protocol.

Protocol 4.9. Simplified authenticated & encrypted data transfer

(1) *Alice*: $E_{Sa}(H(m)) = s$
Alice: $E_{Pb}(m) = c$
(2) *Alice* $\rightarrow$ *Bob*: c, s
Bob: $D_{Sb}(c) = m$
Bob: $D_{Pa}(s) = h2$
Bob: $H(m) =?= h2$

Note since many users may have Alice's public key (*Pa*), if Alice were to sign the message m as opposed to the hash of the message $H(m)$, then users who obtained the signature would be able to obtain the actual message as follows: $E_{Pa}(s) = E_{Pa}(E_{Sa}(m)) = m$. In this protocol the message was encrypted, hence in order to keep the message confidential a hash was used to generate the accompanied signature. Since the hash of the message does not reveal the message, the signature does not reveal any information concerning the message. In both protocols if Bob received the public key of an adversary, instead of Alice's public key, his message would be exposed to the attacker, not Alice. Thus authentication of the public key is a crucial assumption and will be further discussed in Sect. 4.4.1.3.

4.4.1.3 Key Establishment, Certificates, and Protocols

The protocols will now be extended to illustrate conventional public key cryptography providing authentication. Although public keys can be given to anyone (they are not secret), users still need to authenticate them. For example, assume that Alice wishes to speak to Bob. She needs to obtain Bob's authentic public key. She needs proof that indeed she has Bob's real public key and not the public key of some other person or adversary. For this reason a trusted source is required. In conventional public key cryptography, the use of certificates and certificate authorities (CA) serve this purpose. In reality, the certificate contains more information but for now we can assume that its main purpose is to send authenticated copies of a public key. It is also assumed that Alice like other users already has an authentic copy of the CA's public key. In fact they can verify that the public key is authentic because it has an associated signature which can be verified using the CA's public key. Consider the following protocol, where S_{ca} is the secret key of the CA and P_{ca} is the public key of the CA:

Protocol 4.10. Authenticated public key from CA

(1) $CA \rightarrow Alice$: $K_{Pb}, E_{Sca}(K_{Pb})$
    $Alice$: $D_{Pca}(E_{Sca}(K_{Pb})) =?= K_{Pb}$
    $Alice$: $E_{Pb}(m) = c$
(2) $Alice \rightarrow Bob$: c
    Bob: $D_{Sb}(c) = m$

In this approach, all users have the trusted CA's public key, P_{ca}. Alice requests a copy of Bob's public key from the certificate authority. The Certificate Authority sends to Alice a copy of Bob's public key signed by the CA in step (1) of authenticated public key from CA protocol. The digital signature, $E_{Sca}(K_{Pb})$, is concatenated to Bob's public key (note that the key has a reasonable length thus a hash of it is not required when computing the signature). Since Alice has the CA's public key she can verify that the signature is authentic (see =?=) and therefore be assured that she did receive a good copy of Bob's public key. Once Alice has verified Bob's public key in step (1), she uses it to transmit a confidential message to Bob in step (2). Often the message, m, is the session key, and step (2) is referred to as a one pass key transport (Menezes et al. 1996), which allows Alice to establish a secure channel with Bob.

The next protocol will transmit both a session key as well as ciphertext. It is assumed that Alice has Bob's public key and vice versa.

Protocol 4.11. Public key data & key transfer with signature

(1) $Alice$: $E_{Sa}(H(m)) = s$
    $Alice$: $E_{Pb}(k) = e$
    $Alice$: $E_k(m, s) = a$
(2) $Alice \rightarrow Bob$: e, a
    Bob: $D_{Sb}(e) = k$
    Bob: $D_k(a) = m, s$
    Bob: $D_{Pa}(s) =?= H(m)$

In the Public key data and key transfer with signature protocol, Alice signs her message in step (1). She creates and encrypts the session key using Bob's public key and then sends it along with her signed message encrypted with the session key to Bob. Bob receives the transmission and proceeds to first decrypt the session key, k, and then decrypt the signed message. Bob can verify that the signature on the message was in fact from Alice using Alice's public key. This is often referred to as a hybrid scheme since encryption (of the session key) using a public key as well as symmetric key encryption (of the signed message) are both performed.

Section 4.4.2 will examine implementations of public key cryptography based on the integer factorization problem and the discrete logarithm problem. Some mathematical background is covered first followed by the detailed computations.

4.4.2 *Mathematics Behind PKC*

So far we have avoided presenting the mathematical implementation details of the protocols and have only described these using general functions of encryption, decryption, etc. This section will cover the mathematics based upon the integer factorization and discrete logarithm. Some mathematical background will be covered so that readers can follow the implementation. Readers may also refer to the mathematical definitions and terminology listed at the beginning of the chapter.

4.4.2.1 Crypto Based on the Integer Factorization Problem

The integer factorization problem is based on the difficulty of factoring products of a pair of large prime numbers. This problem is stated as: Given an integer $n = pq$, find the prime numbers p and q.

RSA

To set up an RSA scheme, we first need to establish some interesting mathematical properties:

$$n = pq,$$
$$\phi(n) = (p-1)(q-1).$$

The value $\phi(n)$ is known as Euler's function[1] (or the Euler's totient function or phi function). It has the following useful property for any integer k:

$$x^{k\phi(n)+1} = x \bmod n, \quad \text{for any integer } x \in Z_n^*.$$

This property follows from Euler's theorem specifically:

$$x^{\phi(n)} = 1 \bmod n, \quad \text{for any integer } x \in Z_n^*.$$

[1] The Euler function is the number of integers in the interval $[1,n]$ which are relatively prime to n. Further we say integers a and b are relatively prime if $gcd(a,b) = 1$.

Hence the left-hand side of the property above can be expanded and simplified as:

$$x^{k\phi(n)+1} = x^{k\phi(n)}\, x \bmod n = (x^{\phi(n)})^k\, x \bmod n = x \bmod n.$$

In other words, the exponent works with modulo $\phi(n)$ and the resultant value works with *mod n*.

Now assume that Bob's public key consists of (*Pb*,*n*), where *Pb* is a random number between 3 and $\phi(n)$, that has no common factors with $\phi(n)$. The *n* term is called modulus. Bob's private key is *Sb* such that:

$$(Pb)(Sb) = k\phi(n) + 1$$

or in other words $Sb = Pb^{-1}$ $mod\ \phi(n)$ or $(Pb)(Sb)\ mod\ \phi(n) = 1$. It is interesting to note that the key pair can be generated by first selecting two large prime numbers, p and q. Then $n = pq$ and $\phi(n) = (p-1)(q-1)$ are computed. Next a public key is selected *Pa* where $1 < Pa < \phi(n)$ such that $gcd(Pa, \phi(n)) = 1$. Then the private key is computed, *Sa*, such that $1 < Sa < \phi(n)$ where $(Pa)(Sa) = 1\ mod\ \phi(n)$, using the extended euclidean algorithm (Menezes et al. 1996). Now the RSA scheme providing confidential transmission of a message *m* will be given in the Simplified RSA Data Transfer Protocol 4.12 (an RSA version of the earlier Simplified PKC data transfer Protocol 4.7 in Sect. 4.4.1.1) or RSA Encrypt algorithm.

Protocol 4.12. Simplified RSA data transfer (RSA encrypt)

(1) *Alice* → *Bob*: *Pa*, *n*
Bob: $m^{Pa}\ mod\ n = C$
(2) *Bob* → *Alice*: *C*
Alice: $C^{Sa}\ mod\ n = ((m^{Pa})^{Sa})\ mod\ n = m$

In the simplified RSA data transfer protocol, Bob wishes to confidentially send Alice his message, m. We assume that the message is in fact a number in the interval $[0, n-1]$. In step (1) Alice sends Bob her public key, *Pa* and *n*. Again Alice has a public–private key pair, (*Pa*, *Sa*), satisfying $(Pa)(Sa) = k\varphi(n) + 1$. Bob encrypts his message he wishes to send to Alice in the second line of step (1) by performing m^{Pa} *mod n*. Alice decrypts the received ciphertext, C, on the second line of step (2) by performing C^{Sa} *mod n* using her secret key, *Sa*. The RSA scheme decrypts the message since: $((m^{Pa})^{Sa})\ mod\ n = m^{k\phi(n)+1}\ mod\ n = m\ mod\ n = m$.

Next we will illustrate this protocol using a hypothetical small example. Let us consider values $p = 2$ and $q = 5$ and compute $n = 10$ and $\phi(n) = 4$. Let us arbitrarily choose $Pa = 3$. We then solve for a *Sa* where $3Sa = k(4) + 1$, hence we obtain $Sa = 7$ (i.e., $k = 5$). Now that we have generated a pair of public keys, let us consider how these values would be used. Assume that Bob wishes to send a message, *m*, represented as an integer, 8, to Alice. After he receives Alice's public key, $Pa = 3$, as in step (1) of the Simplified RSA Data Transfer Protocol, he computes and sends the ciphertext $C = 8^3\ mod\ 10 = 512\ mod\ 10 = 2$ to Alice in step (2). Alice decrypts C with her private key and obtains *m* by computing $2^7\ mod\ 10 = 128\ mod\ 10 = 8$, thus retrieving the original message *m* whose value is 8.

It is important to note that unlike this simplified example, in practice both *Pa* and *Sa* are very large prime numbers and an attacker must therefore first factor n in order to obtain p,q. This integer factorization problem is what makes RSA very secure since it is very difficult. Next the attacker calculates $\phi(n) = (p-1)(q-1)$ and finally obtains *Sa* from *Pa*, since $Sa = (Pa)^{-1} \; mod \; \phi(n)$.

The simplified RSA encrypted and authenticated data transfer protocol further provides RSA signatures as well as encryption and assumes that both Alice and Bob have an authenticated copy of each other's public keys. It is identical to the general simplified authenticated and encrypted data transfer Protocol 4.9 in Sect. 4.4.1.2 except both the message and concatenated signature are encrypted and it is specifically implemented in RSA.

Protocol 4.13. Simplified RSA encrypted & authenticated data transfer

(1) *Bob*: $s = (H(m))^{Sb} \; mod \; n$
Bob: $(m,s)^{Pa} \; mod \; n = C$
(2) *Bob* → *Alice*: C
Alice: $(C)^{Sa} \; mod \; n = (((m,s)^{Pa})^{Sa}) \; mod \; n = m,s$
Alice: $(s)^{Pb} \; mod \; n =?= H(m)$

Rabin Scheme

The Rabin public-key system is also based on the integer factorization problem, like the RSA system. However, the Rabin system was the first example of a provably secure public-key encryption scheme. It is presented here also since it has possible applications to embedded systems where power is a concern.

Public keys in this scheme consist of n where the corresponding private keys are p and q such that $n = pq$, where p, q are two large and random prime numbers roughly each the same size.

The public-key encryption scheme is given in Protocol 4.14. The message Alice wishes to send to Bob is represented by an integer $0 \leq m \leq (P_b - 1)$. Note that we represent Bob's public key as *Pb* and his private keys as S_b.

Protocol 4.14. Rabin public-key encryption

(1) *Alice*: $m^2 \; mod \; P_b = c$
(2) *Alice* → *Bob*: c

(2.1) *Bob*: $square\_roots(c) = (m_1, m_2, m_3, m_4)$
(2.2) *Bob*: $m = m_1 \; or \; m_2 \; or \; m_3 \; or \; m_4$

In Protocol 4.14 Bob receives the ciphertext and proceeds to take the square root of $m^2 \; mod \; P_b$. He calculates four roots of this value and chooses one of the four messages, specifically $m = m_1 \; or \; m_2 \; or \; m_3 \; or \; m_4$. In practice redundancy is used so that Bob can determine which of the four was sent. For example, he may have agreed with Alice that the last 8-bytes of the plaintext would be replicated. Thus Alice would first replicate the last 8-bytes in her message m before computing step (1) in Protocol 4.14. Then Bob would examine each of the four messages for 8-bytes of redundancy (Menezes et al. 1996) in step (2.2). Details on the square_roots() function can be found in (Menezes et al. 1996), however, we will present the special case algorithm when p and q are chosen such that $p \equiv q \equiv 3(\text{mod } 4)$. The algorithm is shown in Algorithm 4.1.

Algorithm 4.1.

(1) Find a,b such that $ap+bq = 1$ [this can be precomputed by Bob using extended euclidean algorithm, specifically Algorithm 2.107 in Menezes et al. (1996)]
(2) $r = c^{(p+1)/4}\ mod\ p$
(3) $s = c^{(q+1)/4}\ mod\ q$
(4) $x = (aps + bpq)\ mod\ n$
(5) $y = (aps - bqr)\ mod\ n$
(6) $m_1 = x, m_2 = -x\ mod\ n, m_3 = y, m_4 = -y\ mod\ n$

It is important to note the specific modulo integers, where in step (2) and step (3) we are using modulo p and q, respectively, and not modulo n as in step (1) of the protocol (shown as P_b). In solving for the square root we examine *mod p* and *mod q* separately (steps 2 and 3 in Algorithm 4.1) and then proceed to find a solution *mod n* (steps 4 and 5). To briefly illustrate how the Rabin scheme works, we will present two important mathematical properties.

Property 4.1. If we assume all integers are modulo p and if we assume $4|(p + 1)$ (meaning $p + 1 = 4k$, for some k) and c is a quadratic residue, i.e., $m^2 = c\ mod\ p$, then

$$c^{(p-1)/2} \text{ mod } p = 1 \text{ mod } p.$$

Property 4.1 will be used to illustrate the square root function and is used in steps 2 and 3 of Algorithm 4.1.

Property 4.2. Let a, b be integers. If

$$a = 1\, mod\ p = 0\, mod\ q$$

and if

$$b = 1\, mod\ q = 0\, mod\ p$$

then if r,s satisfy the congruence *mod p* and *mod q*, respectively, then

$$z = ar + bs \text{ mod } n$$

z solve the congruence *mod n*.

Property 4.2 is used to calculate the residual of z when divided by n which is the solution we are looking for in the Rabin Scheme protocol steps (2.1) and (2.2).

Property 4.1 example: For example, assume our message is represented by integer $m = 5$, $p = 3$, then $c = m^2\ mod\ n = 25\ mod\ 3 = 1$, thus $1^{(3-1)/2} = 1\ mod\ 3 = 1$.

In order to use these properties and help to illustrate how the Rabin scheme works, we will assume first that unlike the protocol we compute

$$m^2 \bmod p = c_1.$$

In the protocol we used modulo n not modulo p (see step 1 of Protocol 4.14). Consider the following calculation, similar to step (2) of Algorithm 4.1 ($r = c^{(p+1)/4}$ *mod* p), but we assume that all integers are modulo p

$$r_1 = c_1^{(p+1)/2} \bmod p = (c_1)c_1^{(p-1)/2} \bmod p = c_1 \bmod p.$$

Thus applying step (2) in this scenario we would have

$$c_1^{(p+1)/4} \bmod p = \left(c_1^{(p+1)/2}\right)^{1/2} \bmod p = (c_1)^{1/2} \bmod p.$$

Hence we have illustrated that the square roots are calculated for separate modulo p and q in Algorithm 4.1. More details can be found in Menezes et al. (1996) and Rabin (1979). The next section will examine public-key cryptography based on the discrete logarithm problem.

4.4.2.2 Crypto Based on Discrete Logarithm

The discrete logarithm problem is stated as follows: given p a prime number, a generator g of Z_p^*, let $h = g^x$ *mod* p, where $0 \leq x \leq (p-2)$, so the discrete logarithm problem is given p, g, and h find x. In other words computing h from x is easy, whereas computing x from h is not. The security of the protocols which follow are based upon this hard problem. Note that to avoid repetition, all parameters in this section which are exponentiated mod n (such as g, or alpha), are assumed to be generators as just defined.

Diffie–Hellman

With the introduction of public key pairs, the famous Diffie–Hellman key exchange algorithm (Diffie and Hellman 1976) will be discussed next. In particular, this algorithm utilizes the public keys along with random numbers to compute a session key or symmetric key to be used for some period of time for confidential communication between two parties, such as Alice and Bob. Figure 4.4 illustrates this concept again using the lock example. Alice and Bob want to again communicate confidentially. First each of them exchanges their public keys or unlocked locks (Fig. 4.4a). Bob's public key is represented by a shaded lock to differentiate it from Alice's public key. Alice chooses a random number, x, and Bob chooses a different random number, y. They each lock these values and send them to each other. Upon receipt Alice and Bob each unlock their received locks to obtain the random numbers (e.g., y

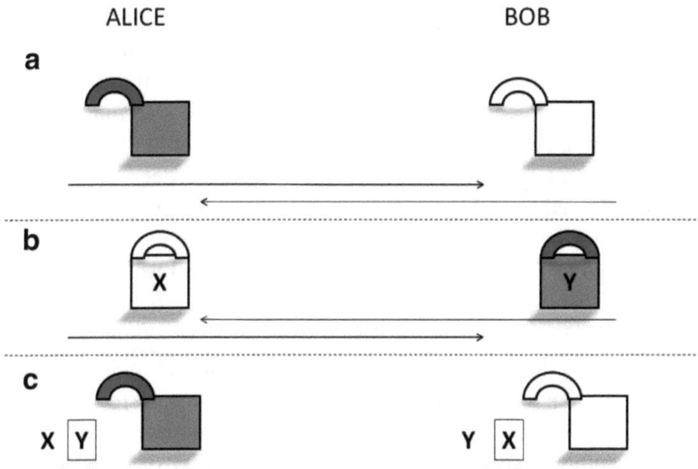

Fig. 4.4 Illustration of Diffie–Hellman protocol using two locks

and x, respectively). Next they each perform a special computation on the received random number (y and x, respectively) and their own random number (x and y, respectively). In order to present the specific mathematical computations using the discrete logarithm public key scheme, some initial mathematical background review will be presented.

A very simple example will be used to illustrate the mathematics in Protocol 4.15. Consider the set $Z_5^* = \{1, 2, 3, 4\}$ and element *3*. Powers of 3 modulo 5 produce: $3^0 = 1$, $3^1 = 3$, 3^2 *mod* $5 = 4$, 3^3 *mod* $5 = 2$. Note that the fourth power is 1, 3^4 *mod* $5 = 1$ hence only four powers are needed to produce all elements, specifically $g^x, 0 \le x \le (p-2)$ for Z_p^*. Thus 3 is said to be a generator.

The Diffie–Hellman scheme is based upon the hardness of this discrete logarithmic problem and it will be described next. In this example, Alice wants to communicate with Bob securely so she wishes to establish a session key. Alice chooses a random number Xa uniformly from the integers $0, 1, 2, \ldots, p-2$ and keeps it secret. Bob does the same choosing Xb. The protocol proceeds as shown in Protocol 4.15.

Protocol 4.15. Simplified Diffie–Hellman key exchange

(1) $Alice \rightarrow Bob$: $Ya = g^{Xa} \; mod \; p$
    Bob: $Ya^{Xb} = g^{XaXb} \; mod \; p$
(2) $Bob \rightarrow Alice$: $Yb = g^{Xb} \; mod \; p$
    $Alice$: $Yb^{Xa} = g^{XaXb} \; mod \; p$

Thus Alice and Bob both create the same secret session key of value $g^{XaXb} \; mod \; q$. They now have a shared secret key which has been established. The keys Xa and $g^{Xa} \; mod \; p$ are referred to as Alice's private and public key, respectively. This is true for Bob as well with regards to keys Xb and $g^{Xb} \; mod \; p$. This is a secure scheme since the problem of solving for *Xa* or *Xb*, given *Yb* or *Ya*, respectively, is a hard problem, namely the discrete logarithm problem. However, this protocol still has a number of problems. First it is not safe from the man-in-the-middle attack. For example, if all communications are intercepted by Eve, she can send her public key and establishes a session key with Alice and Bob without them knowing. This means all communications between Alice and Bob can be decrypted by Eve. Hence Alice or Bob still need to authenticate the public keys. To continue the very simple example, consider $Xa = 2$, and $Xb = 3$ for $p = 5$ where a generator as discussed earlier is 3. In step (1) Alice sends Bob $(3)^2 \ mod \ 5 = 4$ and Bob computes $(4)^3 \ mod \ 5 = 4$. In step (2) Bob sends Alice $(3)^3 \ mod \ 5 = 2$ and Alice computes $(2)^2 \ mod \ 5 = 4$. Now they both share the secret, 4, and can use this as a temporary session key.

The authenticated Diffie–Hellman protocol provides an authenticated version of the Diffie–Hellman key exchange protocol. It is also referred to as the station to station protocol or STS protocol. It provides both mutual entity authentication as well as key authentication. It also supports anonymity of Alice and Bob from eavesdroppers. The STS protocol uses the discrete exponential key authentication, where Alice's signature on m is denoted by $S_{Sa}(m) = (H(m))^{Sa} \ mod \ n$ (and similarly for Bob using Sb), where $H(m)$ is a one-way hash function on m where $H(m) < n$. Again assume given p, a generator g of Z^*_p, let $0 \leq x \leq (p-2)$ and $0 \leq y \leq (p-2)$ again be chosen by Alice and Bob, respectively. The STS protocol proceeds as follows:

Protocol 4.16. Station to station

(1) *Alice* $\rightarrow$ *Bob*: $g^x \ mod \ p$
Bob: $g^{xy} \ mod \ p = k$
Bob: $(H(g^y, g^x))^{Sb} \ mod \ n = S_{Sb}(g^y, g^x) = s_1$
Bob: $E_k(s_1) = c_1$
(2) *Bob* $\rightarrow$ *Alice*: $g^y \ mod \ p, c_1$
Alice: $(g^y)^x \ mod \ p = k$
Alice: $D_k(c_1) = s_1$
Alice: $(s_1)^{Pb} = ((H(g^y, g^x))^{Sb})^{Pb} \ mod \ n =?= H(g^y, g^x)$
Alice: $S_{Sa}(g^y, g^x) = (H(g^y||g^x))^{Sa} \ mod \ n = s_2$
Alice: $E_k(s_2) = c_2$
(3) *Alice* $\rightarrow$ *Bob*: c_2
Bob: $D_k(c_2) = s_2$
Bob: $(s_2)^{Pa} = ((H(g^y||g^x))^{Sa})^{Pa} \ mod \ n =?= H(g^y||g^x)$

In the above example, Alice and Bob both pick random numbers, x,y, respectively, and proceed with the Diffie–Hellman key exchange to create a session key k [e.g., Bob calculated k implicitly in step (2) and Alice computed k in step (3)]. In step 2 Bob signs the concatenation of both exponentials with his private key and sends this encrypted to Alice. Alice does the same in step (4). Alice checks Bob's signature and once confirmed, she then sends her encrypted and signed exponentials to Bob. Bob finally confirms her signature thus completing the establishment of the session key, k. In all cases of the protocol above, we assume that verification succeeds

(=? = is evaluated to =) and hence the computations proceed. If not, generally each application can decide what to do with data whose sender cannot be verified.

Another interesting scheme for public key security management in phones was described in (Diffie 1988). Each secure phone had a public/private pair of keys, authenticated by the key management facility. For each phone session, the caller and callee phone establishes an unique encryption key, using the Diffie–Hellman key exchange protocol, to be used only for the current phone session. This establishes an encrypted channel but not authenticated channel. The caller and callee exchange their public key certificates. They each check the signature on the certificates and extract the public key. The pair of secure phones then performs a challenge–response test by signing test messages and checking these signatures with the acquired public keys. Now the channel has been authenticated as well and the phone conversation can proceed between the two users on each end of the secure phones. The El-Gamal public key system will be described next.

El-Gamal Public Key System

The El-Gamal public key systems security is based on the hardness of the discrete logarithm problem as well as the Diffie–Hellman problem. It is the basis for the DSA standard which will be described immediately after.

Alice's public key parameters are (p, α, α^a), where p is the prime number, $P_a = \alpha^a$ is Alice's public key and $S_a = a$ is Alice's private key. Bob wishes to send message m $(m < p)$ encrypted to Alice and has been given Alice's public key parameters. Bob can use Protocol 4.17 (using a random integer $k < (p-1)$) in order to confidentially send Alice his message.

Protocol 4.17. El-Gamal encryption

(1) *Bob*: $(\alpha^k)\ mod\ p = \gamma$

    (i) *Bob*: $m(P_a)^k\ mod\ p = \partial$

(2) *Bob* → *Alice*: γ, ∂

    (i) *Alice*: $((\gamma)^{-S_a})(\partial)\ mod\ p = m$

This encryption works because Alice's computation on the second line in (2)(i) expands to:

$$= ((\gamma)^{-S_a})(m(P_a)^k) mod\ p,$$

by substituting public and private key values we get

$$\begin{aligned}
&= ((\gamma)^{-a})(m(\alpha^a)^k) mod\ p\ \alpha \\
&= (((\alpha^k))^{-a})(m(\alpha^a)^k) mod\ p \\
&= \alpha^{-ak} m \alpha^{ak} mod\ p \\
&= m
\end{aligned}$$

Since $(\gamma)^{p-1-a} mod\ p = (\gamma)^{-a} mod\ p = \alpha^{-ak} mod\ p.\alpha$

An important standard for digital signatures is known as the digital signature algorithm or DSA. It is a variant of the El-Gamal signature system. The mathematics is given below and further details are provided in FIPS186 (2009). Alice is given the following: m is message to be signed, S_a is private key, $P_a = (g^{Sa}\ mod\ p)$ is Alice's public key, p is prime, $2^{L-1} < p < 2^L$, L is a multiple of 64, $512 \leq L \leq 1{,}024$, q is a prime divisor of $(p - 1)$, $2^{159} < q < 2^{160}$, g has order $(q\ mod\ p)$ (or g is a generator of the unique cyclic group of order q in Z_p^*, which more specifically can be generated by $g = h^{((p-1)/q)}\ mod\ p$, where $1 < h < (p - 1)$ and $g > 1$).

Protocol 4.18. DSA signature generation and verification protocol

(i) *Alice*: $r = (g^k\ mod\ p)\ mod\ q$
(ii) *Alice*: $s = (k^{-1}(H(m) + S_a r))\ mod\ q$

(1) *Alice* → *Bob*: m, r, s

(i) *Bob*: $w = (s)^{-1}\ mod\ q$
(ii) *Bob*: $u1 = ((H(m))w)\ mod\ q$
(iii) *Bob*: $u2 = ((r)w)\ mod\ q$
(iv) *Bob*: $v = (((g)^{u1}(P_a)^{u2}\ mod\ p)\ mod\ q$
(v) *Bob*: $v =?= r$

In step (i) of Protocol 4.17, Alice selects a random value, k, where $(k^{-1}k)\ mod\ q = 1, 0 < k^{-1} < q$, and which is different for each message, $0 < k < q$. She then computes r and s, which is the signature for message m where each r,s is 160-bits long, each an integer in the interval $[1, q - 1]$. The signature is sent to Bob along with (unencrypted or encrypted message) m. The signature verification is performed by Bob to ensure that the message sent came from Alice.

We will show how the DSA scheme works by starting with the definition of the signature Alice has sent or s:

$$s = \left(k^{-1}(H(m) + S_a r)\right) \bmod q.$$

It can be rewritten as (by multiplying both side by k and moving terms around)

$$H(m) = (sk - S_a r) \bmod q,$$

then multiplying by w, we get

$$wH(m) + S_a rw = wsk \bmod q$$

which becomes

$$u1 + S_a u2 = k \bmod q$$

by substituting variables from first 3 lines of DSA verification, 1(i) to 1(iii). Next we raise the generator to the powers above as follows

$$g^{u1+Sa\,u2} = g^k \bmod p \bmod q$$
$$g^{u1}(g^{Sa})^{u2} = g^k \bmod p \bmod q,$$
$$g^{u1}(P_a)^{u2} = g^k \bmod p \bmod q,$$

which finally becomes

$$v = r.$$

DSA is a standard for digital signatures and is a variant of the ElGamal scheme. The security of DSA is based on the hardness of the logarithm problem in Z_p^* and the logarithm problem in the cyclic subgroup of order q. More details can be found in FIPS186 (2009).

Protocols form the basis for using security in many embedded systems. An important criterion for protocols is the amount of computation each device has to perform. For example, when a portable mobile device communicates with a base station, a protocol which involves light computations on the mobile device and moves heavier computations to the base station is ideal. This way the energy dissipation of the mobile device can be minimized. Other possible considerations in choosing a protocol for an embedded system includes how much memory is required, how much data needs to be transmitted (this is a heavy source of energy dissipation), and what kind of trusted memory is required to store secret elements of the protocol such as keys. Chapter 5 will discuss elliptic curve cryptosystems which involve an efficient form of public key cryptography. Elliptic curve protocols will also be covered.

References

Branstad DK (1987) Considerations for security in the OSI architecture. IEEE Netw Mag 1(2): 34–39

Denning DE, Sacco GM (1981) Timestamps in key distribution protocols. Commun ACM 24: 533–536

Diffie W (1988) The first ten years of public key cryptography. Proc IEEE 76(5):560–577

Diffie W, Hellman ME (1976) New directions in cryptography. IEEE Trans Inform Theory IT-22 6:644–654

FIPS186-3 (2009) Digital signature standard (DSS) Federal information processing standards publication. National Institute of Standards and Technology, Gaithersburg, MD

Menezes A, van Oorschot P, Vanstone S (1996) Handbook of applied cryptography. CRC, Boca Raton, FL

Rabin MO (1979) Digitalized signatures and public-key functions as intractable as factorization. http://publications.csail.mit.edu/lcs/pubs/pdf/MIT-LCS-TR-212.pdf MIT/LCS/TR-212:1–16

Rivest R, Shamir A, Adleman L (1978) A method for obtaining digital signatures and public-key cryptosystems. Commun ACM 21(2):120–126. http://people.csail.mit.edu/rivest/Rsapaper.pdf

Schneier B (1996) Applied cryptography. Wiley, New York

Singh S (1999) The code book the science of secrecy from ancient egypt to quantum cryptography. Anchor Books, New York, NY

Smith RE (2002) Authentication from passwords to public keys. Addison-Wesley, Reading, MA

Chapter 5
Elliptic Curve Protocols

It is possible to write endlessly on Elliptic Curves...
from [1] in Miller (1986)

An entire chapter is dedicated to elliptic curves since there is great hope that advances in this area will bring higher levels of security to many embedded devices. Compared to previous public key systems, elliptic curve cryptography (ECC) promises equivalent security strength with smaller key sizes. Since key size impacts storage and in general has been shown to impact performance and thus energy dissipation, these elliptic curves are of great interest to embedded systems. The notion of using elliptic curves in cryptography is generally agreed to have emerged from two famous papers: those of Miller (1986) and Koblitz (1987).

This chapter will provide only a brief introduction to elliptic curves to enable engineers and scientists without significant knowledge of cryptography to gain an appreciation for elliptic curves as well as general understanding of the involved computations. Thus, we start out by addressing high-level algorithms first. The detailed lower level mathematics as well as the general elliptic curve protocols follows. This allows the reader to be introduced to the fundamental mathematical computations of elliptic curves and finally how they are used in standardized protocols. A brief overview of the mathematical background required to understand basic elliptic curve protocols is also presented wherever necessary; however, more in-depth discussion can be found in Hankerson et al. (2000, 2004) as well as in Cohen and Frey (2006) and Stinson (2005).

In Sect. 4.4.2.2 we stated the discrete logarithm problem (DLP) as follows:

Given g, h, and n, DLP is to find x, where
$h = g^x, 0 \leq x \leq (n-1), g \in G$, *$G$ a group, g an element of order n*

Analogous to this problem is the elliptic curve DLP (ECDLP), which can be stated as follows:

Given E, F_q, n, P, and Q, the ECDLP is to find x where
E an elliptic curve defined over F_q (set of integers modulo a prime q, $q > 3$, a finite field of order q)
P a point on E of prime order n
$Q = xP$, *where* $1 \leq x \leq (n-1)$, *Q a point on E*

C.H. Gebotys, *Security in Embedded Devices*, Embedded Systems,
DOI 10.1007/978-1-4419-1530-6_5,

The hardness of the EC DLP is the basis for elliptic curve protocols. Elliptic curves are generally well known for their ability to provide digital signatures.

Examining the security strength of elliptic curves, as discussed in Chap. 3, we compare the time it takes to attack elliptic curves of a given key size to attacks on other cryptographic approaches. Systems with equivalent times for attack are considered to have equal security strength. The more efficient of the two systems is denoted by the smaller key size, since key size is used to vary the security strength of a given cryptographic algorithm. It is well known that for a given secure system, greater security strength can be achieved through the use of larger keys. In performing this analysis on elliptic curves it has been determined that an elliptic curve cryptographic system with key size 160 bits is equivalent to a RSA system using 1,024-bit keys. Other comparisons include 512-bit ECC as equivalent to a 15,360-bit RSA system (Hankerson et al. 2004).

Elliptic curves have the interesting property that if you take two points on the curve and add them, you again get another point on the curve. In general, from a cryptographic point of view, the difficulty of determining which two points were added together to obtain the resulting summation point is extremely useful. Thus, ECC is based upon points located on elliptic curves defined over a field, which we will shortly introduce. The basic operations on the points on the curve are addition and doubling. Typically we refer to P, which is a point on an elliptic curve defined over a particular field. In ECC, k is some integer value defined using over 160 bits, typically referred to as the secret key. The multiplication of the secret key, k, with the point on the elliptic curve, P, is referred to as point multiplication or scalar multiplication. When k is the secret key of a user, then the scalar multiplication, specifically kP defines the corresponding user's public key. The mapping from the public key to the secret key is a hard problem and it is known as the elliptic curve discrete logarithm problem or ECDLP. Since there are no multiplication or division operations in this domain, the only way to compute kP over a curve is to perform a series of point doublings and summations. Apart from state-of-the art attacks [such as the Pollard Rho attack (Pollard 1978)], in general the brute force way of finding k, given kP and all domain parameters, is to try an exhaustive search for k by computing an exhaustive number of point additions and doublings until the answer you get is equivalent to kP. For a 160-bit value of k one would perform 2^{160} guesses of k, each time computing kP, thus illustrating the hardness of the problem. In ECC there are choices for fields, curves, points on the curve, point representation, methods for computing kP, and methods for computing each point double or point addition. This chapter will first discuss the high-level computations involved in ECC, known as scalar multiplications, and then address the detailed mathematics involved in field arithmetic used to perform point doubling and point addition.

The terminology and definition of some mathematical terms that will be used in this chapter are provided below for a quick reference. The specific elliptic curve definitions and terminology are also provided in a second separate table. Additional mathematical definitions that may be useful are those listed in the previous chapters.

In general, there are different levels within ECC as shown in Fig. 5.1. At the highest level there are the ECC protocols and high-level algorithms. Below this level there is scalar multiplication. Next are the point addition and point doubling

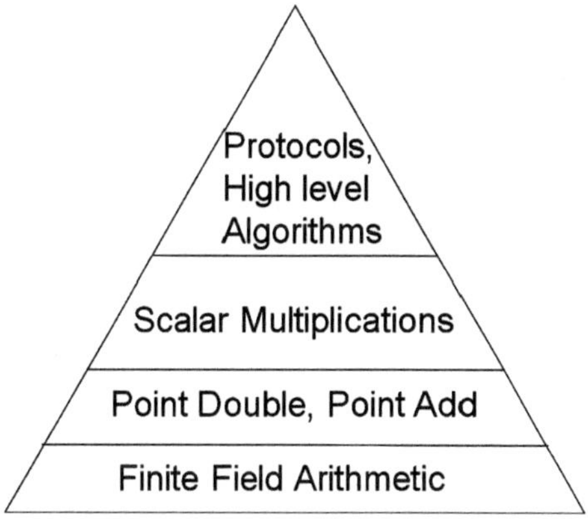

Fig. 5.1 Levels of elliptic curve cryptography

algorithms. Finally the bottom level is composed of the field arithmetic. Readers who are interested only in a high-level understanding of elliptic curves should read general high-level elliptic curve computations in Sect. 5.1 and EC protocols in Sect. 5.3, whereas readers who wish for a more in-depth understanding of the field mathematics involved should read all sections of this chapter and many references provided at the end of this chapter.

General terminology and definitions	
Integer	Represented by a set of coefficients subscripted by the base, e.g., an integer in base b is represented m digits specifically, $(a_{m-1}\ a_{m-2}\ldots a_1a_0)_b$. The conversion of this integer to base 10 is formed by $(a_{m-1}b^{m-1} + a_{m-2}b^{m-2} + \ldots + a_2b^2 + a_1b^1 + a_0)_{10}$
Group	A group consists of a set, G, and operations, *, which satisfy a number of rules. For example, group operations are associative [e.g., $(a^*b)^*c = a^*(b^*c)$ for operation * on elements a, b, c in the set]; they have an identity element (e.g., $a^*1 = a$), and each element from the set has an inverse [e.g., $a^*i = 1$, where i is inverse of a, typically denoted as a^{-1}]. An Abelian group is a group that is also commutative (e.g., $a^*b = b^*a$)
Ring	A ring $(G, +, \times)$ is a group, G, which has two operations, addition, + [where group $(G, +)$ forms an Abelian group with additive identity of 0], and multiplication, (with associative, distributive property and multiplicative identity of 1). The ring is commutative if it also has the commutative property (e.g.,$a \times b = b \times a$). Z_n with addition, multiplication performed modulo n is a commutative ring
Field	A field is a commutative ring where every nonzero element is invertible. Inversion is shown using superscript –1. For example, the inversion of a is $(a)^{-1}$, defined as $(a)^{-1}\ mod\ p = c$, where $a(c)\ mod\ p = 1$.
F_q	A finite field of order q is specifically a commutative ring in which all nonzero elements have multiplicative inverses.

(continued)

(continued)

General terminology and definitions	
	Z_q is a finite field (under operations of addition and multiplication modulo q) if and only if q is prime (in which case we refer to F_q as a prime field). For prime fields, F_q, all operations are performed using normal arithmetic (addition, multiplication) except results are modulo q. Finite fields may also be referred to as Galois fields
$F_q{}^*$	The nonzero elements of F_q form a multiplicative group denoted as F_q^* (F_q^* is a multiplicative group if all elements are invertible, e.g., for all $u \in Z_q^*$, $(u u^{-1}) = 1 \mod n$)
Operations on $F_q{}^*$	$g, h \in F_q{}^*$ $(g)(h)$ g^{-1} g/h g^x
$F_p{}^m$	An extension finite field (binary finite field for $p = 2$ case)
$F_2{}^m$	Binary finite field represented by 2^m elements, each binary polynomial of degree less than m over Z_2
	$f(x)$ any irreducible binary polynomial of degree m over Z_2 (aka the reduction polynomial) $a(x)$ A binary polynomial of degree less than m such as: $a\,(x) = a_{m-1}x^{m-1} + a_{m-2}x^{m-2} + \ldots + a_2x^2 + a_1x^1 + a_0,\ a_i \in Z_2$ for simplicity represented by its coefficient vector (aka polynomial basis representation): $(a_{m-1}a_{m-2}\ldots a_0)$ $a(x) + b(x)$ Polynomial addition (is equivalently addition modulo 2, which is the exclusive or of coefficient vectors): $(a_{m-1} \oplus b_{m-1})\ (a_{m-2} \oplus b_{m-2}) \ldots (a_0 \oplus b_0)$ $a(x)\,b(x)$ Multiplication is modulo $f(x)$ (multiplication is performed normally but followed by modular reduction by the polynomial $f(x)$) $(a(x))^2$ Squaring is also performed modulo $f(x)$ but becomes $\left((a_{m-1})^2\,(a_{m-2})^2 \ldots (a_0)^2\right)$

Elliptic curve terminology and definitions	
$E(F_q)$	Set of all points on the elliptic curve over F_q, specifically: F_q is a set of integers modulo a prime q ($q > 3$) $E(F_q)$ is set of all points (x, y), $x, y \in F_q$, which satisfy equation of elliptic curve over F_q: $y^2 = x^3 + ax + b$ together with the point at infinity (∞, ∞), where $a, b \in F_q$, $(4a^3 + 27b^2) \mod q \neq 0$ (nonequality requirement for curve to be elliptic). All the points that are on this curve, along with the point at infinity, form an Abelian group, specifically $(E(F_q), +)$, where the point at infinity is the additive identity. This is the Weierstra β equation for an elliptic curve. (special curves: if q is a Mersenne-like prime)

(continued)

(continued)

Elliptic curve terminology and definitions	
$E(F_2{}^m)$	Set of all points on the elliptic curve over $F_2{}^m$, specifically: $E(F_2{}^m)$ is set of all points (x, y), $x, y \in F_2{}^m$, which satisfy equation of elliptic curve over $F_2{}^m$: $y^2 + xy = x^3 + ax^2 + b$ together with the point at infinity (∞, ∞), where $a, b \in F_2{}^m, b \neq 0$ (nonequality requirement for curve to be elliptic). All the points that are on this curve, along with the point at infinity, form an Abelian group, specifically $(E(F_2{}^m), +)$, where point at infinity is the additive identity. (special curves: if m is composite (composite fields);$a, b \in F_2$, aka binary anomalous curve or Koblitz curve)
$P \in E(F_q)$ [or $P \in E(F_2{}^m)$]	Point $P = (x, y)$ on the elliptic curve $x, y \in F_q$ [or $x, y \in F_2{}^m$]
Operations on $E(F_q)$ or $E(F_2{}^m)$	$P, Q \in E(F_q)$ $P + Q$ $-P$ $P - Q$ xP
Cyclic subgroup of $E(F_q)$	For any point G on $E(F_q)$, the cyclic subgroup is $(0G, 1G, 2G, 3G, \ldots)$

5.1 High-Level Elliptic Curve Computations

This section will briefly introduce scalar multiplication, which can be performed with a point on an elliptic curve over a field. The detailed low-level mathematics behind ECC will be presented in Sect. 5.2. However, in this section we will discuss elliptic curves at the high level without providing details so that the reader can get an appreciation for high-level ECC protocols and the types of computations involved. In particular, we will focus on scalar or point multiplication. This is the focus of ECC and at the heart of all protocols, which will be described in Sect. 5.3. It is also the sole computation known for requiring significant amounts of computing effort. At the end of this section we will talk about ongoing efforts to improve the performance at this high level of computation.

Specifically for two points, P and Q on the curve we can define the smallest set of operations to be only addition. Of course, we can build other operations such as subtracting (since we can also form the negative of any point), doubling, tripling, quintupling, etc. Point multiplication, also referred to as scalar multiplication (specifically the computation of kP, where P and kP are both points on the elliptic curve and k is the secret key), is used in session key generation, signature generation, signature verification (for authenticity), and so on. In ECC, there is no multiplication operation; hence, the only method for computing kP is by doubling and adding (or summing). In general, this is why elliptic curves are so secure since the problem of finding k given P and kP is a very hard problem.

Algorithms for performing scalar multiplication or specifically the computation of *kP* for a point P on the elliptic curve will be discussed next. Scalar multiplication is also referred to as point multiplication. These algorithms are typically divided into two types. There are algorithms that work for any P (discussed here) and those for a specific value of P (or window methods to be discussed in the next section, where precomputed tables using P are utilized). For example, let us assume that k is equal to 11 (in our base 10 system) or using a binary representation $(00001011)_2$. An algorithm for computing the scalar multiplication, when $k = (k_{n-1}\ k_{n-2}\ldots k_0)_2$ is in binary form, is given in Algorithm 5.1, where the most significant bit $(n-1)$ of k is a "1" (which in our case, using k equal to 11, n = 4):

$$kP = \sum_{i=0}^{i=n-1} k_i 2^i P, \quad k_i \in \{0, 1\}. \tag{5.1}$$

Algorithm 5.1. Double and add point multiplication $kP|k_i \in \{0,1\}$

1. $Q = P$
2. For $i = n-2$ to 0
   a. $Q \leftarrow 2Q$
   b. If $k_i = 1$ then $Q \leftarrow Q + P$
3. Return Q

Other forms or bases are possible for representing k and computing *kP*. For example, consider using the basis set $\{0, \pm 1\}$. The introduction of negative coefficients works due to the well-known relationship (Koc 1994) $2^{i+j-1} + 2^{i+j-2} + \ldots + 2^i = 2^{i+j} - 2^i$.

For example, the value $(1100)_2 = 12$ can be represented by $(1\ 0 -1\ 0\ 0)_2 = 16 - 4 = 12$, which is a result of using $i = 2, j = 2$. As another example consider $k = (0000\ 1011)_2 = (11)_{10}$, which can be transformed several times as $(0\ 1\ 1\ 0\ -1)_2$ with $i = 0, j = 2$ for replacing bit positions 0 and 1, and finally to $(1\ 0\ -1\ 0\ -1)_2$ with $i = 2, j = 2$ for replacing bit positions 2 and 3 (calculated as $(1)2^4 + (0)2^3 + (-1)2^2 + (0)2^1 + (-1)2^0 = 16-4-1 = 11$). In EC mathematics the efficiency of point subtraction is typically assumed to be equivalent to the efficiency of a point addition (see Sect. 5.2 for details). An algorithm for computing $kP = \sum_i k_i 2^i P,\ k_i \in \{0, \pm 1\}$ is shown in Algorithm 5.2.

Algorithm 5.2. Left to right point multiplication $kP|k_i \in \{0, \pm 1\}$

1. $Q = P$
2. For $i = n-2$ to 0
   a. $Q \leftarrow 2Q$
   b. If $k_i = 1$ then $Q \leftarrow Q + P$
   c. If $k_i = -1$ then $Q \leftarrow Q - P$
3. Return Q

Of course, this algorithm also works if k is strictly in a binary representation with no 1 digits. To further illustrate this left to right algorithm, the computations performed specifically for calculating the point kP, where $k = (1\ 0\ -1\ 0\ -1)_2 = (0b)_{16}$, are shown below in Fig. 5.2.

There are many variations of the double and add algorithm, which have differing properties. An interesting variation of the double and add algorithm is the Montgomery Ladder, which is presented in Algorithm 5.3. It has the property of computing both a double and add during each loop iteration. Although this seems to be energy wasteful, it turns out to be useful for resisting a side channel attack (see Chap. 8). This algorithm is further illustrated in Fig. 5.3.

Algorithm 5.3. Montgomery ladder

1. $R_0 = P, R_1 = 2P$
2. For $i = n - 2$ to 0
    a. $b \leftarrow 1 - k_i$
    b. $R_b \leftarrow R_b + R_{kj}$
    c. $R_{kj} \leftarrow 2R_{kj}$
3. Return R_0

Algorithm	Iteration	Computation
1. Q=P 2. For i = n-2 to 0 a. Q←2Q b. If k_i = 1 then Q←Q+P c. If k_i = -1 then Q←Q-P 3. Return Q	*n = 5*	Q=P *(line 1.)*
	i=3, *k_i=0*	Q←(2P) *(line 2.a.)*
	i=2, *k_i=-1*	Q←2(2P) *(line 2.a.)* Q←2(2P)-P *(line 2.d.)*
	i=1, *k_i=0*	Q←2[2(2P)-P] *(line 2.a.)*
	i=0, *k_i=-1*	Q←2(2[2(2P)-P]) *(line 2.a.)* Q←2(2[2(2P)-P])-P *(line 2.d.)*
		Exits

Fig. 5.2 Double and add example with $k = (1\ 0\ -1\ 0\ -1)_2$

Algorithm	Iteration	Computation
1. R_0=P, R_1=2P 2. For i = n-2 to 0 a. b←1- k_i b. R_b←R_b+R_{kj} c. R_{kj}← 2R_{kj} 3. Return R_0	*k=(1011)$_2$* *n= 4*	R_0=P, R_1=2P *(line 1.)*
	i=2, *k_i=0,* b←1	R_1=2P+P *(line 2.b.)* R_0=2P *(line 2.c.)*
	i=1, *k_i=1,* b←0	R_0=2P+3P *(line 2.b.)* R_1=2(3P) *(line 2.c.)*
	i=0, *k_i=1*	R_0=5P+6P *(line 2.b.)* R_1=2(6P) *(line 2.c.)*

Fig. 5.3 Montgomery ladder example with $k = (1011)_2$

The next section will examine other forms of the general scalar multiplication algorithm, which aim to improve the performance. They specifically focus on the number representation of k as well as precomputation approaches for improving performance.

5.1.1 Performance Improvements

This section will describe various approaches aimed at improving the performance of the scalar multiplication task. Since point doubling is typically more efficient than point adding (see Sect. 5.3), one way of optimizing computations is to try to reduce the number of point additions. The nonadjacent form (NAF) for representing k is an example of this. In NAF, the representation of k is a series of digits from the set $\{0, +1, -1\}$ such that there are no two sequential nonzero elements. However more important the NAF of a number has the fewest number of nonzero elements for any signed digit representation of a number. Thus, fewer nonzero elements mean fewer point additions and therefore more efficiency (assuming that the number of doublings does not increase significantly, thus offsetting the gains from the reduction of point additions). Although it has been shown that in general there are an infinite number of representations for numbers using digits $\{0, \pm 1\}$ [see Prodinger (2000) for further interesting discussions], it turns out that the NAF of an integer is in fact unique. Furthermore, there are efficient algorithms for computing the NAF of a number, such as those listed in Hankerson et al. (2000).

In many ECC protocols, the base point, P, is fixed. In other words the protocol will be executed several times, each time with the same base point. In these cases it is extremely efficient to precompute several point multiples and store them to be used during all relevant point multiplication computations. In some instances even if the base point changes, it may still be worthwhile to perform precomputations in order to speed up the point multiplication. The precomputation approach is referred to as a window method. It is not worthwhile to store the even multiples of the base point since they can all be computed using the efficient point doubling algorithm. However, it does make sense to store the odd multiples of the base point since these otherwise would require the inefficient point addition. For example, let us assume that the precomputed points are $3P$, $5P$, 7P. In order to compute $11P$ we can use these precomputed points as follows: $2(5P) + P = 11P$, by loading (5P), doubling it, and adding P to it. Thus, this computation requires only one point doubling and one point addition compared to the computation for the usual binary representation, $(11)_{10} = (1011)_2$, requiring three doubles and two additions (or alternatively as shown in the last section using NAF, $(11)_{10} = (10\,-1\,0\,-1)_{naf}$, requiring four doubles and two additions). This technique precomputes odd multiples of P and stores them in a table and calculates kP using $kP = \sum_i 2^i(k_i P)$, where $k_i P$ is stored for $|k_i| < 2^{w-1}$ for a window of size w (known as window w-NAF or wNAF). Note that negative values do not need to be stored since they can easily be obtained from the stored positive values as will be shown in Sect. 5.2. For example, consider the scalar $k = (10\ 01\ 10\ 00\ 00\ 11)_2$. Using a window of width 2, the scalar is recoded as (2 1 2 0 0 3) using six windows. The scalar value of k is split into pieces each of size w digits long. A variant of the window method is called the sliding window method (Cohen and Frey 2006), and it allows the algorithm to skip over series of zeros. For example, the sliding window would recode our example for $k = (10\ 01\ 10\ 00\ 00\ 11)_2$ as $(1\ 00\ 3\ 00000\ 3)_2$, thus requiring fewer additions as well. There are other window algorithms as well such as the fractional width-w NAF algorithm [where a more flexible number of points is stored in the precomputation table; see Moller (2001)], etc. Progress on performance-optimized precomputation of points has also been made (Longa and Gebotys 2009b). An example of a window-based NAF point multiplication is given in Algorithm 5.4.

Algorithm 5.4. Left to right window NAF method point multiplication

1. $Q = P$
2. For $i = n - 2$ to 0
   a. $Q \leftarrow 2Q$
   b. If $k_i \neq 0$
      i. If $k_i > 0$ then $Q \leftarrow Q + P_{ki}$
      ii. else $Q \leftarrow Q - P_{ki}$
3. Return Q

For example, consider the binary scalar $k = (1\,0 - 10 - 1)_2 = (11)_{10}$. Using a window of three bits we can transform this into $NAF(k) = (1\,0\,0\,0 - 5)_2$. Using Algorithm 5.4 we iterate through the loops as $Q = 2P$, $Q = 2(2P)$, $Q = 2(2(2P))$, $Q = 2(2(2(2P))) - 5P$, which finally gives $Q = 11P$.

Another technique that may improve the performance of point multiplications is to study bases other than base 2. For example, in base 3 one would use a triple and add algorithm (Cohen and Frey 2006) analogous to the left to right point multiplication or double and add algorithm. For example, using a base 3 representation for a number may lead to a faster implementation if the computation of $3P$ is faster than the equivalent base 2 computation of $2P + P$ using a double and add approach (Ciet et al. 2006). Recently multibase scalar multiplication has been researched in attempts to further improve the performance of elliptic curve computations. Specifically double-base number system (DBNS) introduced by Dimitrov et al. (1997) and then generalized to multibases by Longa (2007) and Mishra and Dimitrov (2007) were developed. A general multibase number representation for an integer, k, where $B = \{b_1, b_2, \ldots b_l\}$ is the set of bases, is as follows (Mishra and Dimitrov 2007):

$$k = \sum_{j=1}^{m} s_j \left(\prod_{i=1}^{l} b_i^{e_{ji}} \right), s_j \in \{\pm 1\}.$$

For example, using base set $B = \{2, 3\}$ we can represent the integer $k = 23$ as $k = 2^1 3^2 + 2^0 3^1 + 2^1 3^0$, or $k = 2^4 3^0 + 2^1 3^1 + 2^0 3^0$, etc. Depending upon which representation is used, a different number of point triplings, doublings, and additions would be required. Initially, double-base representations for k were found using a greedy approach, which is costly in terms of memory (Dimitrov et al. 2005; Doche and Imbert 2006). However, more efficient and memory-friendly algorithms have been developed, which efficiently find a multibase chain for a given window width and set of bases, which is unique for each integer (Longa 2007), later refined, and extended to fractional windows in Longa and Gebotys (2009a).

Whether or not it is faster depends upon a number of factors including the field computations required for each point operation, type of curve, choice of coordinate system, which bases are used, etc. Section 5.2 will provide some details on these lower level computations required and the factors which impact their efficiency.

5.2 The Mathematics Behind Elliptic Curves

This section will first examine the mathematics of the elliptic curve including points that exist on the curve. Next it will examine the low-level computations involved in point doubling, point adding, and point negation. Details of different point representations are finally discussed with emphasis on performance.

5.2.1 The Curve Over a Field

The elliptic curve and its properties as well as point computations on the curve will be outlined in this section. The objective of this section is to introduce elliptic curve mathematics in a very simplistic manner without significant complexity. However, readers wishing for a more in-depth treatment should refer to Cohen and Frey (2006) as well as Hankerson et al. (2004).

As mentioned at the beginning of this chapter, elliptic curves support the notion of adding two points on the curve to obtain another point on the curve. To define this addition operation on field elements it is necessary to also define an identity element. The identity element is called the point at infinity. Before we discuss the point at infinity, let us examine points on an elliptic curve. The addition rules works as follows: (1) draw a line between the two points on the curve that you are going to add, (2) the negative of the third point where the line intersects the curve is the point representing the addition of the first two points. Now getting back to the point at infinity, if we add two points where one point is (x, y) and the other point is $(x, -y)$, when we draw a line through the two lines, it in fact becomes a vertical line. In order to support our addition rule, we say that this third point is in fact the point at infinity, (∞, ∞). Though it is not very useful in practice, in theory the point at infinity is important for defining our addition rule and thus supports the notion of a group.

In general the finite field (also known as the Galois field) over an elliptic curve forms a group. The group elements are points, such as $P = (x, y)$, on the curve plus the point at infinity, (∞, ∞). In its simplest form it allows one to extend the real number line into a real projective line where increasing positive numbers extend to the same infinity as increasing negative numbers. Thus, it joins two ends of the real line and is referred to as a real projective line. In ECC this concept provides the projective version of the elliptic curve so that the addition operation can be supported. The point at infinity is known as the identity element since $P+(\infty,\infty) = P$ and $P + (-P) = (\infty,\infty)$.

A general elliptic curve is defined in cryptography as a smooth curve E over a field K described by the Weierstraß equation:

$$E : y^2 + axy + by = x^3 + cx^2 + dx + e \quad a, b, c, d, e \in K. \tag{5.2}$$

The field K is more generally represented as $K = F_{p^m}$. Using this representation we can change p and m to create two different kinds of fields, which are studied in ECC. These two types of fields are the prime fields (where m =1, thus the field F_p) and the general extension fields (where $m \geq 2$). Binary fields (where $p^m = 2^m$) are a special case of the general extension fields. In this chapter we will study these two common fields where the elliptic curve is defined over a prime field and a binary field.

5.2.1.1 Prime Fields

In prime fields, F_p, the field elements are integers from 0 to $(p - 1)$ and field operations are performed *modulo p*. In prime fields the general form of elliptic curve from (5.2) is used except $a = b = c = 0$. Hence, an example of a general elliptic curve (the *Weierstraß* curve), E, over a prime field F_p [a set of integers, from 0 to $(p - 1)$, modulo a prime $(p > 3)$] can be written as:

$$E : y^2 = x^3 + ax + b \quad a, b \in F_p, \tag{5.3}$$

where a, b [equivalently d, e from (5.2) but given new names here for simplicity] are also defined as integers modulo p such that the following equation holds: $4a^3 + 27b^2 \neq 0 (\text{mod}\, p), \quad a, b \in F_p$. This last nonzero requirement is added to ensure that the elliptic curve does not intersect itself or cross.

For illustration purposes, consider the elliptic curve in (5.3) with $a = b = 1$ defined over the finite field modulo prime $p = 17$. The EC equation is $y^2 = x^3 + x + 1$. Now let us determine some points on the curve. By solving the left and right hand sides of the curve for various values of x and y we can generate a table, from which we can extract points. In Table 5.1 we can see the left hand side and the right hand side values of the curve equation for various values of x and y. By matching the left and right hand side values that are equivalent, we can determine the corresponding x and y values that represent a point on the curve.

Examining Table 5.1, we can see that a valid point is (4,1) since when $x = 4$ we have the right hand side of equation equal to 1, specifically $(4^3+4+1)\ mod\ 17 = 1$, since we have $4^3+4+1 = 69 = 4(17)+1$; the sixth row fourth column of Table 5.1.

Table 5.1 Evaluating elliptic curve equation (5.3) for $a = b = 1$, $p = 17$

y	Left hand side of curve = (y^2) mod 17	x	Right hand side of curve = $(x^3 + x + 1)$ mod 17
0	0^2 *mod* 17 = 0	0	$(0^3 + 0 + 1)$ *mod* 17 = 1
1	1^2 *mod* 17 = 1	1	$(1^3 + 1 + 1)$ *mod* 17 = 3
2	2^2 *mod* 17 = 4	2	$(2^3 + 2 + 1)$ *mod* 17 = 11
3	3^2 *mod* 17 = 9	3	$(3^3 + 3 + 1)$ *mod* 17 = 14
4	4^2 *mod* 17 = 16	4	$(4^3 + 4 + 1)$ *mod* 17 = 1
5	5^2 *mod* 17 = 8	5	$(5^3 + 5 + 1)$ *mod* 17 = 12
6	6^2 *mod* 17 = 2	6	$(6^3 + 6 + 1)$ *mod* 17 = 2
7	7^2 *mod* 17 = 15	7	$(7^3 + 7 + 1)$ *mod* 17 = 11
8	8^2 *mod* 17 = 13	8	$(8^3 + 8 + 1)$ *mod* 17 = 11
9	9^2 *mod* 17 = 13	9	$(9^3 + 9 + 1)$ *mod* 17 = 8
10	10^2 *mod* 17 = 15	10	$(10^3 + 10 + 1)$ *mod* 17 = 8
11	11^2 *mod* 17 = 2	11	$(11^3 + 11 + 1)$ *mod* 17 = 0
12	12^2 *mod* 17 = 8	12	$(12^3 + 12 + 1)$ *mod* 17 = 7
13	13^2 *mod* 17 = 16	13	$(13^3 + 13 + 1)$ *mod* 17 = 1
14	14^2 *mod* 17 = 9	14	$(14^3 + 14 + 1)$ *mod* 17 = 5
15	15^2 *mod* 17 = 4	15	$(15^3 + 15 + 1)$ *mod* 17 = 8
16	16^2 *mod* 17 = 1	16	$(16^3 + 16 + 1)$ *mod* 17 = 16

When $y = 1$ the left hand side of the equation on the third row and second column is equal to 1, specifically 1^2 *mod* $17 = 1$. Since the left hand side is equal to the right hand side, the pair $(x, y) = (4, 1)$ represents a valid point on the curve. Other points on the curve found in this table are (0,1), (0,16), (4,1), (4,16), (6,6), (6,11), (9,5), (9,12), (10,5), (10,12), (11,0), (13,1), (13,16), (15,5), (15,12), (16,4), and (16,13).

Hasse theorem states that the number of points on the elliptic curve over finite field F_p is bounded by $(p + 1 \pm 2p^{0.5})$ (Hankerson et al. 2004). The exact number of points can also be computed exactly in polynomial time using Schoof's algorithm (Hankerson et al. 2004). Next we will discuss the elliptic curve defined over binary fields, illustrating points on this curve.

5.2.1.2 Binary Fields

Elliptic curves can also be defined over binary fields, F_{2^m}. In these binary fields the elements are not integers as in the case of prime fields. Instead the elements are binary polynomials (or polynomials with binary coefficients) of degree less than m. For example, $(x + 1)$ represents an element of the field. These binary polynomials can also be represented in their binary numeric form (c_3, c_2, c_1, c_0), representing $c_3x^3 + c_2x^2 + c_1x^1 + c_0x^0$, where $c_i \in \{0, 1\}$. For example, in the field F_{2^4}, where $m = 4$, the element $(x + 1)$ could be represented in binary form as (0011). The operations on this field include addition and multiplication. Addition is performed using the exclusive-or operation. For example, $(x + 1) + (x^3 + x) = (x^3 + 1)$

since the exclusive-or cancels out the x (e.g., $x + x = 0$, similarly to $1 + 1 = 0$ in arithmetic modulo 2, specifically exclusive-oring 1 with 1 is 0). In the binary representation this is easier to see since equivalently exclusive-oring follows $(0011) + (1010) = (1001)$, which is equivalent to $(x^3 + 1)$ in polynomial representation. Multiplication can be performed similarly except the resulting value is modularly reduced by the irreducible polynomial as we will shortly show. An irreducible polynomial is one that cannot be factored. For example, for the binary field F_{2^4} there are three irreducible polynomials, which are $f(x) = x^4 + x + 1$, $f(x) = x^4 + x^3 + 1$, and $f(x) = x^4 + x^3 + x^2 + x + 1$ (Hankerson et al. 2004). For illustration purposes we will use both polynomial as well as binary representation in this section.

Consider the elliptic curve, E, defined over the binary field [defined from (5.2) with coefficients $a = 1, b = c = 0$, where d, e are equivalently shown in (5.4) as a, b] given as follows:

$$E : y^2 + xy = x^3 + ax^2 + b \quad a, b \epsilon F_{2^m}, b \neq 0. \tag{5.4}$$

Lets also consider the binary field F_{2^4}, where $m = 4$ in (5.4), represented by the irreducible polynomial $f(x) = x^4 + x + 1$ and $a = \alpha^4$ [in polynomial form using alpha so there is no confusion with the x variable of the curve equation, see discussion which follows *or* $a = (0011)$] and $b = 1[or\ b = (0001)]$. Hence, the elliptic curve is shown below in (5.5) with $\alpha = x[or\ (0010)]$ used to generate F_2^4.

$$y^2 + xy = x^3 + \alpha^4 x^2 + 1. \tag{5.5}$$

We can show that α is a generator for the field represented by the polynomial $f(x)$. From Chap. 4 recall that powers of the generator create all elements in the field. For example, by computing $\alpha, \alpha^2, \alpha^3$, etc., we would observe that the numbers generated repeat themselves from α^{15}. For example, the values for all powers of in binary and polynomial form are as follows:

$$\begin{array}{ll}
\alpha^1 = (0010) = x, & \alpha^2 = (0100) = x^2, \\
\alpha^3 = (1000) = x^3, & \alpha^4 = (0011) = x + 1, \\
\alpha^5 = (0110) = x^2 + x, & \alpha^6 = (1100) = x^3 + x^2, \\
\alpha^7 = (1011) = x^3 + x + 1, & \alpha^8 = (0101) = x^2 + 1, \\
\alpha^9 = (1010) = x^3 + x, & \alpha^{10} = (0111) = x^2 + x + 1, \\
\alpha^{11} = (1110) = x^3 + x^2 + x, & \alpha^{12} = (1111) = x^3 + x^2 + x + 1, \\
\alpha^{13} = (1101) = x^3 + x^2 + 1, & \alpha^{14} = (1001) = x^3 + 1, \\
\alpha^{15} = (0001) = 1. &
\end{array}$$

Let us illustrate how these values are attained with an example. Consider computing the value of α^5. It is computed as follows by first substituting $\alpha = x$ into α^5 to get $x^5\ mod\ (x^4 + x + 1)$. The modular reduction is equivalent to the remainder after dividing x^5 by $(x^4 + x + 1)$. The division shown below obtains a remainder of $(x^2 + x)$. Thus, $\alpha^5 = x^5 \; mod \; (x^4 + x + 1) = (x^2 + x) = (0110)$.

$$
\begin{array}{r|l}
 & \quad\quad x \\
\hline
x^4 + x + 1 & x^5 \\
 & x^5 + x^2 + x \\
\cline{2-2}
 & \quad\quad x^2 + x
\end{array}
$$

Now let us get back to points on the elliptic curve. Consider the point $(x, y) = (\alpha^3, \alpha^8) = ((1000), (0101))$. Let us show that this point is in fact on the curve $y^2 + xy = x^3 + \alpha^4 x^2 + 1$ defined over the field F_{2^4} represented by the irreducible polynomial $f(x) = x^4 + x + 1$. By substituting the point into the left hand side of the elliptic curve equation we can show that its value is equivalent to the right hand side value. Substituting the values of x and y from the point $(x, y) = (\alpha^3, \alpha^8)$ into the left hand side of the curve becomes:

$$y^2 + xy = (\alpha^8)^2 + (\alpha^3)(\alpha^8).$$

Next we substitute the equivalent polynomial values for α^3 and α^8 into the equation and reduce it (recalling that $x^2 + x^2 = 0$).

$$
\begin{aligned}
&= \left(x^2 + 1\right)^2 + \left(x^3\right)\left(x^2 + 1\right) \\
&= x^4 + x^2 + x^2 + 1 + x^5 + x^3 \\
&= x^4 + 1 + x^5 + x^3
\end{aligned}
$$

Now we have to reduce this to an element in the field by performing modular reduction with the polynomial $f(x) = x^4 + x + 1$.

$$
\begin{aligned}
&= x^5 + x^4 + x^3 + 1 \; modulo \; (x^4 + x + 1) \\
&= x^3 + x^2
\end{aligned}
$$

Again the modulo computation that takes us from the second last line to the bottom line is the remainder from the polynomial division computation given below:

$$
\begin{array}{r|l}
 & \quad\quad\quad\quad\quad\quad\quad\quad x + 1 \\
\hline
x^4 + x + 1 & x^5 + x^4 + x^3 \quad\quad\quad\quad\quad + 1 \\
 & x^5 \quad\quad\quad\quad + x^2 + x \\
\cline{2-2}
 & \quad\quad x^4 + x^3 + x^2 + x + 1 \\
 & \quad\quad x^4 \quad\quad\quad\quad\quad + x + 1 \\
\cline{2-2}
 & \quad\quad\quad\quad x^3 + x^2
\end{array}
$$

We can now substitute into the right hand side of the curve and using similar computations we obtain the same value as shown below.

$$\begin{aligned}
&x^3 + \alpha^4 x^2 + 1 \\
&= (\alpha^3)^3 + (\alpha^4)(\alpha^3)^2 + 1 \\
&= \left(x^3\right)^3 + (x + 1)\left(x^3\right)^2 + 1 \\
&= x^9 + x^7 + x^6 + 1 \; modulo \; (x^4 + x + 1) \\
&= x^3 + x^2
\end{aligned}$$

Again the modulo reduction comes from the remainder of the polynomial division shown below.

```
                     x5 +     x3    + x  +1
x4 +x+1 | x9 + x7  + x6               + 1
          x9       + x6 + x5
          ---------------------
               x7       + x5            + 1
               x7            + x4 + x3
               ----------------------
                    x5 + x4 + x3        + 1
                    x5            + x2 + x
                    ---------------------
                         x4 + x3+ x2 + x + 1
                         x4           + x + 1
                         -------------------
                              x3+ x2
```

As shown both the left and right hand side reduce to $x^3 + x^2$ *or* (1100). Since they are equal we have proof that the point $(\alpha^3, \alpha^8) = ((1000), (0101))$ lies on the curve $y^2 + xy = x^3 + \alpha^4 x^2 + 1$ over the binary field F_{2^4} represented by the irreducible polynomial $f(x) = x^4 + x + 1$.

The field computation details will be discussed in the next section.

5.2.2 *Point Computations*

In order to compute with points on the elliptic curve one has to have equations for computing point addition and point doubling. These equations will be presented in this section, though they can also be derived from the roots of cubic equations (Rosing 1999) or more formally from the geometry of elliptic curves (Cohen and Frey 2006). The equations will be illustrated with examples. However, for implementation details Sect. 5.2.3 will cover specific approaches for optimizing field computations for performance. We will continue to describe the two cases, prime fields and binary fields. In this section we will start with prime fields since the mathematics is easier to understand.

5.2.2.1 Point Computations Over Prime Field

In prime fields the negative of a point $P = (x, y)$ is equal to $-P = (x, -y)$. The equation for adding two points over the EC: $y^2 = x^3 + ax + b$ over F_p [curve from (5.3)], specifically adding $P = (x_1, y_1)$ and $Q = (x_2, y_2)$, to obtain $P + Q = (x_3, y_3)$, $P \neq (\infty, \infty)$ is given in (5.6).

$$\lambda = \begin{cases} \dfrac{3{x_1}^2 + a}{2y_1}, P = Q \\ \dfrac{y_2 - y_1}{x_2 - x_1}, P \neq Q \end{cases},$$
$$x_3 = \lambda^2 - x_1 - x_2,$$
$$y_3 = \lambda(x_1 - x_3) - y_1. \tag{5.6}$$

From (5.6), it can be seen that point doubling requires two squarings $(\lambda^2, {x_1}^2)$, two multiplications (one to create y_3 and one to create λ), and one inversion $((x_2 - x_1)^{-1})$. Point addition requires one squaring (λ^2), two multiplications (one to create y_3 and one to create λ), and one inversion $((x_2 - x_1)^{-1})$.

For example, the curve, $y^2 = x^3 + x + 1$ is defined over a prime field F_{17} ($a = b = 1, p = 17$) as introduced in Sect. 5.2.1. Point $P = (4, 1)$ was shown to exist on this curve. Let us now compute the doubling of this point, $2P = (x_3, y_3)$. Applying the equation where $2P = P + P$, we obtain $\lambda = (3(4)^2 + 1)/(2(1))\ mod\ 17 = 16$ and $x_3 = ((16)^2 - 4 - 4)\ mod\ 17 = 10$ and $y_3 = (16)(4 - 10) - 1\ mod\ 17 = 5$. Hence, $2P = (10, 5)$, which we have previously verified to also lie on the elliptic curve as expected. To compute $3P$ we can now add P with $2P$. Again using the equations we compute $\lambda = (5 - 1)/(10 - 4)\ mod\ 17 = 12$, and $x_3 = (12)^2 -$ 10-4 $mod\ 17 = 11$ *and* $y_3 = 12(4 - 11) - 1\ mod\ 17 = 0$. Hence, $3P = (11, 0)$, which is also a point on the elliptic curve. We can continue to double and add points on the curve using the formula.

5.2.2.2 Point Computations Over Binary Field

This section will discuss binary fields, starting with point addition for elliptic curves. In binary fields the negative of a point $P = (x, y)$ is equal to $-P = (x, x + y)$. The point addition of $P = (x_1, y_1)$ and $Q = (x_2, y_2)$ over binary fields is computed as $P + Q = (x_3, y_3)$ defined by (5.7) where $P \neq \pm Q$.

$$\lambda = \frac{y_1 + y_2}{x_1 + x_2}.$$
$$x_3 = (\lambda)^2 + (\lambda) + (x_1 + x_2) + a.$$
$$y_3 = (\lambda)(x_1 + x_3) + y_1 + x_3. \tag{5.7}$$

Point doubling formulas ($P \neq -P$) are shown in (5.8), where $P = (x_1, y_1)$ and $2P = (x_{12}, y_{12})$.

$$x_{12} = (x_1)^2 + \left(\frac{b}{(x_1)^2}\right).$$

$$y_{12} = (x_1)^2 + \left(x_1 + \frac{y_1}{x_1}\right)(x_3) + x_3. \tag{5.8}$$

From (5.7) we can see that point addition requires one squaring, two multiplications, and one inversion. Point doubling in (5.8) requires two squarings, two multiplications, and one inversion.

Now let us illustrate point doubling and point addition over binary fields. We will consider the elliptic curve previously used for illustration purposes, specifically $y^2 + xy = x^3 + \alpha^4 x^2 + 1$ over the binary field F_{2^4} represented by the irreducible polynomial $f(x) = x^4 + x + 1$ [where the standard curve equation has $a = \alpha^4 = (0011)$ and $b = 1 = (0001)$]. We will illustrate point doubling and then point addition by computing with $P = (x, y) = (\alpha^3, \alpha^8) = ((1000), (0101))$. First we will compute $2P$ using (5.8) as shown below in the point doubling example.

Example 5.1. x-Coordinate point doubling of $P = (\alpha^3, \alpha^8)$

1. $x_{12} = (\alpha^3)^2 + (1/(\alpha^3)^2)$
2. $= \alpha^6 + (1/\alpha^6)$
3. $= \alpha^6 + \alpha^9$
4. $= \alpha^5$

In line 1 of the point doubling example, we substitute in the point values, specifically x_1 and curve parameter $b = 1$. In transforming step 2 into step 3 we have to compute the multiplicative inverse of α^6 to convert $(1/\alpha^6)$ into the form α^w. We do this by finding w such that $(\alpha^w)(\alpha^6) = 1 \ mod \ (x^4 + x + 1)$ In general we can use the Extended Euclidean Algorithm for polynomials to find w (Cohen and Frey 2006). However, in this example we will just show you how we verify that as shown in line 3 of the point doubling example, $w = 9$ The multiplication of $(\alpha^9)(\alpha^6)$ is given below in binary form where $\alpha^9 = (1010)$ and $\alpha^6 = (1100)$ from Sect. 5.2.1.2.

```
      1100
  ×   1010
  --------
      0000
     1100
    0000
   1100
  --------
   1111000   which is equal to x^6 + x^5 + x^4 + x^3
```

Alternatively one can perform the multiplication using the polynomial representation where $(\alpha^9)(\alpha^6) = (x^3+x)(x^3+x^2) = x^6+x^5+x^4+x^3$. The reduction of the result of the multiplication modulo the polynomial ($mod\ (x^4+x+1)$) is given below.

$$
\begin{array}{r|l}
 & \qquad\qquad\qquad\qquad x^2 + x + 1 \\ \hline
x^4+x+1 & x^6 + x^5 + x^4 + x^3 \\
 & \underline{x^6 \qquad\qquad\quad + x^3 + x^2} \\
 & \quad x^5 + x^4 \qquad\quad + x^2 \\
 & \quad \underline{x^5 \qquad\qquad\quad + x^2 + x} \\
 & \qquad\quad x^4 \qquad\qquad\quad + x \\
 & \qquad\quad \underline{x^4 \qquad\qquad\quad + x + 1} \\
 & \qquad\qquad\qquad\qquad\qquad\quad 1
\end{array}
$$

Hence, from the remainder of the polynomial division, we have shown that $(\alpha^9)(\alpha^6)\ mod\ (x^4 + x + 1) = 1$. Therefore, $(1/\alpha^6) = \alpha^9$. The next computation is addition or $(\alpha^6 + \alpha^9)$ from line 3 of the point doubling example. This can be performed by converting to binary and performing simple addition, specifically $1100 + 1010 = 0110 = \alpha^5$, which appears in line 4 to end the illustration of the x-coordinate of the point doubling example.

By computing y_{12} in a similar manner, specifically

Example 5.2. y-Coordinate point doubling of $P = (\alpha^3, \alpha^8)$

$$
\begin{aligned}
y_{12} &= (x_1)^2 + (x_1 + y_1/x_1)\,x_3 + x_3 \\
&= (\alpha^3)^2 + (\alpha^3 + \alpha^8/\alpha^3)\alpha^5 + \alpha^5 \\
&= \alpha^6 + (\alpha^3 + \alpha^8(\alpha^{12}))\alpha^5 + \alpha^5 \\
&= \alpha^6 + (\alpha^3 + \alpha^{10})\alpha^5 + \alpha^5 \\
&= \alpha^6 + (\alpha^{12})\alpha^5 + \alpha^5 \\
&= \alpha^6 + \alpha^2 + \alpha^5 \\
&= \alpha^{11}
\end{aligned}
$$

Therefore for $P = (\alpha^3, \alpha^8)$, we have shown how to use the formulas to compute $2P = (\alpha^5, \alpha^{11})$ or point doubling.

Next we will illustrate point addition over binary fields by examining specifically $P + 2P$. Using (5.8) with the previous points $P = (\alpha^3, \alpha^8) = (x_1, y_1)$ and computed $2P = (\alpha^5, \alpha^{11}) = (x_2, y_2)$ one can compute the point $P + 2P = (x_3, y_3)$, as shown in the point addition example.

Example 5.3. Point addition, $(\alpha^3,\ \alpha^8) \pm (\alpha^5,\ \alpha^{11}) = (x_3,\ y_3)$

$$\begin{aligned}
\lambda &= (y_1 + y_2)\,/\,(x_1 + x_2) \\
&= (\alpha^8 + \alpha^{11})/(\alpha^3 + \alpha^5) \\
&= \alpha^7/\alpha^{11} = \alpha^7\alpha^4 \\
&= \alpha^{11}
\end{aligned}$$

$$\begin{aligned}
x_3 &= \lambda^2 + \lambda^2 + (x_1 + x_2) + a \\
&= (\alpha^{11})^2 + \alpha^{11} + (\alpha^3 + \alpha^5) + \alpha^4 \\
&= \alpha^7 + \alpha^{11} + (\alpha^3 + \alpha^5)\ + \alpha^4 \\
&= \alpha^3
\end{aligned}$$

$$\begin{aligned}
y_3 &= \lambda\,(x_1 + x_3) + x_3 + y_1 \\
&= \alpha^{11}(\alpha^3 + \alpha^3) + \alpha^3 + \alpha^8 \\
&= 0 + \alpha^3 + \alpha^8 \\
&= \alpha^{13}
\end{aligned}$$

Hence, $P + 2P = (\alpha^3,\ \alpha^{13})$, which can also be shown to lie on the elliptic curve as one would expect.

Now that we have illustrated the basic computations involved in elliptic curves over binary and prime fields, the next section will present some techniques to improve the performance of field computations including different point representations and bases.

5.2.3 Improving Performance

This section will discuss some techniques to improve performance. In elliptic curve computations, performance can be optimized only by examining improvements at all levels. For example, the high-level improvements in NAF, multibase NAF, windowing techniques, etc. are important and perhaps have the most significant impact on performance. These were discussed in Sect. 5.1.1. This section will provide examples of some improvements at middle and lower levels of design. The second level including curve selection, field selection, coordinate representation, etc. also has an important impact on performance. Finally the lower level implementation of field computation algorithms such as modular reduction will be examined to illustrate how algorithm performance can be improved. This section will only cover architecture-independent optimizations; discussion of parallelism/etc. will be provided in Sect. 5.4. At the middle level we will illustrate how coordinate representation and curve selection can impact performance. However, further details on improving performance can be found in Cohen and Frey (2005) or Hankerson et al. (2004). Elliptic curve codes are available at many Web sites; however, a very good text that includes "C" codes is also available (Rosing 1999).

Table 5.2 Example of field arithmetic comparison

Number of DSP processor cycles[a] per field operation		
Field	163-Bit binary field	192-Bit prime field
Operation	$X^{163}+x^7+x^6+x^3+1$	$x^{192}-x^{64}-1$
Inversion	7,752	23,146
Multiplication	2,812	330
Squaring	135	213

[a]Cycle count for optimized assembly routines of *Starcore 140 VLIW DSP processor* (Motorola/Lucent 1999)

All the computations discussed so far involve what is called affine operations and involve multiprecision arithmetic since they manipulate m-bit numbers, where $m \geq 163$ bits. These operations are very compute intensive. Table 5.2 shows that even optimized assembly implementations in highly parallel processors require a significant number of cycles (Gebotys et al. 2005). The table provides results for the elliptic curve point operations using the elliptic curve $y^2 + xy = x^3 + ax^2 + b$ over 163-bit binary fields using prime polynomial $x^{163} + x^7 + x^6 + x^3 + 1$ (a standard in cryptography). The implementation shown in the second and third columns indicates the cycle times for each field operation. This implementation uses optimized assembly programs for the Star*core SC140 DSP processor, a highly parallel VLIW processor (Star*core 1999). The inversion operation is the slowest even in a parallel implementation. Recall that point addition and doubling, in either binary or prime fields, requires $1I + 2M + 1S$ and $1I + 2M + 2S$, respectively (where I, M, S refer to the number of inversions, multiplications, and squaring, respectively). For current standards with 163-bit point multiplication, one point double would require approximately 14,000 cycles and a point sum 13,617 cycles on the DSP processor. Additional cycles come from optimized field additions, modular reductions, etc., which are generally insignificant compared to major operations listed, multiplication, squaring, and inversion. To provide further illustration of the computational effort in a 163-bit scalar multiplication over this curve (computing kP where k is a 163-bit key), assuming half the bits are nonzero, would require 163 double point operations and 81 sum point operations totaling over three million cycles. At 350 MHz, this DSP processor can execute a typical 163-bit scalar multiplication in approximately 10 ms using affine coordinates. It is interesting to note that the ratio of inversion cycles to field multiplication cycles (I/M) is 7:1 for prime fields and 3:1 for binary fields in this case shown in Table 5.2. In general it is normally assumed that the inversion operation takes longer than seven times the execution of a field multiplication, though in some platforms it may take 24 times the execution time (Seo et al. 2008). For example, prime field implementations on PCs exhibit I/M ratios greater than 30 (Brown et al. 2001; Lim and Hwang 2000). The next section will illustrate how coordinate representation can optimize the performance, by eliminating the need for inversion during point addition and point doubling.

5.2.3.1 Coordinate Representation

There are many choices of coordinate representation and these have a significant impact on performance. Table 5.3 lists a few of these from Cohen and Frey (2006). One method for attempting to speed up the EC computations is to transform the representation of points on the curve to their projective representation, as illustrated in the second row of Table 5.3. This provides a speedup by eliminating the slow inversion operation.

By transforming the points on the elliptic curve, which we have discussed so far in this chapter into another representation, for example, projective representation, we can modify the point addition and doubling equations. For example, the points on the EC in general are known as affine coordinates, such as $P = (x, y)$, defined over the EC: $y^2 = x^3+ax+b$ over F_p. These affine coordinates can be transformed into points of the form $P = (xz, yz, z)$ by introducing a z variable. These are called projective representation of points and they are defined over EC: $y^2z = x^3+axz^2+bz^3$ over F_p for each $z \neq 0$, $z \in F_p$. This can be easily shown by substituting $(x/z, y/z)$ into the x, y values of the original Weierstraß form of the elliptic curve equation, for example, $(y/z)^2 = (x/z)^3 + a(x/z) + b$ and multiplying both sides by z^3. Using this projective representation we obtain new formulae for addition and doubling, again by substitution, which have no inversions. However, at the end of the point multiplication, kP, a final inversion is necessary since mapping back to the affine representation is necessary.

For example, in the second row of Table 5.3, the standard projective coordinates compute using points of the form (x, y, z). Thus, at the beginning of the point multiplication algorithm the affine points are transformed into these projective points by selecting a z value and computing the projective point shown in the second column of the second row (xz, yz, z). The mapping back into the affine point that is executed at the end of the point multiply is performed by computing $(x/z,\ y/z)$ illustrated in the third column of Table 5.3, requiring one inversion. Other transformations (Cohen and Frey 2006) are also possible such as those shown in Table 5.3.

For example, let us illustrate how the choice of coordinate affects the point doubling and addition formulas. Consider the Jacobian coordinates where (x, y, z) represents the affine coordinate $(x/z^2,\ y/z^3)$. Also assume we have an elliptic curve of the form $y^2 + xy = x^3 + ax + b$, defined over F_{2^m}, for $a, b \in F_{2^m}$, $b \neq 0$. By substituting the affine coordinate into the point doubling equations we can obtain new equations for the transformed coordinates. For example, the x coordinate of

Table 5.3 Coordinate representations and transforms

Transform name	From affine	To affine
Standard projective	(xz, yz, z)	$(x/z, y/z)$
Jacobian	(xz^2, yz^3, z)	$(x/z^2, y/z^3)$
Lopez-Dahab	(xz, yz^2, z)	$(x/z, y/z^2)$
Modified Jacobian	(xz^2, yz^3, z, az^4)	$(x/z^2, y/z^3)$
Chudnovsky-Jacobian	$(xz^2, yz^3, z, z^2, z^3)$	$(x/z^2, y/z^3)$

point doubling, where $P = (x_1, y_1)$ using affine coordinates, was given in (5.8) but restated below.

$$2P = (x_{\mathrm{dbl}}, y_{\mathrm{dbl}}), x_{\mathrm{dbl}} = (x_1)^2 + \left(\frac{b}{(x_1)^2}\right).$$

So if we substitute $x_1 = x/(z)^2$ and $y_1 = y/(z)^3$ into this formula we get

$$(x_{\mathrm{dbl}}) = \left(\frac{x}{z^2}\right)^2 + \frac{b}{\left(\frac{x}{z^2}\right)^2} = \frac{x^2}{z^4} + b\frac{z^4}{x^2}.$$

Furthermore, from our doubling formula the y coordinate is as follows

$$y_{\mathrm{dbl}} = x_{\mathrm{dbl}} + y_1 + (x_1 + x_{\mathrm{dbl}})\left(x_1 + \frac{y_1}{x_1}\right).$$

When we substitute $x_1 = x/(z)^2$ and $y_1 = y/(z)^3$ into this formula we get

$$\begin{aligned} y_{\mathrm{dbl}} &= x_{\mathrm{dbl}} + \frac{y}{z^3} + \left(\frac{x}{z^2} + x_{\mathrm{dbl}}\right)\left(\frac{x}{z^2} + \frac{y}{xz}\right) \\ &= x_{\mathrm{dbl}} + \frac{y}{z^3} + \frac{x^2}{z^4} + \frac{y}{z^3} + x_{\mathrm{dbl}}\frac{x}{z^2} + x_{\mathrm{dbl}}\frac{y}{xz} \\ &= x_{\mathrm{dbl}} + \frac{x^2}{z^4} + x_{\mathrm{dbl}}\frac{x}{z^2} + x_{\mathrm{dbl}}\frac{y}{xz}. \end{aligned}$$

Next we can map these double affine units into the projective coordinate by multiplying x_{dbl} and y_{dbl} by $(z_3)^2$ and $(z_3)^3$. Specifically we can choose $z_3 = xz^2$ to compute $x_3 = x_{\mathrm{dbl}}(z_3)^2$, $y_3 = y_{\mathrm{dbl}}(z_3)^3$. By doing this we get the final equations for doubling a point in projective coordinates that do not require any inversions.

$$\begin{aligned} x_3 &= x_{\mathrm{dbl}}\left(xz^2\right)^2 = \left(xz^2\right)^2\left(\frac{x^2}{z^4} + b\frac{z^4}{x^2}\right) \\ &= x^4 + bz^8. \\ y_3 &= y_{\mathrm{dbl}}\left(xz^2\right)^3 \\ &= \left(x^3z^6\right)\left(\frac{x_3}{(xz^2)^2} + \frac{x^2}{z^4} + \frac{x_3}{(xz^2)^2}\left(\frac{x}{z^2}\right) + \frac{x_3}{(xz^2)^2}\frac{y}{xz}\right) \\ &= xz^2x_3 + x^5z^2 + x^2x_3 + zyx_3 \\ &= x_3\left(xz^2 + x^2 + zy\right) + x^5z^2. \\ z_3 &= xz^2. \end{aligned}$$

As one can see there are no longer any inversion operations required. In fact, the total number of operations for point doubling are five multiplications and five squarings [e.g., x^2, $(x^2)^2$, z^2, $(z^2)^2$, $((z^2)^2)^2$].

There are a number of other possible transformations using the prime or binary field, which eliminate the slow inversion operation. An example of some of these transformations and their impact on performance is given in Table 5.4 for prime fields and in Table 5.5 for binary fields (Cohen and Frey 2006), where I, M, S represent the number of inversions, field multiplications, and field squarings. In these tables M includes multiplications by a fixed coefficient. The derived result for point doubling using Jacobian coordinates, which was performed in this section, is shown in row 4 of Table 5.5.

Additionally it is also possible to mix the coordinate representations such as keeping one point in affine coordinates but using the second point in projective coordinates (Cohen et al. 1998).

The selection of appropriate curves also has a large impact on performance. There are many curves and ongoing efforts to find curves whose computations are more efficient. Examples of some curves with improved efficiency are Koblitz curves (Solinas 2000), Jacobian form the curve (Miller 1986), and Edwards curves (Edwards 2007). For example, Koblitz curves (Solinas 2000) use a tau-adic representation (also known as a Frobenius mapping) of a finite field element and avoid the costly point doubling to improve the algorithmic performance. Using this curve, only two finite field squarings are used instead of a point doubling. The Koblitz curve is also known as the ABC curve, or anomalous binary curve, defined as follows:

$$E : y^2 + xy = x^3 + ax^2 + 1 \quad \text{a} = 0 \text{ or } 1.$$

The tau-adic representation combined with the window method leads to an efficient implementation. In this representation, projective coordinates are not always necessary since the point doubling has already been eliminated and the inversion is more efficient than ten multiplications. Typically tau-adic NAF representations are used and the scalar multiplication with the windows method provides improved performance.

Table 5.4 Prime Fields (Cohen and Frey 2006)

Transform	Point sum	Point double
Affine	I+2M+1S	I+2M+2S
Standard Projective	12M+2S	7M+5S
Jacobian	12M+4S	4M+6S
Chudnovsky-Jacobian	11M+3S	5M+6S
Modified Jacobian	13M+6S	4M+4S

Table 5.5 Binary fields (Cohen and Frey 2006)

Transform	Point sum	Point double
Affine	I+2M+S	I+2M+S
Standard projective	16M+2S	8M+4S
Jacobian	16M+3S	5M+5S
Lopez-Dahab	13M+4S	5M+4S

Another example is Edwards curves (Edwards 2007). Edwards curves can be generalized (Bernstein and Lange 2007) as follows:

$$E : x^2 + y^2 = c^2(1 + dx^2y^2),$$

where $cd(1-dc^4) \neq 0$. Using Edwards curves, computations for point doubling and addition require only 3M + 4S and 9M + 1S, which can be compared to the fastest known formulas (with Jacobian coordinates and the elliptic curve with $a = -3$) requiring 3M + 5S and 7M + 4S, respectively (Bernstein and Lange 2007). It is interesting to note that the Edwards curve also has the nice property of completeness. Completeness specifically means that the point doubling and addition formulas do not have any exceptions, such as $P \neq (-P)$, and there are no divisions by zero (for the point at infinity). This is more computationally efficient since division by zero, checking for negative points, and other checking often cause more complications for implementers (Bernstein and Lange 2007). Recommended curves can be found in various standards as well; there are curves of particular interest for resisting side channel analysis (for example where point addition and point doubling have the same set of computations). The next section will provide further mathematical details for optimizing finite field operations.

5.2.3.2 Field Computations

In addition to implementation of field computations using multiprecision algorithms (e.g., computing using several 32-bit words to store elements of the field since they are typically 163-bits long or 192-bits long or more), there are many algorithm optimizations one can perform. Of course, performance improvements likely also can be attained through assembly coding, but this section will discuss platform-independent optimizations. For illustration purposes, the modular reduction algorithm will be used as an example of performance improvements.

In a direct software implementation, modular reduction performed by polynomial division would be quite inefficient. Recall in Sect. 5.2.2.2 the computation $x^6 + x^5 + x^4 + x^3 \ mod \ (x^4 + x + 1) = 1$ was illustrated using polynomial division. This modular reduction succeeded in reducing the 2m-bit wide element $(A\ B) = (0111\ 1000)$ into a m-bit wide element $C = (0001)$. In essence the polynomial division exclusive-ored the irreducible polynomial (10011 in Fig. 5.4) with any bits which are "1" at bit locations beyond x^3 as shown in Fig. 5.4 for $(A\ B) = (0111\ 1000)$, where $C = (0001)$. This computation can be optimized for a fixed value of irreducible polynomial. For example, for the irreducible polynomial, $f(x) = (x^4 + x + 1)$, the modular reduction of $(A\ B)$ can be performed by computing $C = (B + A \ll 1 + A) \ mod\ 2$ as illustrated in Fig. 5.5. For example, by examining the range of terms in $f(x)$, which will cover parts of B, one can write a modular reduction function that will reduce any 2m-bit data into m-bit data for the defined $f(x)$. In Fig. 5.5, the least significant word, B, is exclusive-ored with the most significant word, A, according to the polynomial term "1" from $f(x)$ and with

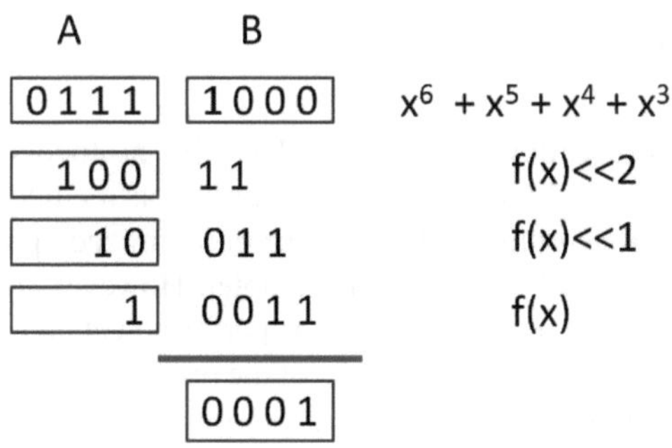

Fig. 5.4 Modular reduction illustration by exclusive-oring with irreducible polynomial

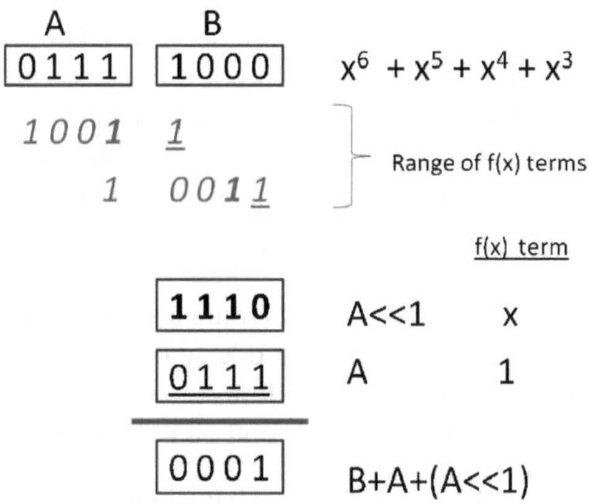

Fig. 5.5 Modular reduction illustration by exclusive-oring with most significant words of data for a fixed reduction polynomial

the one-bit left shift of the most significant word, $(A \ll 1)$, according to the polynomial term "x" from $f(x)$. For example, consider applying this technique to the standardized reduction polynomial $x^{163}+x^7+x^6+x^3+1$. In this case we are reducing (2×163)-bit polynomials into 163-bit polynomials. In practise we are moving from 11 32-bit words, $B[10], \ldots B[0]$, into 6 32-bit words, $C[5], \ldots C[0]$. Note that we only use the three least significant bits of $C[5]$ and six least significant bits of B[10]. Consider the operations performed on the fifth most significant word, $B[4]$. According to the reduction polynomial, it will be affected by bits in $B[9]$. Thus, we can write $C[4] = B[4] + (B[9] \gg 3) + (B[9]) + (B[9] \ll 3) + (B[9] \ll 4) \; mod \; 2$, where computations $B[9] \gg 3$, $B[9]$, $B[9] \ll 3$, $B[9] \ll 4$ represent the effect of terms 1, x^3, x^6, and x^7, respectively. More details can be found in Hankerson et al. (2004).

This technique can also be applied to prime fields, replacing modular 2 addition with regular addition, e.g., addition with carries. There are many algorithms that have been developed for other field arithmetic as well. For example, there are a significant number of field multiplication algorithms. A survey of improvements in lower level field arithmetic algorithms can be found in Cohen and Frey (2006) and Hankerson et al. (2004). The next section will describe elliptic curve protocols where point multiplication is used to provide digital signatures, key establishment as well as encryption.

5.3 EC Protocols

This section will cover the basic EC protocols including ECDSA (ANSI 1999): EC-DH and ECIES. All of these protocols include the time-consuming point multiplication and thus will benefit from optimized approaches, such as window techniques, described earlier in this chapter. However, this section will additionally point out where computations can be performed at an earlier point in time or precomputed in anticipation of future use of the protocols as well. This is important for some embedded systems with limited energy availability or strict time constraints. Encryption, key establishment, and digital signatures using elliptic curves will be illustrated in this section.

The Diffie–Hellman protocol, ECDH (Diffie et al. 1976), will be used to illustrate the application of elliptic curves. Recall from Sect. 4.4.2.2 that Alice wants to communicate with Bob securely. Hence, she must first establish a session key. The objective is for both parties to compute a secret no one else can compute. In the elliptic curve case, they must also first agree on certain elliptic curve parameters such as the finite field, F_q, the elliptic curve, E over F_q or $E(F_q)$, and a base point $P \in E(F_q)$ of order n. These elliptic curve parameters do not need to be secret. Next, Alice chooses a random number x uniformly from the integers $0,1,2,\ldots,n-1$ and keeps it secret. Bob does the same by choosing a random number y. The EC protocol proceeds as follows to establish the shared key, $k = (xy)P$, where xy is the scalar and P is the point on the elliptic curve:

Protocol 5.1. EC Diffie–Hellman key exchange

(1) *Alice* $\rightarrow$ *Bob*: xP
    Bob: $k = y(xP)$
(2) *Bob* $\rightarrow$ *Alice*: yP
    Alice: $k = x(yP)$

In Protocol 5.1 an attacker sees only xP and yP. Furthermore, the attacker cannot obtain x from xP nor y from yP, since this is the ECDLP that is hard. Therefore, the attacker cannot compute k.

The protocol also allows Alice and Bob to select a random value at an earlier point in time and precompute one point multiplication each. For example, Alice can compute and store x and xP, whereas Bob can also compute and store both y and yP. This will minimize the computations required during a future session, which instead of requiring two point multiplications each, now requires only one point multiplication each.

An extension of the EC Diffie–Hellman protocol has been standardized. It is called ECMQV and has been standardized in ANSI X9.63 (ANSI 2001), IEEE 1363–2000, ISO/IEC 15946–3 (Hankerson et al. 2004). The two pass version will be briefly described. Assume Alice has private key equal to (S_a) and public key equal to (P_a). Similarly Bob has private key equal to (S_b) and public key equal to (P_b). Note that the private key and public keys are related as follows: $P_b = (S_b P)$ and $P_a = (S_a P)$. Note that unlike other texts in public key cryptography, the private secret key will remain in upper case notation to be consistent with other notation in this book. Assume Alice and Bob have each other's public keys (authenticated from a trusted source) and they also share the elliptic curve parameters. The protocol proceeds as follows:

Protocol 5.2. ECMQV session key establishment protocol

(1) *Alice* → *Bob*: $X_a = r_a P$

(i) *Bob*: $r_b P = Y_b$
(ii) *Bob*: $K = H([(r_b + half(Y_b)S_b) \; mod \; n][(X_a) + half(X_a)(P_a)])$

(2) *Bob* → Alice: $Y_b = r_b P$

(iii) *Alice*: $K = H([(r_a + half(X_a)S_a) \; mod \; n][(Y_b) + half(Y_b)(P_b)])$

Alice chooses a random value r_a, where $1 \leq r_a \leq n - 1$, where P is a point on E of prime order n. On line (1), Alice computes $r_a P$. This is referred to as Alice's ephemeral public key, P. Note that Alice will use a different value, r_a, for each execution of the protocol (thus ephemeral). Also the ephemeral public key is different from her actual public key, $P_a = S_a P$. Alice sends Bob her ephemeral public key specifically this value $X_a = r_a P$, where P is a point on the elliptic curve. Bob also chooses a random value r_b, $1 \leq r_b \leq n - 1$ and sends Alice the value $Y_b = r_b P$ (which is Bob's ephemeral public key). The *half*() function generally can be viewed as taking at most the bottom half of the bits, but starting with the first nonzero "1" bit (defined as $half(X) = (x \bmod 2^{\lceil f/2 \rceil}) + 2^{\lceil f/2 \rceil})$ where $f = \lfloor \log_2 n \rfloor + 1$, or f is bitlength of n). Readers are referred to Hankerson et al. (2004) or the standards for further details. The function $H()$ is the hash function, which further processes the shared secret into the final shared key, K. The random numbers, r_a and r_b, are one-time random numbers used to help establish a session key. Let us now prove that both Alice and Bob end up with the same session key K. Taking Bob's computation in 1(ii) of Protocol 5.2, we can reduce it as follows:

$$
\begin{aligned}
&= H\left([(r_b + half(Y_b)\, S_b) \;\; mod\, n]\, [(X_a) + half(X_a)\, (P_a)]\right)\\
&= H\left([(r_b + half(Y_b)\, S_b) \;\; mod\, n]\, [(r_a) + half(X_a)\, (S_a)]\; P\right)\\
&= H\left([x_b]\, [x_a]\; P\right).
\end{aligned}
$$

We substitute in the definition of Alice's public key ($S_a P$) for P_a. Then we factor out the point P to obtain the second line above. Finally we define the computation of the first pair of brackets and second pair of brackets as x_b ($x_b = [(r_b + half(Y_b)S_b) \; mod \; n]$) and x_a ($x_a = [(r_a) + half(X_a)(S_a)]$), respectively. Now taking Alice's computation, it also can be reduced similarly.

$$
\begin{aligned}
&= H\left([(r_a + half(X_a)\, S_a) \;\; mod\, n]\, [(Y_b) + half(Y_b)\, (P_b)]\right)\\
&= H\left([(r_a + half(X_a)\, S_a) \;\; mod\, n]\, [(r_b) + half(Y_b)\, (S_b)]\; P\right)\\
&= H\left([x_a]\, [x_b]\; P\right).
\end{aligned}
$$

Hence, both Alice and Bob share the same session key. It is interesting to note that even if an attacker found out Alice's value of S_a, and observed the value of $r_a P$ *and* $r_b P$, the attacker could still not determine the value of the session key K since they do not have access still to r_a. Furthermore, Alice is always convinced that the value $r_b P$ came from Bob since he is the only one who has r_b, thus computing the same session key as Alice has computed. This provides implicit key authentication. A three-pass version of ECMQV is also part of the standard and it provides an explicit key authentication.

Precomputation is also possible with the ECMQV protocol, Protocol 5.2. In particular, Bob can precompute $(r_b + \mathit{half}(Y_b)S_b)\ \mathit{mod}\ n$ and $r_b P = Y_b$. He stores both values ahead of time. Similarly Alice can precompute $[(r_a + \mathit{half}(X_a)S_a)\ \mathit{mod}\ n]$ and $X_a = r_a P$. This leaves just one point multiplication each during the future ECMQV session.

The next protocol to be described is the EC encryption protocol, also a standard in ANSI X9.63 (ANSI 2001). The protocol allows Alice to send Bob an encrypted message. Here, we also assume that both Alice and Bob share the EC parameters and have authenticated copies of public keys. The protocol uses a function referred to as *KDF*(), which is a general key derivation function, which for example may be a hash function.

Protocol 5.3. ECIES encryption

(i) *Alice*: $r_a P = X_a$
(ii) *Alice*: $H(r_a P_b) = K$
(iii) *Alice*: $KDF(K) = (k1, k2)$
(iv) *Alice*: $E_{k1}(m) = c$
(v) *Alice*: $MAC_{k2}(E_{k1}(m)) = s$

(1) *Alice* → *Bob*: X_a,c,s

(vi) *Bob*: $H(S_b X_a) = K$
(vii) *Bob*: $KDF(K) = (k1, k2)$
(viii) *Bob*: $MAC_{k2}(c) =?= s$
(ix) *Bob*: $D_{k1}(c) = m$

In the ECIES encryption protocol, there is only one transmission (line (1)) from Alice to Bob of the encrypted message similar to the general public key encryption protocol previously presented in Sect. 4.4, Protocol 4.7. However, there are more computations, which provide added security. For example, on line (i) Alice computes her ephemeral public key. She then computes the point multiplication, $r_a P_b$, and then hashes it to obtain the value K on line (*ii*). Using a key derivation function (agreed upon with Bob) on line (*iii*), she further processes K into a value from which she splits into two quantities, $k1$ and $k2$ (the upper half and lower half of the output from *KDF*()). Finally on lines (*iv*) and (*v*), she encrypts the message with the key $k1$ to produce ciphertext and then she computes the MAC using key $k2$ of the ciphertext. She sends her ephemeral public key along with the cipher text and

MAC to Bob. Bob hashes Alice's ephemeral public key (X_a) along with his private key (S_b). This computation on line (*vi*) produces the value K as shown by the computation that follows.

$$S_b X_a = S_b r_a P = r_a (S_b P) = r_a P_b.$$

The resultant value of this computation $r_a P_b$ is the value that Alice computed and hashed in line (*ii*) to get K. Once Bob has computed K, he uses the key derivation function to obtain $k1$ and $k2$ on line (*vii*). He then checks the MAC on line (*viii*) and decrypts the message on line (*ix*). The ECIES encryption Protocol 5.7 also allows Alice to generate $k1$ and $k2$ ahead of time and store them together with X_a for the future when a message is ready for encryption. However, there are no precomputations possible for Bob. He must hash the result of his point multiplication ($S_b X_a$) in 1(*vi*) before he can decrypt or authenticate the encrypted message from Alice.

Another interesting encryption approach that will be further discussed in Sect. 5.4 is ECMV or elliptic curve Menezes Vanstone (Stinson 2005). Given an elliptic curve $E : y^2 = x^3 + ax + b \bmod p$, choose a point on the curve, P, and a random integer, r_a where $2 \le r_a \le n - 1$, where P is a point on E of prime order n. Again assume Bob's private key is S_b and his public key is $S_b P = P_b$, the complete public key with EC parameters is $(E,\ p,\ P,\ P_b)$. The ECMV encryption is given below in Protocol 5.4.

Protocol 5.4. ECMV encryption protocol for message (x_1, x_2) transport

(1) *Alice*: $r_a P_b = (c_1, c_2)$

(i) *Alice*: $X = r_a P$
(ii) *Alice*: $Y_1 = c_1 x_1 \ mod\ p$
(iii) *Alice*: $Y_2 = c_2 x_2 \ mod\ p$

(2) *Alice* → *Bob*: (X, Y_1, Y_2)

(i) *Bob*: $S_b X = (c_1, c_2)$
(ii) *Bob*: $c_{1-}{}^{1} Y_1 \ mod\ p = x_1$
(iii) *Bob*: $c_{2-}{}^{1} Y_2 \ mod\ p = x_2$

The ECMV protocol allows the message, represented as a pair of numbers (x_1, x_2) to be encrypted and sent to the recipient using the recipient's public key. Alice chooses a random number in step (1), $2 \le r_a \le n - 1$, and computes $r_a P$ where P is a point on E of prime order n. She then computes with her message and sends Bob the three values shown in Protocol 5.8 step (2). Bob uses his secret key, S_b, to compute (c_1, c_2) and computes its inverse to further obtain the message (x_1, x_2). Details of Bob's computations in step (2), which allow him to obtain the message, are shown below.

$$\begin{aligned}
S_b X &= S_b r_a P = r_a S_b P = r_a P_b = (c_1, c_2),\\
c_{1-}{}^{1} Y_1 \ mod\ p &= c_{1-}{}^{1} c_1 x_1 \ mod\ p = x_1,\\
c_{2-}{}^{1} Y_2 \ mod\ p &= c_{2-}{}^{1} c_2 x_2 \ mod\ p = x_2.
\end{aligned}$$

The use of EC includes an important computation, that of digital signature generation. The EC digital signature algorithm can be used to authenticate the origin of data, to provide data integrity as well as nonrepudiation. It has been standardized by ANSI X9.62 (ANSI 1999), FIPS 186–2 (FIPS 2000), as well as IEEE 1363. Again this algorithm assumes that both Alice and Bob share the EC parameters as well as share authenticated copies of each other's public keys. The ECDSA algorithm consists of a signature generation and a signature verification. For transmission from Alice to Bob of a signed message, Alice will generate a signature, s, and append it to the (encrypted) message, m, whereas Bob will verify her signature. The signature consists of a part r and a part s, each 160-bit-long integers. The algorithm for signature generation and verification are presented next.

Algorithm 5.5. ECDSA signature generation and verification

(i) *Alice*: $r_a P = (x_a, y_a)$
(ii) *Alice*: $x_a \ mod \, n = r$
(iii) *Alice*: $(r_a)^{-1} \, [H(m) + S_a r] \ mod \, n = s$

(1) *Bob* m, r, s

[ECDSA signature verification]

(i) *Bob*: $w = (s)^{-1} \ mod \, n$
(ii) *Bob*: $(H(m)w) \ mod \, n = u_1$
(iii) *Bob*: $(rw) \ mod \, n = u_2$
(iv) *Bob*: $(u_1 P + u_2 P_a) = (x_c, y_c)$
(v) *Bob*: $x_c \ mod \, n = ? = r$

To generate a signature Alice produces her ephemeral public key in line (i) and takes the x-coordinate modulo n in line (*ii*) as part of her signature. She then computes the second part of her signature in line (*iii*) by computing with the hash of the message, her secret key and her secret ephemeral key to produce s. The signature (r, s) is sent to Bob along with the (encrypted) message. To illustrate how this scheme works we start with Alice's definition of s. We multiply the base point by s to get the following expression for sP (where s is defined in line (*iii*) of ECDSA 5.5):

$$sP = \left((r_a)^{-1} \, [H\,(m) + S_a r] \ mod \, n \right) P.$$

Now we multiply both sides by (r_a) and $(s)^1$ to get:

$$r_a P = \left((s)^{-1} \, [H\,(m) + S_a r] \ mod \, n \right) P.$$

Finally we take this expression for $r_a P$ and manipulate it to show that it is equal to the computed value in line (*iv*).

$$
\begin{aligned}
r_a P &= \left((s)^{-1} \left[H(m) + S_a r \right] \, mod \, n \right) P \\
&= (s)^{-1} H(m) P + (s)^{-1} S_a r P \\
&= (s)^{-1} H(m) P + (s)^{-1} r (P_a) \\
&= u_1 P + u_2 P_a .
\end{aligned}
$$

Below we distribute point P through the computation in the third line above, and substitute (P_a) *for* $(S_a P)$ in the fourth line. Finally we substitute the definitions of u_1 and u_2 from lines (*ii*) and (*iii*) of the protocol to show that Bob actually has calculated the x and y coordinates of point $(r_a P)$ in line (*iv*) that Alice had computed in line (i). Specifically $(x_c, y_c) =? = (x_a, y_a)$ or in line (*v*) it will check that $x_c \; mod \; n$ is equal to $x_a \; mod \; n$, which Alice computed in line (*ii*). In Protocol 5.5 ECDSA Alice can compute the point multiplication (her ephemeral public key) and store $(r_a)^{-1}$ and $(S_a r)$ ahead of time. Bob cannot perform any precomputations, apart from the standard lookup tables with precomputed values of base point P to help improve the performance of his point multiplication.

5.4 Embedded Implications

Elliptic curve systems are suitable for some but not all embedded systems. As Sect. 5.3 has illustrated a large number of multiplications using multiprecision arithmetic which tends to dominate the computations. Hence it is not surprising that DSP processors (Guajardo et al. 2001; Gebotys 2004), as well as graphics processing units and special-purpose architectures implemented with FPGAs (Kim et al. 2004; Guneysu and Paar 2008) have been studied for elliptic curve and public key cryptographic systems in general. Also there are several studies of efficiency of ECC for platforms such as PDAs (Dabholkar and Yow 2004). Measured energy of a PDA for various public key systems is shown in Table 5.6. However, as explained in Chap. 1, a careful analysis of security requirements and security strengths should always be performed before deciding on which cryptographic algorithms are to be used.

Since elliptic curve systems require significant amounts of computation, it is important to examine all methods for saving energy, latency, memory accessing, etc.

Table 5.6 Measured energy of PDA from Potlapally et al. (2003)

Algorithm	Key size	Key generation (mJ)	Key exchange (mJ)
DH	1,024	875.96	1,046.5
ECDH	163	276.70	163.5
DH	512	202.56	159.6
			(sign/verify)
RSA	1,024	270.13	546.5/15.97
DSA	1,024	293.20	313.6/338.02
ECDSA	163	226.65	134.2/196.23

Precomputation, lookup tables, pipelining, and parallelism from field operations up to point multiplication should all be examined. In Sect. 5.3 we discussed precomputational aspects after every protocol to examine how the latency of the protocol could be minimized. Precomputation for point multiplication as well as field operations was discussed to examine how lookup tables can be used to store precomputed values to speed implementation. Additional research has examined parallelism and pipelining aspects of more detailed computations, for example, pipelining of EC scalar multiplication in Mishra (2006) and parallelization of ECC in Jarvinen and Skytta (2008).

An interesting low-power architecture for ECC public key crytography was fabricated in Goodman and Chandrakasan (2001). Their architecture allowed low power and scalable security by using a bitslice approach and powering down slices of the circuit, which were not needed. It estimated that this architecture at 0.7 V at 3 MHz would consume at most 525 μW using a 0.25-μm technology, supporting 8–1,024-bit ECC. There have been significant amounts of research since that time examining FPGA implementations, specific ECC (co)processor design as well as hardware–software codesign of ECC processors. In most cases only performance and area measurements are provided, however some researchers have reported energy estimations. For example, an ECC processor (Batina et al. 2007) estimated to require less than 13 μW at 200 KHz was designed with 0.13-μm technology supporting up to 83-bit ECC. Another research group (Luo et al. 2008) developed a 226-bit ECC architecture with estimated 6.6 μW at 1.28 MHz with 0.18-μm technology.

Nevertheless, it remains debatable whether they are suitable for ultraenergy constrained embedded systems due to their large number of computations. Recent studies indicate that there is potential in this area (Gaubatz et al. 2004; Kaps 2006). In a sensor network application, key distribution is an important concern. Thus, public key encryption is one method for enabling key distribution when nodes in the network do not yet have shared symmetric keys. For example public key encryption can be used to distribute keys for cheaper encryption using symmetric means. Alternatively nodes may encrypt data with the base station's public key in order to facilitate confidentiality.

ASIC hardware implementations using the TMSC 0.13-μm standard cell library were created for Rabin's scheme from Chap. 4 and the ECMV scheme of public key encryption from 5.3. Table 5.7 (Kaps 2006) illustrates these research results obtained from their standard cell designs simulated using back-annotated wire capacitances with PrimePower. Packets of 30 bytes were simulated for the tabulated results. From the tabulated results ECMV offers efficient support for small message sizes and low overhead only requiring two packets. The Rabin scheme offered comparable encryption to ECMV. This research also indicated that another approach using Ntru's public key system (Hoffstein et al. 1998) offered the lowest average power consumption but required five packets, making it inefficient in devices where transmission power was dominant. In wireless sensor nodes the power for transmitting a single bit is roughly equivalent to the power for executing around 1,000 instructions (Kaps 2006), hence transmission power is very significant. It is interesting to note that other security research for sensor wireless networks have examined symmetric security approaches (see Chap. 6).

Table 5.7 Public key encryption simulated chip results (Kaps 2006)

Algorithm	Details	Rabin	ECMV
	Size		
	Message payload	<512 bits	>200 bits
	Ciphertext (packets)	512 bits (3)	400 bits (2)
Encryption	Time per message	2.88 ms	817.7 ms
	Avg. power	148.18 μW	394.4 μW
	Energy per message	426.76 μJ	322.5 μJ
Decryption	Time per message	1.089 s	411.5 ms
	Avg. power	191.5 μW	394.4 μW
	Energy per message	208.6 μJ	162.3 μJ

Other researchers have examined public key systems (Wolkerstorfer 2005), ECC in particular, for RFID applications, again a low power, low-energy application. A 196-bit ECC architecture was simulated to show a 170-μW/MHz power (225-MHz clock frequency maximum, area of 0.35 mm^2) using 0.18-μm technology, which they showed scales to 55-μW/MHz power using 90-nm technology with 0.09 mm^2 area and 600-MHz clock rate. They remark that the 0.18-μm technology supports RFID requirements if used at maximum clock rate.

The next chapter will provide a brief look at properties of ciphers as well as present some details of some standard and lightweight ciphers important for many embedded systems.

References

ANSI X9.62 (1999) Public key cryptography for the financial services industry: the elliptic curve digital signature algorithm (ECDSA). American National Standards Institute, Washington, DC

ANSI X9.63 (2001) Public key cryptography for the financial services industry: key agreement and key transport using elliptic curve cryptography. American National Standards Institute, Washington, DC

Batina L, Mentens N, Sakiyama K, Preneel B, Verbauwhede I (2007) Public-key cryptography on the top of a needle. IEEE symposium on circuits and systems:1831–1834. DOI: 10.1109/IS-CAS.2007.378270

Bernstein DJ, Lange T (2007) Faster addition and doubling on elliptic curves. LNCS 4833:29–50

Brown M, D Hankerson, J López, A Menezes (2001) Software implementation of elliptic curve cryptography over prime fields. Cryptographers' track at the San Francisco RSA conference 2001. Lecture Notes in Computer Science (Topics in Cryptology – CT-RSA 2001), Springer 2020:250–265

Ciet M, Joye M, Lauter K, Montgomery PL (2006) Trading inversion for multiplications in elliptic curve cryptography. Designs Codes Cryptogr 39:189–206

Cohen H, Frey G (2006) Handbook of elliptic and hyperelliptic curve cryptography. Chapman & Hall/CRC, Boca Raton, FL

Cohen H, Miyaji A, Ono T (1998) Efficient elliptic curve exponentiation using mixed coordinates. Asiacrypt 1998, LNCS 1514:51–65

Dabholkar A, Yow KC (2004) Efficient implementation of elliptic curve cryptography (ECC) for PDA (PDAs). Wireless Personal Commun 29:233–246

Diffie W, Hellman ME (1976) New directions in cryptography. IEEE Trans Inform Theory IT 22(6):644–654

Dimitrov et al. (1997) Theory and applications for a double-base number system. ARITH 1997, p 44

Dimitrov V, Imbert L, Mishra PK (2005) Efficient and secure elliptic curve point multiplication using double-base chains. LNCS 3788:59–78

Doche C, Imbert L (2006) Extended double-base number system with applications to elliptic curve cryptography. Indocrypt 2006, LNCS 4329:335–348

Edwards HM (2007) A normal form for elliptic curves. Bull Am Math Soc 44(3):393–422 http://www.ams.org/bull/2007-44-03/S0273-0979-07-01153-6/S0273-0979-07-01153-6.pdf. Accessed June 2009

FIPS 186–2 (2000) Digital signature standard (DSS) federal information processing standards publication 186–2. National Institute of Standards and Technology, Gaithersburg, MD

Gaubatz G, Kaps J-P, Sunar B (2004) Public key cryptography in sensor networks – revisited. First European Workshop on Security in Ad-Hoc and Sensor Networks, ESAS 2004, Heidelberg, Germany

Gebotys CH (2004) Design of secure cryptography against the threat of power-attacks in DSP-embedded processors. ACM Trans Embedded Comput Syst 3(1):92–113

Guajardo J, Blumel R, Krieger U, Paar C (2001) Efficient implementation of elliptic curve cryptosystems on the TI MSP 430x33x family of microcontrollers. LNCS 1992:365–382

Güneysu T, Paar C (2008) Ultra high performance ECC over NIST primes on commercial FPGAs. CHES2008, LNCS 5154:62–78

Hankerson D, Hernandez J, Menezes A (2000/2001) Software implementation of elliptic curve cryptography over prime fields (white paper. http://www.certicom.com)

Hankerson D, Hernandez J, Menezes A (2000) Software implementation of elliptic curve cryptography over binary fields (white paper. http://www.certicom.com) CHES 2000 LNCS, vol 1965. Springer, Berlin, pp 243–267

Hankerson D, Menezes A, Vanstone S (2004) Guide to elliptic curve cryptography. Springer, New York

Hoffstein J, Pipher J, Silverman JH (1998) NTRU: a ring-based public key cryptosystem. Algorithmic number theory (ANTS III), LNCS. Springer, Berlin 1423:267–288

Jarvinen K, Skytta J (2008) Parallelization of high-speed processors for elliptic curve cryptography. IEEE Trans VLSI 16(9):1162–1175

Kaps J-P (2006) Cryptography for ultra-low power devices. PhD dissertation. http://www.crypto.wpi.edu/Publications/Documents/phd_kaps.pdf

Kim HW, Wollinger T, Choi YJ, Chung K, Paar C (2004) Hyperelliptic curve coprocessors on a FPGA. LNCS 3325:36–374

Koblitz N (1987) Elliptic curve cryptosystems. Math Comput 48(177):203–209

Koc CK (1994) High-speed RSA implementation. Report online ftp://ftp.rsasecurity.com/pub/pdfs/tr201.pdf

Lim CH, Hwang HS (2000) Fast implementation of elliptic curve arithmetic in GF(p^m). PKC2000, LNCS 1751:405–421

Longa P (2007) Accelerating the scalar multiplication on elliptic curve cryptosystsems over prime fields. Masters thesis, University of Ottawa, Ottawa

Longa P, Gebotys C (2009a) Fast multibase methods and other several optimizations for elliptic curve scalar multiplication. PKC LNCS 5443:443–462

Longa P, Gebotys CH (2009b) Novel precomputation schemes for elliptic curve crypgosystems. ACNS LNCS 5536:71–88

Luo P, Wang X, Feng J, Xu Y (2008) Low-power hardware implementation of ECC processor suitable for low-cost RFID tags. ICSICT Intl Conf on Solid State and IC Technology:1681–1684. DOI: 10.1109/ICSICT.2008.4734876

Miller VS (1986) Use of elliptic curves in cryptography. LNCS 218:417–426

Mishra PK, Dimitrov V (2007) Efficient quintupling formulas for elliptic curves and efficient scalar multiplication. LNCS 4779:390–406

Mishra PM (2006) Pipelined computation of scalar multiplication in elliptic curve cryptosystems (extended version). IEEE Trans Comput 55(8):1000–1010

Möller B (2001) Algorithms for multi-exponentiation. SAC2001, LNCS 2259:165–180

Motorola/Lucent (1999) Star*core 140 DSP core reference manual. Motorola, Lucent, Freescale Semiconductor

Pollard JM (1978) Monte Carlo methods for index computation mod p. Math Comput 32(143):918–924

Potlapally R et al. (2003) Analyzing the energy consumption of security protocols. ISLPED, 771 doi:10.1109/LPE.2003.1231830:30–35

Prodinger H (2000) On binary representations of integers with digits -1, 0, 1. INTEGERS (electronic journal of combinatorial number theory), vol 0. http://www.emis.de/journals/INTEGERS/vol0.html

RFC4050 Using the elliptic curve signature algorithm (ECDSA). http://www.faqs.org/rfcs/rfc4050.html

Rosing M (1999) Elliptic curve cryptography. Manning, Greenwich, CT

Solinas JA (2000) Efficient arithmetic on Koblitz curves. Designs Codes Cryptogr 19:195–249

Stinson D (2005) Cryptography theory and practise. Discrete mathematics and its application, vol 36. Chapman & Hall/CRC, Boca Raton, FL

Wolkerstorfer J (2005) Scaling ECC Hardware to a minimum. CRASH presentation at http://www.cosic.esat.kuleuven.be/crash/ScalingECCtoMin.pdf

Chapter 6
Symmetric Key Protocols Including Ciphers

...if Mary's cipher was strong enough... a life hung on the strength of a cipher

(Singh 1999)

The objectives of cipher design as well as a brief description of some common ciphers, which have become standards, will be covered in this chapter. In addition to highlighting two general categories of ciphers, key streaming and block ciphers, some ciphers that have interesting properties relevant for embedded systems are also covered. Additionally various modes of cipher operation as well as schemes providing not only confidentiality but also authentication are covered. Finally hardware as well as software implementation notes are provided in each section.

6.1 The Theory of a Cipher

Historically substitution ciphers (such as the Caesar cipher) and transposition ciphers (such as the Scytale) were some of the earliest ciphers to be used. However, as discussed in Chap. 2, these were not very secure. Nevertheless, they form an important basis for many ciphers used today, which repeatedly combine substitution and transposition. These ciphers are referred to as product ciphers. They combine two or more transformations such that the cipher is more secure than it would be using any one transformation alone. The most famous product cipher was the Enigma as discussed in Chap. 2.

In general, cipher design is based upon two main principles, confusion and diffusion. These principles relate back to Shannon's communication theory in 1949 on secrecy systems (Shannon 1949). Confusion attempts to decorrelate the plaintext from the ciphertext and key, by using substitution techniques. Ciphers typically use substitution in a complex manner by modifying the substitution based upon bit values in the key or plaintext. The relationship between plaintext and ciphertext/key is hidden with the confusion principle. However, substitution is not sufficient to thwart attacks. Diffusion is also required in order to spread out the redundancy of the plaintext over the ciphertext. The diffusion principle spreads the influence of

C.H. Gebotys, *Security in Embedded Devices*, Embedded Systems,
DOI 10.1007/978-1-4419-1530-6_6,

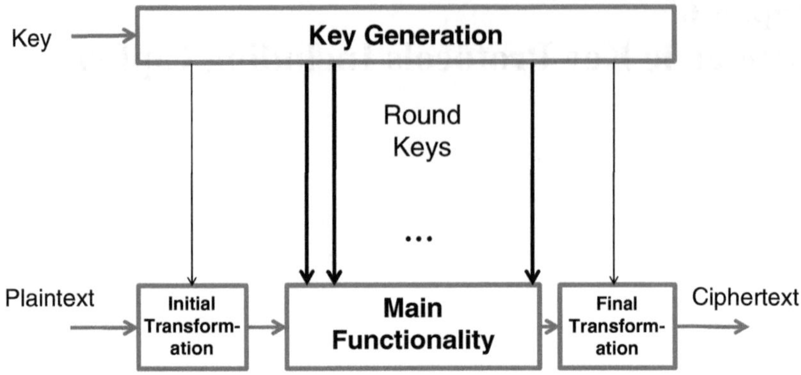

Fig. 6.1 Components of a general block cipher

plaintext or key bits over as much of the ciphertext as possible (Schneier 1996). Typically diffusion is created through permutations. Block ciphers, where blocks of data are input and a resulting block of ciphertext is output, typically use both substitution as well as diffusion. Stream ciphers, where data are input bit serially and ciphertext is also generated bit serially, typically were developed only with substitution in mind but with feedback structures diffusion can also be introduced. As a general rule diffusion alone is easily cracked (Schneier 1996).

As block cipher implementations go, confusion and diffusion are generally implemented with substitution and permutation, respectively. The substitution involves nonlinear operations. Most ciphers are composed of four different parts, namely the key generation (which accepts the master key as input), the initial transformation (which accepts the plaintext as input), the main functionality, and the final transformation (which outputs the ciphertext), shown in Fig. 6.1. For reasons of efficient implementation the main functionality of most block ciphers is iterative, where a particular set of operations called a round are performed repeatedly over a number of iterations. The encryption key is typically input to a key generation routine, which produces a set of round keys to be used in each round of the cipher. For generality there may be some part of the keys used in the initial and final transformation as shown.

Most iterative block ciphers fall under the class of Feistel networks. This was published by H. Feistel in 1973 (Feistel 1973). A Feistel network is defined as taking a block of length n, dividing it into two halves, the left half, L_i, and right half, R_i, and iterating as shown in (6.1).

$$\begin{aligned} L_i &= R_{i-1}, \\ R_i &= L_{i-1} \oplus f(R_{i-1}, K_i). \end{aligned} \tag{6.1}$$

The function $f()$ is referred to as a round function and K_i is called the round key. The Feistel function defined in (6.1) is illustrated in Fig. 6.2. The Feistel network is of great importance in cryptography due to its property that guarantees that a cipher using this technique is reversible. In other words the decryption can use the same

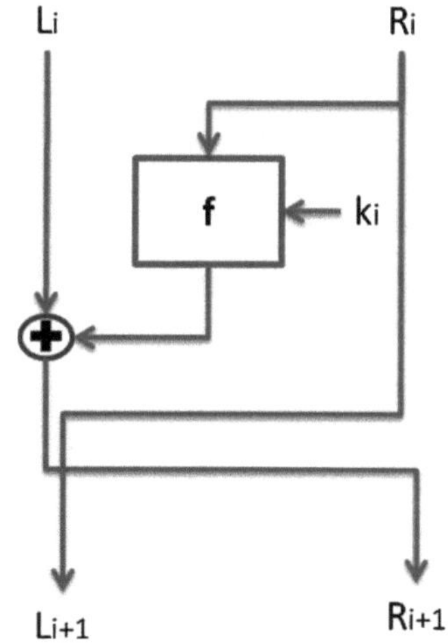

Fig. 6.2 Feistel round

functionality as shown in (6.1) with the same round function, etc. Hence, designers do not have to implement a different decryption algorithm. This property can be more easily seen from the relationship in (6.2), which holds true since the operation uses an exclusive-or.

$$L_{i-1} \oplus f(R_{i-1}, K_i) \oplus f(R_{i-1}, K_i) = L_{i-1}. \tag{6.2}$$

A particular Feistel cipher referred to as DES will be used to further illustrate a general block cipher. DES was a standard effective from 1977, but was replaced by triple DES or 3DES, which is also referred to as TDEA in the standard NIST SP800–67 (Barker 2008). DES is, however, a component within 3DES. In 2001 AES (FIPS-197 2001) emerged as the new encryption standard in order to address increased security needs. AES will be discussed in Sect. 6.4.

DES is an iterated block Feistel cipher with 64-bit block inputs and 56-bit key size. For example, the DES definition of the function f from Fig. 6.2 is shown in Fig. 6.3a). The initial transformation is independent of the key and performs a fixed 64-bit to 64-bit permutation. DES also performs an expansion permutation (see top box in Fig. 6.3a) before exclusive-or with the round key. The expansion permutation maps a 32-bit input into a 48-bit output. It permutes and repeats bits. This is responsible for the cipher avalanche affect where one bit affects two substitutes. The substitution and permutation functions are shown as S-box and P-box, respectively. Both the S-box and P-box take 32-bit inputs and generate 32-bit outputs. The S-box does this using eight 6-bit to 4-bit lookup tables.

DES will again be used as an example of how round keys may be generated. The DES key starts off as a 64-bit key. First a 56-bit key is extracted from the 64-bits, by ignoring every eighth bit. It is interesting to note that the eighth bit is typically used as a parity bit to ensure that the key is error free. The 56-bits are then mapped into 16 unique 48-bit round keys one for each of 16 rounds. To generate the round

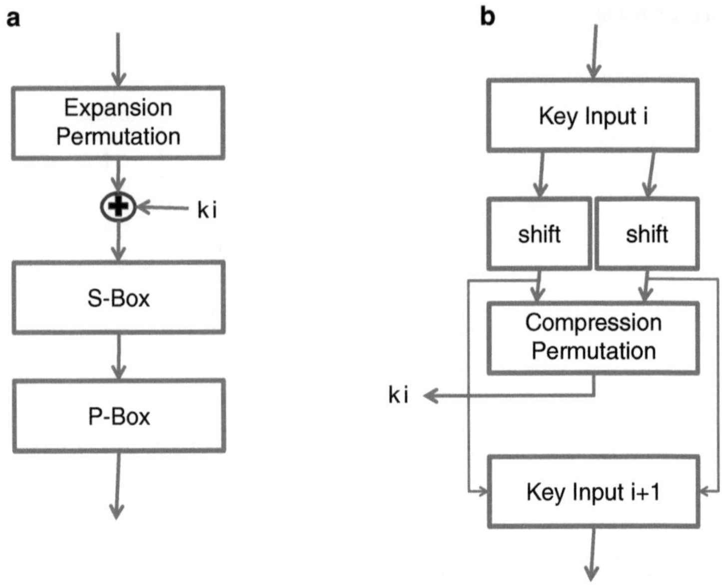

Fig. 6.3 DES function f in (**a**) and key generation function in (**b**)

keys, the 56-bit key is divided into two 28-bit subkeys. According to which round key is being computed, the subkeys are circularly shifted left by one or two bits (the bit shift per round is defined according to this vector {1,1, 2,2, 2,2, 2,2, 1,2, 2,2, 2,2, 2,1}). After the shifting of both subkeys, 48-bits out of the 56-bits total are selected for the particular round key. This shift and select round key repeats to generate the next key. This process is referred to as a compression permutation, and Fig. 6.3b illustrates the ith iteration of this key generation process.

In general, ciphers should be strong irregardless of which keys are used. For example, there should be no weak keys, though in general if they are few in number and known then the algorithm could easily test for them and refuse to use them. The strength of a Feistel network generally is determined by the strength of its substitution function. The substitution function is often referred to as the S-box. The S-box maps an m-bit input to an n-bit output. The number of input bits is more important than the number of output bits. In general, the S-box is a Boolean function, which should have the following properties (Schneier 1996) in order to be secure:

- Far from linear or affine
- Balance of zeros and ones
- No correlations between combinations of bits
- Output bits should behave independent of any input bit
- Satisfy as much as possible a strict avalanche criteria, namely half of the output bits change when one input bit changes

Stream ciphers also have a set of principles for secure design (Schneier 1996) such as

- Long period, no repetitions
- Large linear complexity
- Statistical criteria
- Confusion and diffusion properties
- Nonlinear criteria

The security of ciphers should depend only upon total secrecy of the key and not the obscurity or secrecy of the algorithm. This is known as Kerckhoff's principle. The actual cipher algorithms are typically well published and standardized.

The remainder of this chapter will describe some encryption algorithms, discuss modes of operation, and examine embedded system implications.

6.2 Block Ciphers

A block ciphers is an invertible function also known as a bijection function for a fixed key (where a function f is a bijective if and only if its inverse f^{-1} is a function). Bijection means that the function has a one-to-one correspondence between the inputs and the outputs (so the inputs applied in one direction of the function can be obtained from applying the reverse function on the resulting output). Block ciphers can be stated as a mapping from n-bits to n-bits along with a key input. The iterated block ciphers have a number of parameters including number of rounds, block size in bits, bit size of each round key, and bit size of the master key, which is used to derive the set of round keys. Each round function is also a bijection.

Offering increased security, a whitening technique is sometimes used. This approach was introduced first in DESX (Schneier 1996) and is shown in (6.3). The values k_1, k_2 are random values often referred to as whitening values, whereas P, k, and C are the plaintext, key, and ciphertext.

$$\mathrm{DESX}_{k,k_1,k_2}(P) = k_2 \oplus \mathrm{DES}_k(k_1 \oplus P) = C. \tag{6.3}$$

This whitening technique can also be used for any general encryption algorithm. For example, let W_1 and W_2 represent whitening values, also known as key material since it must remain confidential to the party of interest. They can be used along with an encryption algorithm as shown in Protocol 6.1 and further illustrated in Fig. 6.4. The whitening technique increases the security from $2^{(n+m/p)}$ for $W_1 = W_2$ to $2^{(n+m+1)}$ when $W_1 \neq W_2$ assuming brute force attack is the only known attack possible on this cipher (where n, m, and p are the key size, block size, and number of known plaintexts, respectively).

Protocol 6.1. Cipher whitening

$$\begin{aligned} \mathit{Alice} \rightarrow \mathit{Bob} &: W_2 \oplus E_k(P \oplus W_1) = C \\ \mathit{Bob} &: W_1 \oplus D_k(C \oplus W_2) = P \end{aligned}$$

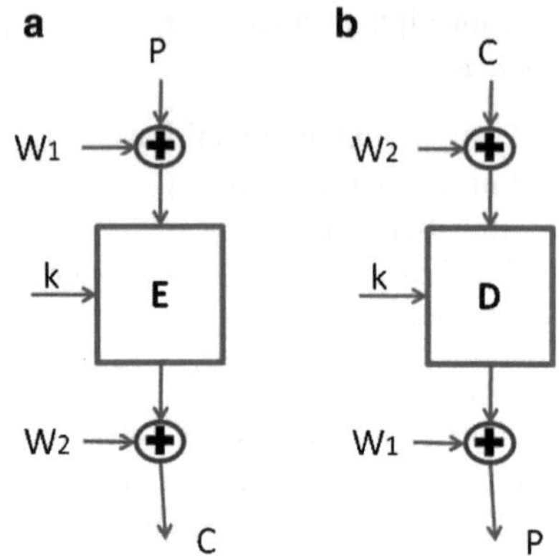

Fig. 6.4 Example of whitening used with encryption (**a**) and decryption (**b**)

Ciphers may also be combined in order to increase security in some cases. For example, one can encrypt data using one key and encrypt the resulting ciphertext with another cipher using a different and independent key. Decryption in this case follows the reverse order. However, it is not advisable to reencrypt data more than once with the same key and same cipher. Block cipher modes also have an impact on the combination of ciphers, and these will be discussed in Sect. 6.4. Triple DES, 3DES or TDEA (Barker 2008), is an example of encrypt–decrypt–encrypt type of cipher mode. The 3DES cipher is shown in protocol 6.2, where $DES_k()$ is the DES encryption and $DES_k^{-1}()$ is the DES decryption function, with key k.

Protocol 6.2. Triple DES

$$Alice \rightarrow Bob : DES_{k3}(DES_{k2}^{-1}(DES_{k1}(P))) = C$$
$$Bob : DES_{k1}^{-1}(DES_{k2}(DES_{k3}^{-1}(C))) = P$$

6.2.1 Rijndael: The Advanced Encryption Standard

AES became a standard in 2001 when NIST chose it over 15 competing cipher designs after 5 years of evaluation. The standard is described in FIPS-197 (FIPS-197 2001). It is based upon Rijndael, which was developed by two Belgian cryptographers, J. Daemen and V. Rijmen. It accepts 128-bit block sizes, which can be processed with 128, 192, or 256-bit key sizes with 10, 12, or 14 rounds, respectively, depending upon the size of the key. Rijndael can handle other block sizes and key sizes; however, these are not supported in the standard.

In general, AES is composed of several rounds where each round involves the four processes shown in Fig. 6.5. The original key is transformed into a set of 128-bit round keys, where each round key is responsible for functioning with each round. For example the first round key is the same as the original key. The details of key expansion or generation can be found in the standard (FIPS-197 2001). The 128-bit data input is transformed according to the algorithm and the value of secret key

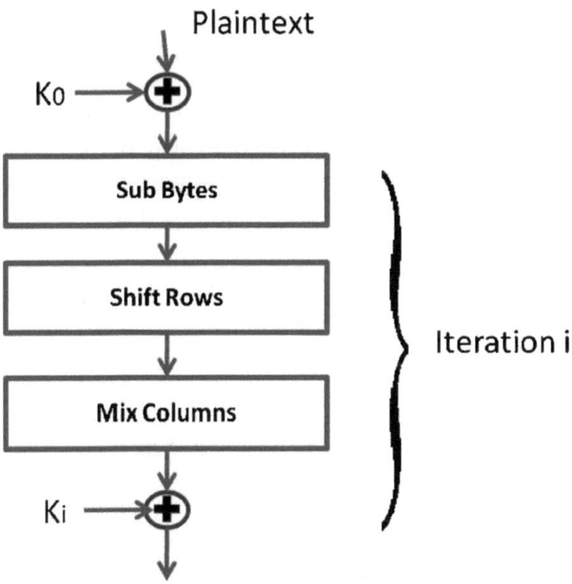

Fig. 6.5 Parts of AES

into a 128-bit output. Typically within AES these transformed data are referred to as a 4×4 array of bytes, known as state. For example, a byte within the state is often referred to as $s_{r,c}$ where r, c refers to the row number ($r = 0, 1, 2, 3$) and column number ($c = 0, 1, 2, 3$) of the byte within the 4×4 array. Rows of this array often are referred to as w_i, for the ith row. The byte substitution function (sub bytes in Fig. 6.5) typically is implemented with a lookup table or substitution table also called S-box. The S-box performs a nonlinear byte substitution. The shift rows function shifts each row of the state array by a specified amount. Finally mix column function mixes the data within each column of state. The S-box, shift rows, and mix column functions can be combined into one lookup table (8-bit input and 32-bit output), which performs both substitution as well as permutation. Section 6.2.1.4 will describe this implementation since provides very fast software implementation when memory is not constrained. The final stage of a round takes the state and exclusive-ors it with the specific 128-bit round key.

Each function within a round of AES is further defined below. For FPGA implementation details, refer to Sect. 10.3.

6.2.1.1 Sub Bytes Function

The Sub Bytes function takes each byte of state and transforms it through a S-box implemented as a lookup table with an 8-bit input and an 8-bit output. Hence, it is simply substituted with one of 256 values from the table. The S-box lookup table is provided in Fig. 6.7 of the standard (FIPS-197 2001).

Mathematically the table represents two transformations. One is the multiplicative inverse and the other is the affine transformation. These two transforms could also be implemented as a combination of a lookup table (byte in and byte out) and an affine transformation. Clearly it is not as efficient as one S-box lookup table. Nevertheless, it may be of interest as discussed in Chap. 10 for reliable security. The two steps are shown below in (6.4), where the multiplicative inverse (*MultInverse*()) of a byte a is denoted as a^{-1} and the output of the Sub Bytes function is *affine*(a^{-1}).

$$
\begin{aligned}
MultInverse(a) &= a^{-1}. \\
affine(a^{-1}) &= \begin{bmatrix} 1\,1\,1\,1\,1\,0\,0\,0 \\ 0\,1\,1\,1\,1\,1\,0\,0 \\ 0\,0\,1\,1\,1\,1\,1\,0 \\ 0\,0\,0\,1\,1\,1\,1\,1 \\ 1\,0\,0\,0\,1\,1\,1\,1 \\ 1\,1\,0\,0\,0\,1\,1\,1 \\ 1\,1\,1\,0\,0\,0\,1\,1 \\ 1\,1\,1\,1\,0\,0\,0\,1 \end{bmatrix} a^{-1} \oplus \begin{bmatrix} 0 \\ 1 \\ 1 \\ 0 \\ 0 \\ 0 \\ 1 \\ 1 \end{bmatrix} \\
&= \begin{bmatrix} a_7^{-1} \oplus a_6^{-1} \oplus a_5^{-1} \oplus a_4^{-1} \oplus a_3^{-1} \oplus (0) \\ a_6^{-1} \oplus a_5^{-1} \oplus a_4^{-1} \oplus a_3^{-1} \oplus a_2^{-1} \oplus (1) \\ a_5^{-1} \oplus a_4^{-1} \oplus a_3^{-1} \oplus a_2^{-1} \oplus a_1^{-1} \oplus (1) \\ a_4^{-1} \oplus a_3^{-1} \oplus a_2^{-1} \oplus a_1^{-1} \oplus a_0^{-1} \oplus (0) \\ a_3^{-1} \oplus a_2^{-1} \oplus a_1^{-1} \oplus a_0^{-1} \oplus a_7^{-1} \oplus (0) \\ a_2^{-1} \oplus a_1^{-1} \oplus a_0^{-1} \oplus a_7^{-1} \oplus a_6^{-1} \oplus (0) \\ a_1^{-1} \oplus a_0^{-1} \oplus a_7^{-1} \oplus a_6^{-1} \oplus a_5^{-1} \oplus (1) \\ a_0^{-1} \oplus a_7^{-1} \oplus a_6^{-1} \oplus a_5^{-1} \oplus a_4^{-1} \oplus (1) \end{bmatrix} \\
&= \begin{bmatrix} A_7 \\ A_6 \\ A_5 \\ A_4 \\ A_3 \\ A_2 \\ A_1 \\ A_0 \end{bmatrix}. \qquad (6.4)
\end{aligned}
$$

6.2.1.2 Shift Row Function

The shift row function has the simplest functionality. Each row is shifted a fixed number of bits specifically, row 0 stays the same, row 1 is circularly shifted to the right by 1, row 2 is circularly shifted to the right by 2, and row 3 is circularly shifted to the right by 3.

6.2.1.3 Mix Column Function

The mix columns transformation is a function of each column of a state [where each column $(a\ b\ c\ d)$ has a polynomial representation: $dx^3 + cx^2 + bx + a$] and produces the corresponding output column. It multiplies each column by the constant polynomial in (6.5).

$$\{03\}x^3 + \{01\}x^2 + \{01\}x + \{02\} \bmod (x^4 + 1). \tag{6.5}$$

All the polynomial coefficients are in $GF(2^8)$. The resulting column of the state matrix is shown in (6.6) in its polynomial presentation, and in (6.7) in the matrix multiplication form.

$$\text{result} = (dx^3 + cx^2 + bx + a)(3x^3 + 1x^2 + 1x + 2) \bmod (x^4 + 1). \tag{6.6}$$

In the resulting column $[a'\ b'\ c'\ d']^T$ (where $[]^T$ represents the transpose of the vector), defined in (6.7), each 8-bit row depends on all the rows of the input column $[a\ b\ c\ d]^T$.

$$\begin{bmatrix} a' \\ b' \\ c' \\ d' \end{bmatrix} = \begin{bmatrix} 02 & 03 & 01 & 01 \\ 01 & 02 & 03 & 01 \\ 01 & 01 & 02 & 03 \\ 03 & 01 & 01 & 02 \end{bmatrix} \begin{bmatrix} a \\ b \\ c \\ d \end{bmatrix} = \begin{bmatrix} 2a \oplus 3b \oplus c \oplus d \\ a \oplus 2b \oplus 3c \oplus d \\ a \oplus b \oplus 2c \oplus 3d \\ 3a \oplus b \oplus c \oplus 2d \end{bmatrix}. \tag{6.7}$$

The multiplication by 2 is shown as $2a$ in row 1 of (6.7) where a is a byte. This multiplication by 2 is equivalent to a shift to the left followed by a modulo reduction by a polynomial (which was covered in Sect. 5.2.3.2; see Fig. 5.3). For example, in Fig. 6.6 the byte is represented by $z_7, \ldots, z_0$. It is shifted to the left by one bit and

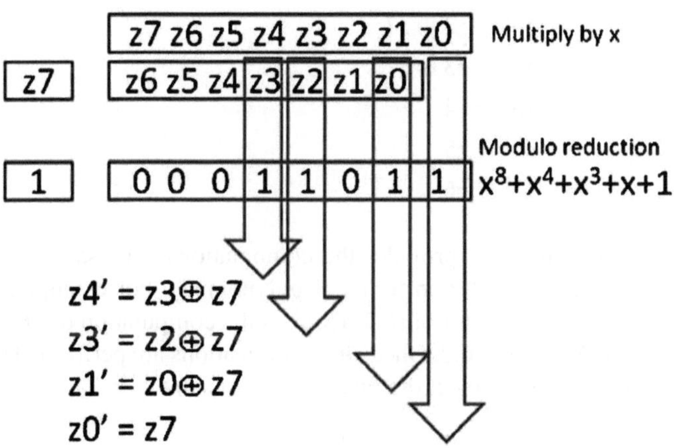

Fig. 6.6 Derivation of multiply by 2 or x (or *xtime*()) with modulo reduction

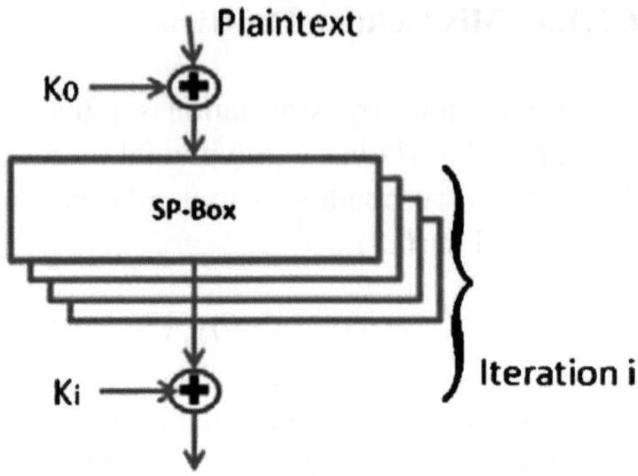

Fig. 6.7 AES table method using four SP-boxes

reduced by polynomial $m(x) = x^8 + x^4 + x^3 + x + 1$ in $GF(2^8)$. Algorithm (6.1) illustrates the result of this multiplication by 2 for each of 8-bits of state. This computation uses three exclusive-or operations. For example, the computation of $2a$ in row 1 column 1of the resulting column vector in (6.7) is denoted by *xtime*(a), where byte a is circularly shifted to the left and the bit $a7$ is exclusive-ored with the new bit positions 1, 3, and 4. For example, using (6.7) together with algorithms 6.1 and 6.2 applied to $2a$ and $3b$, respectively, we get,

$$a' = 2a \oplus 3b \oplus c \oplus d, \text{ so for } i = 0,$$
$$a'_0 = (a_7) \oplus (b_0 \oplus b_7) \oplus c_0 \oplus d_0.$$

Algorithm 6.1. Given $z = [z7\ z6\ z5\ z4\ z3\ z2\ z1\ z0]$, *Compute* $z' = 2z = xtime(z)$

$$z0' = z7$$
$$z1' = z0 \oplus z7$$
$$z2' = z1$$
$$z3' = z2 \oplus z7$$
$$z4' = z3 \oplus z7$$
$$z5' = z4$$
$$z6' = z5$$
$$z7' = z6$$

Consequently, Algorithm 6.2 provides the computations necessary for the multiplication by 3, or *xtime*(z) $\oplus$ z in terms of polynomials. For example, the row 1 column 1 of the resulting vector of (6.7) requires the computation of $3b$, which can be obtained using Algorithm 6.2. Once all multiplications are performed the vectors are exclusive-ored together to get the final column of state in (6.7).

Algorithm 6.2. Given $z = [z7\ z6\ z5\ z4\ z3\ z2\ z1\ z0]$, compute $z' = 3z = xtime(z) \oplus z$

$$\begin{aligned}
z0' &= z0 \oplus z7\\
z1' &= z1 \oplus z0 \oplus z7\\
z2' &= z2 \oplus z1\\
z3' &= z3 \oplus z2 \oplus z7\\
z4' &= z4 \oplus z3 \oplus z7\\
z5' &= z5 \oplus z4\\
z6' &= z6 \oplus z5\\
z7' &= z7 \oplus z6
\end{aligned}$$

The next Sect. 6.2.1.4 describes an AES implementation for 32-bit processors. Details on a FPGA implementation of AES, with emphasis on SEU-resistance, are also provided in Sect. 10.3.

6.2.1.4 High-Speed AES Software Implementation

In the table method of Rijndael (described in Gladman (2003), lookup tables perform both substitution as well as permutations, for fast implementation on 32-bit processors. These tables have 8-bit inputs and 32-bit outputs and are referred to as SP-boxes in the equations of this section. This is unlike the standard AES S-boxes described in Sect. 6.2.1, which have 8-bit inputs and 8-bit outputs. The four data outputs accessed from the SP-boxes are exclusive-ored together and then exclusive-ored with the round key within one round as illustrated in Figs. 6.7 and 6.8. This is performed four times within each round as shown in (6.8). Equation (6.7) performs the initial add-round-key operation on plaintext inputs $p0$, $p1$, $p2$, $p3$ with the initial set of four round keys, $rk0$, $rk1$, $rk2$, and $rk3$. Equation (6.8) partially define the Rijndael encryption process, where $SPbox0, \ldots, SPbox3$ represent the four SP-boxes and subscripts $B3$, $B2$, $B1$, $B0$ represent the most to least significant bytes of the 32-bit word. Equation (6.9) illustrates part of round 2 and finally (6.10) illustrates part of round 10, which uses a different lookup table, shown as *SPbox* 4 in

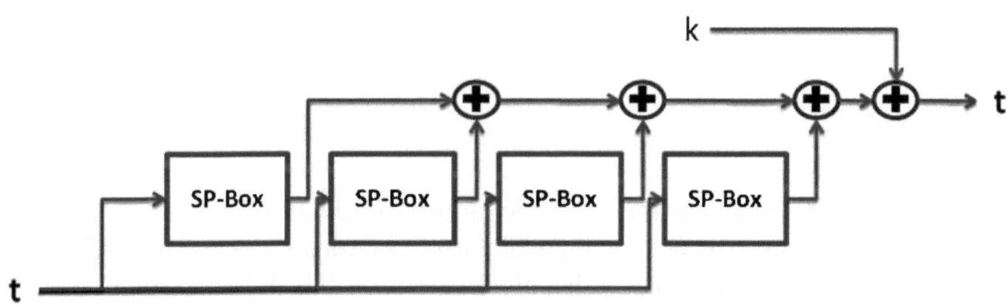

Fig. 6.8 AES table method showing exclusive or computations

round 10. In the algorithm below the looping is unrolled, and so the values $t0 \ldots t3$ are assigned in one round from computations with $s0 \ldots s3$ and vica versa in the next round. Similar code is also provided in Sect. 8.7.1.

$$\begin{aligned} s0 &= p0 \oplus rk0;\; s1 = p1 \oplus rk1;\\ s2 &= p2 \oplus rk2;\; s3 = p3 \oplus rk3; \end{aligned} \tag{6.7}$$

$$\begin{aligned} &//round1:\\ &\quad t0 = SPbox0[s0_{B3}] \oplus SPbox1[s1_{B2}]\\ &\qquad \oplus SPbox2[s2_{B1}] \oplus SPbox3[s3_{B0}] \oplus rk4;\\ &\quad t1 = SPbox0[s1_{B3}] \oplus SPbox1[s2_{B2}]\\ &\qquad \oplus SPbox2[s3_{B1}] \oplus SPbox3[s0_{B0}] \oplus rk5;\\ &\quad t2 = SPbox0[s2_{B3}] \oplus SPbox1[s3_{B2}]\\ &\qquad \oplus SPbox2[s0_{B1}] \oplus SPbox3[s1_{B0}] \oplus rk6;\\ &\quad t3 = SPbox0[s3_{B3}] \oplus SPbox1[s0_{B2}]\\ &\qquad \oplus SPbox2[s1_{B1}] \oplus SPbox3[s2_{B0}] \oplus rk7; \end{aligned} \tag{6.8}$$

$$\begin{aligned} &//round2:\\ &\quad s0 = SPbox0[t0_{B3}] \oplus SPbox1[t1_{B2}]\\ &\qquad \oplus SPbox2[t2_{B1}] \oplus SPbox3[t3_{B0}] \oplus rk8;\\ &\quad \ldots\\ &\quad s3 = SPbox0[t3_{B3}] \oplus SPbox1[t0_{B2}]\\ &\qquad \oplus SPbox2[t1_{B1}] \oplus SPbox3[t2_{B0}] \oplus rk11;\\ &\quad \ldots \end{aligned} \tag{6.9}$$

$$\begin{aligned} &//round10:\\ &\quad \ldots\\ &\quad s3 = SPbox4[t3_{B3}]_{B3} \oplus SPbox4[t0_{B2}]_{B2}\\ &\qquad \oplus SPbox4[t1_{B1}]_{B1} \oplus SPbox4[t2_{B0}]_{B0} \oplus rk43; \end{aligned} \tag{6.10}$$

The table method is illustrated in Figs. 6.7 and 6.8, where K_i in Fig. 6.7 is equivalent to k in Fig. 6.8, and the output value t in Fig. 6.8 becomes the input value t for the next round.

6.2.2 CAST

CAST is a Feistel cipher and it is available for use worldwide on a royalty-free basis for commercial as well as noncommercial purposes (Adams 1997). The CAST-128 uses 128-bit maximum key size with a block size of 64-bits. It has 16 rounds (or 12 rounds if the key size is 80-bits or less) and uses rotations, exclusive-ors, addition, and subtraction modulo 2^{32} in the round function. It uses three different types of round functions and within each there are 8×32 S-boxes. Often this cipher is referred to as CAST5-m where m is the number of bits of the key, e.g., CAST5–128 for a 128-bit key. The CAST algorithm is described in Algorithm 6.3. The key generation algorithm, *GenKeys(k)* in (6.3), takes the input key and generates pairs of round keys using four key generation S-boxes. In the pair of round keys there is a 32-bit masking key, K_i^m, and a 5-bit rotation key, K_i^r. Further details on key generation can be found in Adams (1997).

Algorithm 6.3. CAST5 – encrypt

Input:
plaintext = P
key = k
Output:
ciphertext = $c_1, \ldots, c_{64}$
$(K_i^m,\ K_i^r) = GenKeys(k)$
$(L_0, R_0) = P$
$For\ i = 1\ to\ 16\{L_i = R_{i-1};\ R_i = L_{i-1} \oplus f_i(R_{i-1},\ K_i^m,\ K_i^r)\}$
$(c_1, \ldots, c_{64}) = (L_{16}, R_{16})$

The CAST algorithm divides the data into two 32-bit left and right words *(Lo,Ro)* using a Feistel arrangement. The round function $f_i()$ is defined in Algorithm (6.4). The 32-bit value of I is separated into bytes and input to one of four different S-boxes, $S_1(), S_2(), S_3(), S_4()$. The most significant byte of I is I_3, whereas the least significant byte of I is defined as I_0. The shift operation shifts the value to the left by a number of bits determined by the value of K_r.

Algorithm 6.4. CAST5 – round function $f_i(R, K_m, K_r)$ definition

Input:
Data input = R
Masking key = K_m
Rotation key = K_r
Output:
data = F
$if\ i = 1, 4, 7, 10, 13, 16$

$$I = ((K_m + R) \ll K_r)$$
$$F = ((S_1(I_3) \oplus S_2(I_2) - S_3(I_1)) + S_4(I_0)$$

$if\ i = 2, 5, 8, 11, 14$

$$I = ((K_m \oplus R) \ll K_r)$$
$$F = ((S_1(I_3) - S_2(I_2) + S_3(I_1)) \oplus S_4(I_0)$$

$if\ i = 3, 6, 9, 12, 15$

$$I = ((K_m - R) \ll K_r)$$
$$F = ((S_1(I_3) + S_2(I_2) \oplus S_3(I_1)) - S_4(I_0)$$

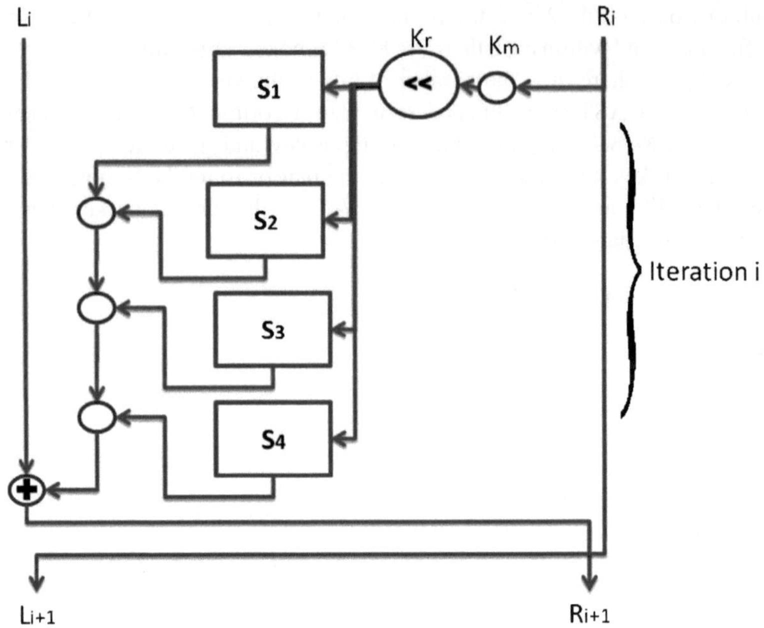

Fig. 6.9 An example of an iteration of CAST

The CAST cipher is further illustrated in Fig. 6.9 where the small circles indicate either an exclusive-or, addition or subtraction according to which round is being performed. For example, in round 1 the masking key is added to the data and then shifted according to the value of K_i^r in Algorithm 6.3. The output of S_1 and S_2 are exclusive-ored and the output of S_3 is then subtracted from this result. Finally the output from S_4 is added to obtain the final output, which is exclusive-ored with the left data shown with the exclusive-or symbol in Fig. 6.9.

6.2.3 TEA

TEA is a lightweight encryption algorithm (Wheeler and Needham 1994) originally designed for high-speed easy implementation in software. The algorithm is illustrated in Algorithm 6.5 where + is addition and all variables are 32-bits. A 128-bit key is used along with a 64-bit plaintext input.

Algorithm 6.5. TEA – encrypt

Input:
plaintext = d_1, d_2
key = k_0, k_1, k_2, k_3

Output:
ciphertext $= d_1, d_2$
$s = 0$
$c = 0x9e3779b9$
$For\ n = 32\ to\ 1\{$
$s = s + c$
$d_1 = d_1 + (d_2 \ll 4) + k_0 \oplus d_2 + s \oplus (d_2 \gg 5) + k_1$
$d_2 = d_2 + (d_1 \ll 4) + k_2 \oplus d_1 + s \oplus (d_1 \gg 5) + k_3$
$\}$

TEA provides nonlinearity by alternating addition and exclusive-or operations. The decryption is shown in TEA decryption Algorithm 6.6.

Algorithm 6.6. TEA – decrypt

Input:
ciphertext $= d_1, d_2$
key $= k_0, k_1, k_2, k_3$
Output:
Plaintext $= d_1, d_2$
$c = $ 0x9e3779b9
$s = (c \ll 5)$
$For\ i = 1\ to\ 32\{$
$d_2 = d_2 - (d_1 \ll 4) + k_2 \oplus d_1 + s \oplus (d_1 \gg 5) + k_3$
$d_1 = d_1 - (d_2 \ll 4) + k_0 \oplus d_2 + s \oplus (d_2 \gg 5) + k_1$
$s = s - c$
$\}$

TEA is considered a Feistel operation, more easily seen from the structure shown in Fig. 6.10 where half of each software algorithm iteration is considered a general Feistel round. In this figure, the function f is defined as (6.11)

$$f(d, k_i, k_j, s) = (d \ll 4) + k_i \oplus d + s \oplus (d \gg 5) + k_j. \quad (6.11)$$

6.2.4 HIGHT

HIGHT was introduced in 2006 as a cipher suitable for low-resource devices (Hong et al. 2006). Hardware implementations are also very efficient and thus of interest for embedded systems (Lim et al. 2009). HIGHT initially transforms the 64-bit plaintext using the whitening technique introduced earlier in Sect. 6.2. The round keys are generated from an initial 128-bit key. The 32 iterative round functions are performed and at the end of the cipher another stage of whitening occurs. Operations within HIGHT are very simple including addition, subtraction both modulo 2^8, exclusive-or and rotation.

Fig. 6.10 Feistel structure of TEA

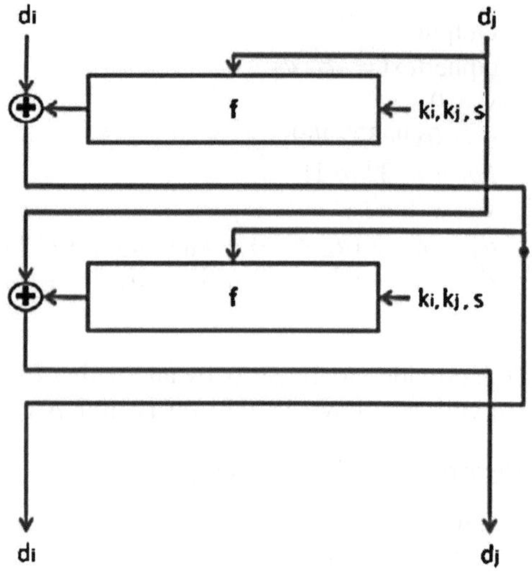

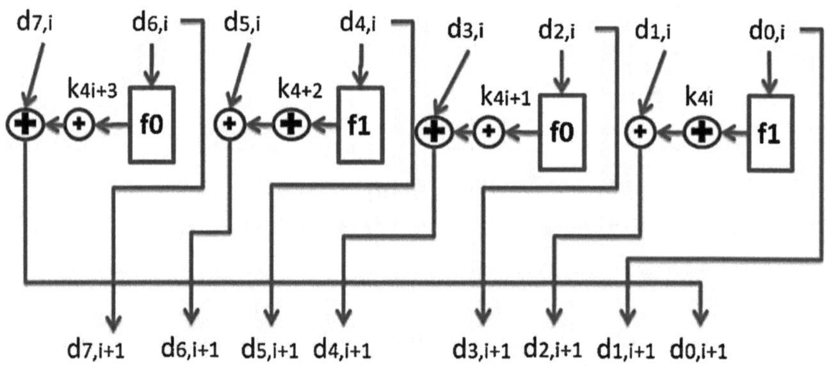

Fig. 6.11 HIGHT encryption round function

HIGHT can be described using variables representing bytes as shown in Algorithm 6.7. In Fig. 6.11 the smaller circled + is arithmetic addition, whereas the larger circled + is an exclusive-or function. The round function that performs the loop body iteration in Algorithm 6.7 is illustrated in this figure.

Algorithm 6.7. HIGHT – encrypt (byte variables)

Input:
plaintext = $d_{7,0}, \ldots d_{0,0}$
round keys = $k_0, \ldots k_{127}$
whitening keys = $w_0, \ldots w_7$
Output:
ciphertext = $c_7, \ldots c_0$
$d_{6,0} = d_{6,0} \oplus w_3$
$d_{4,0} = d_{4,0} + w_2$

$d_{2,0} = d_{2,0} \oplus w_1$
$d_{0,0} = d_{0,0} + w_0$
For $i = 0$ *to* $31\{$
$d_{0,i+1} = d_{7,i} \oplus (f0(d_{6,i}) + k_{4i-3})$
$d_{1,i+1} = d_{0,i}$
$d_{2,i+1} = d_{1,i} + (f1(d_{0,i}) \oplus k_{4i-2})$
$d_{3,i+1} = d_{2,i}$
$d_{4,i+1} = d_{3,i} \oplus (f0(d_{2,i}) + k_{4i-1})$
$d_{5,i+1} = d_{4,i}$
$d_{6,i+1} = d_{5,i} + (f1(d_{4,i}) \oplus k_{4i})$
$\}$
$c_7 = d_{32,0}$
$c_6 = d_{32,7} \oplus w_7$
$c_5 = d_{32,6}$
$c_4 = d_{32,5} + w_6$
$c_3 = d_{32,4}$
$c_2 = d_{32,3} \oplus w_5$
$c_1 = d_{32,2}$
$c_0 = d_{32,1} + w_4$

The whitening keys in Algorithm 6.7 as well as the round keys are generated from the input 128-bit key for HIGHT. Also the two functions are defined according to (6.12) and (6.13).

$$f0(x) = (x \lll 1) \oplus (x \lll 2) \oplus (x \lll 7). \tag{6.12}$$

$$f1(x) = (x \lll 3) \oplus (x \lll 4) \oplus (x \lll 6). \tag{6.13}$$

Decryption as in most cases involves the reverse computations on the cipher text.

6.2.5 PRESENT Cipher

PRESENT (Bogdanov et al. 2007) is a substitution-permutation network cipher, which accepts 64-bit block input and a 80-bit or 128-bit key. Its major focus of design was efficiency, so it was based on the most hardware-efficient cipher of the AES finalists, Serpent (Anderson et al. 1998). It uses only one 4-bit to 4-bit S-box, which is used 16 times in parallel in each of 31 rounds. Although the algorithm for PRESENT shows round keys being generated in *GenRoundKeys*, they can in fact be generated in each round on the fly.

Algorithm 6.8. PRESENT – encrypt

Input:
plaintext $= P$
key $= k$

Output:
ciphertext $= C$
$(k_1, \ldots, k_{32}) = GenRoundKeys(k)$
$state = P$
$For\ i = 1\ to\ 31\{$
$Addroundkey(state, k_i)$
$Sbox(state)$
$Permutation(state)$
$\}$
$Addroundkey(state, k_{32})$

Algorithm 6.9. PRESENT – round key generation

Input:
Key Register $= K$
Output:
Key Register $= K$
Round key $= R_k$
$R_k = (K_{79}..K_{16})$
$(K_{79}...K_0) = (K_{18}, K_{17}, \ldots K_{20}, K_{19})$
$(K_{79}..K_{76}) = S\ (K_{79}..K_{76})$
$(K_{19}..K_{15}) = rcntr \oplus (K_{19}..K_{15})$

Each round key is equivalent to the most significant 64-bits of the key register. The key register holds the initial 80 or 128-bit key, and at the end of each round it is updated. The key update rotates the key register to the left by 61-bits. The resulting four most significant bits are passed through the S-box. Finally the round counter value is exclusive-ored with bits 19 through 15 of the key register. This round key generation is shown in Algorithm 6.9.

6.3 Stream Ciphers

Most regard stream ciphers as more efficient and faster from a hardware point of view than block ciphers. It is believed that their use in commercial applications is widespread (Robshaw 2007). The standardized ciphers include A5/3 or Kasumi GSM ciphers, RC4 (part of the IEEE 802.11b WEP standard), and the UMTS/3GPP f8 function. There is also an eSTREAM project whose objectives are to identify efficient hardware as well as software stream ciphers for academic attention. Currently the hardware candidates are Trivium, Grain, and Mickey and the software candidates are HC-128, Rabbit, Salsa20, and SOSEMANUK (Bjorstad 2008). This section will review the principles of stream ciphers, examine some techniques for building stream ciphers from LFSRs as well as block ciphers, and finally examine some specific stream ciphers, both well known, RC4, and lightweight, Grain.

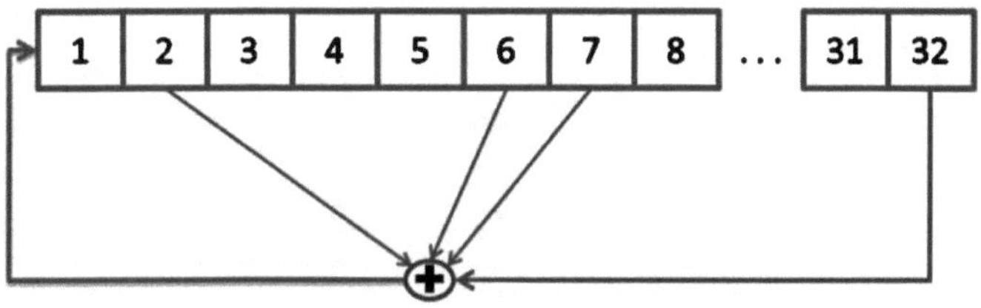

Fig. 6.12 LFSR representing prime polynomial $x^{32} + x^7 + x^6 + x^2 + 1$

6.3.1 Stream Cipher Principles

Linear feedback shift registers (LFSR) of maximal length are the most popular basis for stream ciphers. LFSRs are guaranteed to be of maximal length if their representation in terms of a polynomial is in fact a primitive polynomial (or irreducible). An irreducible polynomial of degree n is one that divides $x^{2^{n-1}} + 1$ but not $x^d + 1$ for any d that divides $2^n - 1$. Tables of primitive polynomials are available in Schneier (1996) and other places.

In Fig. 6.12 the LFSR has four taps, bit 32, 7, 6, and 2 (for labeling bits 1 through 32 left to right as shown in the figure), where a tap is an output of a bit register, which is input to an exclusive-or in addition to the serial shift register functionality. The highest order in the polynomial denotes the bit in the LFSR, which outputs the pseudo-random pattern and feeds back into the exclusive-or together with the taps. The output of the exclusive-or feeds into the bit "1" of the LFSR. This is also called a Fibonacci configuration of a LFSR. Each tap represents the superscript of a term in the polynomial together with one or the zeroth superscript. Figure 6.12 can be described therefore as a polynomial shown in (6.14).

$$x^{32} + x^7 + x^6 + x^2 + 1. \tag{6.14}$$

Some researchers prefer to define the LFSR using an update function (Hell et al. 2005, 2006). The update function unlike the polynomial representation uses the concept of state. It is sometimes preferred to avoid any confusion as to whether the polynomial is describing a left to right numbering or right to left numbering of bits for a LFSR, which shifts to left or right, or right to left, etc. Consider again the LFSR in Fig. 6.12 but this time represented as a series of states shown in Fig. 6.13. The flipflops holding state are now renumbered as shown. In this case each clock cycle generates the next set of state, consecutively numbered. The first set of state is $(s_1, \ldots, s_{32})$. The second set of state is $(s_{33}, \ldots, s_{64})$. Using this state concept one can define the LFSR using an update function instead of a polynomial. For the example we are considering (Figs. 6.12 and 6.13) the update function can be defined as shown in (6.15), where i is multiples of 32, e.g., $i = 0, 32, \ldots$.

$$s_{32+i} = s_i + s_{i+25} + s_{i+26} + s_{i+30}. \tag{6.15}$$

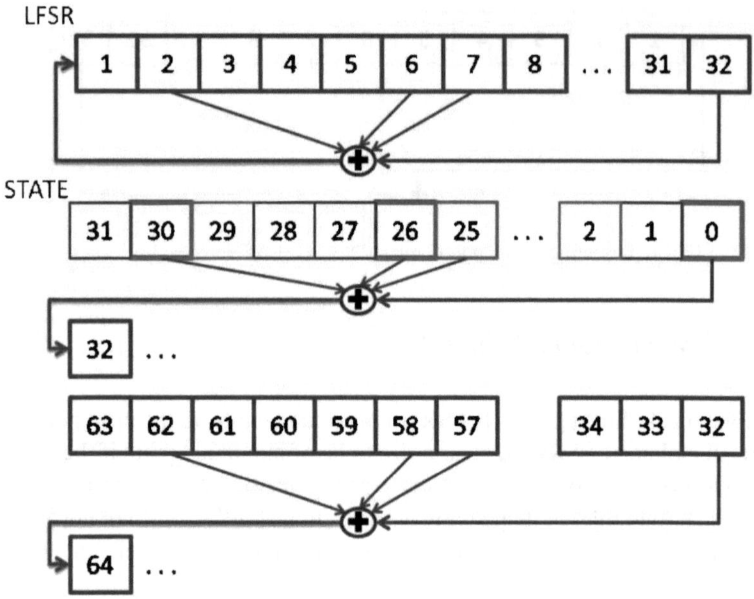

Fig. 6.13 Example of a LFSR and its state representation

The update function numbers state left to right in the same direction of shifting. For example, consider $i = 0$, we have $s_{32} = s_0 + s_{25} + s_{26} + s_{30}$, examining the state in Fig. 6.13; we see that one cycle later s_{32} in the second row will be input to the next row indicating the next state. Then 31 cycles later this value will shift from box 63 to box 32. Thus, $i = 32$ will generate $s_{64} = s_{32} + s_{57} + s_{58} + s_{62}$, which is shown at the bottom of the figure. This is analogous to hardware where the second row represents the bits stored in the flipflops at time $t + 1$ (one clock cycle later), which is defined by the bits stored in the flipflops at time t.

Maximal length means that it will cycle through $2^n - 1$ values before repeating. In general, the fewer the taps the sparser and less secure is the polynomial. The more the taps the denser and more secure the polynomial will be. Security indicates how difficult it is to determine the taps of the LFSR given the series of outputs generated by the LFSR. In general for a n-bit LFSR, it only takes $2n$ sequential outputs to determine the LFSR tap arrangement using the Berklekamp–Massey algorithm (Meyer and Tuchman 1972, 1979). Often linear complexity profiles are used to measure the linear complexity of the stream ciphers. It is interesting to note that low linear complexity indicates security weakness, but high linear complexity does not indicate high security. Thus, outputs from LFSRs are typically highly correlated and alone do not offer significant security, though they offer great efficiency in hardware.

An alternate arrangement of a LFSR is called a Galois LFSR and an example is given in Fig. 6.14. Here the output bit is combined with several input bits to generate new input bits. This LFSR scheme is easier to implement in software and offers parallelism. This scheme was also discussed in Chap. 3 with regards to oscillators.

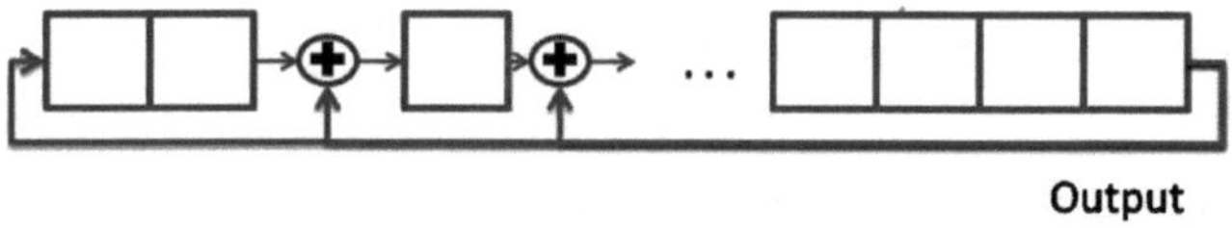

Fig. 6.14 Example of a Galois LFSR

In the simplest implementation of a LFSR in a stream cipher, the output bit is exclusive-ored with the plaintext bit to generate the ciphertext bit one cycle at a time. As discussed earlier, LFSR schemes do not offer sufficient security for a cipher. Typically several different LFSRs are used and combined in some nonlinear way to generate a cipher. However, it is very difficult to generate a secure cipher using this approach. Correlation attacks are often used to break these types of ciphers. For example, in some cases, over 70% of the cipher outputs come from one of the LFSRs, thus the details of this LFSR can often be determined using correlations. Examples of correlation attacks on these types of ciphers can be found in Schneier (1996). Further discussion of other types of pseudo-random number generators such as linear congruential generators can also be found in Stinson (2006).

6.3.2 RC4

RC4 is an old variable key size stream cipher, which is still widely used in many software applications (Fluhrer et al. 2002) though no longer recommended in new applications. It was developed in 1987 by Ron Rivest for RSA Data Security, but is still used today in SSL and file encryption (RSA Web site). The RC4 algorithm is shown in Algorithm 6.10. It uses 8-bit input S-box with 8-bit outputs. Two indices i and j are generated iteratively and then used to identify which two locations in the S-box to swap. The S-box entries are swapped. The final index into the S-box is calculated, t, as a function of the two S-box entries. This 8-bit value, S_t, is exclusive-ored with the plaintext to generate the ciphertext.

Algorithm 6.10. RC4 – encryption

Input:
Plaintext $= p$
Output:
Ciphertext $= c$
$i = (i + 1)\ mod\ 256$
$j = (j + S_i)\ mod\ 256$
$h = S_i$
$S_i = S_j$
$S_j = h$
$t = (S_i + S_j)\ mod\ 256$
$k = S_t$
$c = p \oplus k$

Algorithm 6.11. RC4 – S-box generation

Input:
Key(or seed) $= K_i, i = 0, \ldots 255$
Output:
S-box $= S_i, i = 0, \ldots 255$
$For\ i = 0\ to\ 255\{$
$S_i = i\}$
$j = 0$
$For\ i = 0\ to\ 255\{$
$j = (j + S_i + K_i)\ mod\ 256$
$h = S_i$
$S_i = S_j$
$S_j = h\}$

The S-box is easy to generate as well using Algorithm 6.11. An initial array of key bytes or seeds is used in this algorithm. The initial entries in the S-box are the indices, 0–255. Next indices are generated using the S-box values, and the initial key bytes and the two values of the S-box are swapped iteratively in a loop. There is also a key setup phase, which is more difficult than the encryption.

6.3.3 Grain

GRAIN is composed of only two components, a pair of two feedback shift registers and the nonlinear combiner or filter. Figure 6.15 illustrates the stream cipher.

The LFSR and NLFSR (nonlinear feedback shift register) are both polynomials of degree 80. The LFSR is defined by the polynomial in (6.16) and uses 6-bits exclusive-ored together to define the input bit.

$$x^{80} + x^{67} + x^{57} + x^{42} + x^{29} + x^{18} + 1. \tag{6.16}$$

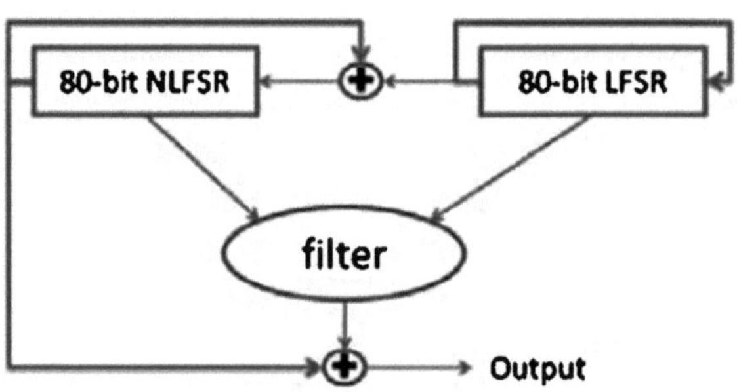

Fig. 6.15 Grain stream cipher structure

The equivalent update function (Hell et al. 2005) is given in (6.17).

$$s_{i+80} = s_{i+62} + s_{i+51} + s_{i+38} + s_{i+23} + s_{i+13} + s_i. \quad (6.17)$$

The feedback function polynomial of the NLFSR is quite complex and it is shown in (6.18).

$$\begin{aligned} &1 + x_{18} + x_{20} + x_{28} + x_{35} + x_{43} + x_{47} + x_{52} + x_{59} + x_{66} + x_{71} + x_{80} \\ &+ x_{17}x_{20} + x_{43}x_{47} + x_{65}x_{71} + x_{20}x_{28}x_{35} + x_{47}x_{52}x_{59} + x_{17}x_{35}x_{52}x_{71} \\ &+ x_{20}x_{28}x_{43}x_{47} + x_{17}x_{20}x_{59}x_{65} + x_{17}x_{20}x_{28}x_{35}x_{43} + x_{47}x_{52}x_{59}x_{65}x_{71} \\ &+ x_{28}x_{35}x_{43}x_{47}x_{52}x_{59} \end{aligned} \quad (6.18)$$

Both these polynomials are defined in Hell et al. (2005), with update functions as well. The filter function, $h()$, uses nonlinear constructs and combines five bits output from the two feedback shift registers. One bit is from the NLFSR and the other four bits are taken from the LFSR. The filter function is defined by $h(x_0, x_1, x_2, x_3, x_4)$ and is given in (6.19).

$$\begin{aligned} h(x_0, x_1, x_2, x_3, x_4) = {} & x_1 + x_4 + x_0x_3 + x_2x_3 + x_3x_4 + x_0x_1x_2 \\ & + x_0x_2x_3 + x_0x_2x_4 + x_1x_2x_4 + x_2x_3x_4 \end{aligned} \quad (6.19)$$

The output function exclusive-ors seven bits from the NLFSR and exclusive-ors them to the output of the filter function. A key and an initialization vector are used to fill the feedback shift registers at the start of the cipher and it is run for 160 cycles. The output from the output function is fedback to the input of both feedback registers. This helps to scramble the contents of the feedback shift registers. After this key initialization, this feedback is disconnected and the output from the cipher is used to generate ciphertext (through exclusive-or with the plaintext).

6.4 Cipher Modes of Operation

There are different ways of combining calls to a cipher when the plaintext size is greater than the input plaintext block size of the cipher. This section will discuss five general different modes of operation for block ciphers, which form the standard discussed in Dworkin (2001). These modes are referred to as electronic codebook or ECB, cipher block chaining or CBC, cipher feedback mode or CFB, output feedback mode or OFB, and counter mode CTR. However, in addition to these modes, there are other modes whose details can be found in Schneier (1996).

When the plaintext is smaller than the input block size of the cipher, *padding* is performed such as adding a single "1" bit followed by as many "0" bits as needed. However, when the plaintext is longer, the plaintext is then divided up into blocks, where each block length is equal to the length of the input to the cipher. Padding may then be applied to the final block unless the plaintext is some integer multiple of the cipher block length. In addition to padding, many cipher modes require an *initialization vector* or IV. This value must be unique for each message over the lifetime of the key, or in other words the same IV cannot be used for two messages. Typically the initialization vector can be any value or random value, and it can be transmitted in the clear. In some modes of operation, the IV must be unpredictable; hence, a random value can be used or sent through a cipher. In some cases, such as counter mode, which will be discussed shortly, there is more than one input vector to be generated, referred to as *counter values* or *counter blocks* instead of an initialization vector. Normally the counter values are generated using an initial value, which is then incremented to generate the next counter value, which is incremented and so on. However, it is often the case that these counter values must be unique for each message. Hence, an alternative method for generating counter values is to assign each message a $b/2$-bit nonce (where the specific cipher uses block sizes of b-bits). Then the counter values would be defined successively with the most significant $b/2$-bits equal to the nonce and the least significant $b/2$-bits equal to the block number within the message. Since each message will have a different nonce, then the counter values will always be different as long as there are less than or equal to $2^{b/2}$ blocks and messages over the lifetime of the key.

The Electronic Code book or ECB mode is shown in Fig. 6.16. It is the simplest mode possible since each ciphertext block $(C_i, \ldots C_n)$ is strictly dependent upon the input plaintext block (in addition to the key and cipher) and no other plaintext or ciphertext. It is called code book since a block of plaintext will always encrypt to the same ciphertext; hence, it is analogous to looking up codes in a "table." However, for this reason it is not a secure mode and should not be used on its own. The encryption and decryption can be performed in parallel.

Cipher block chaining mode or CBC mode chains the plaintext with the previous ciphertext before input to the cipher as shown in Fig. 6.17. An initialization vector, *IV* shown in Fig. 6.17, is required in order to perform the first exclusive-or or chaining. Unfortunately, this mode does not support parallelism directly. Decryption uses the inverse cipher function, applied to the first ciphertext block, and the resulting

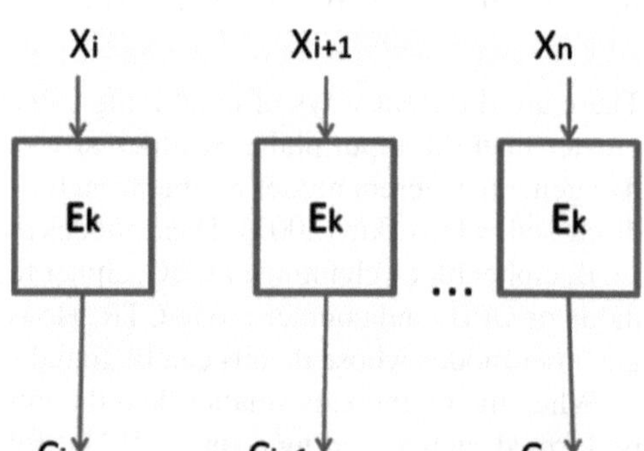

Fig. 6.16 Electronic code book (ECB) mode

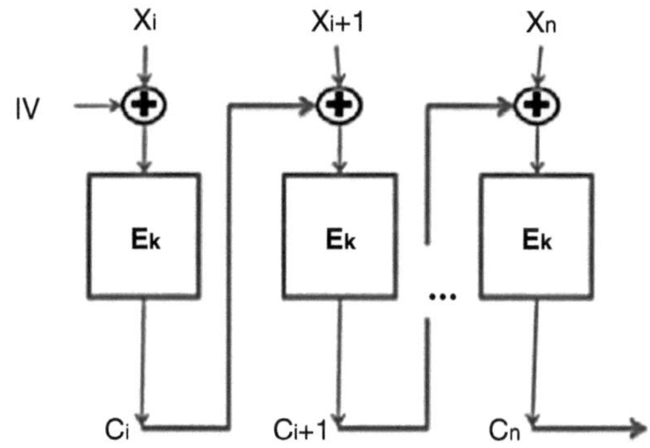

Fig. 6.17 Cipher block chaining (CBC) mode

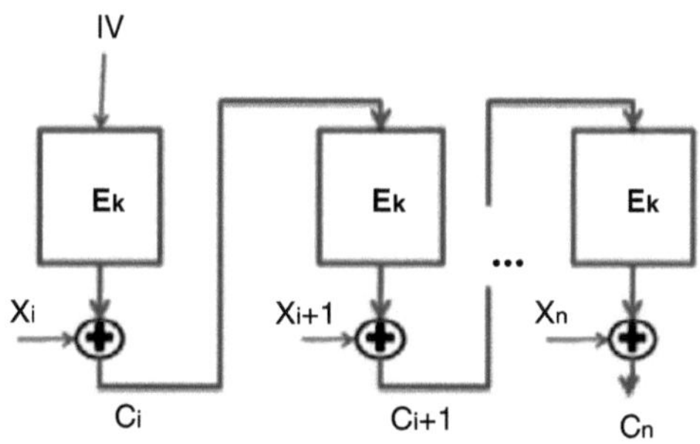

Fig. 6.18 CFB mode

output block is exclusive-ored with the initialization vector to obtain the first plaintext block. This continues except that the previous ciphertext block is used in place of the initialization vector.

The cipher feedback (CFB) mode is shown in a simplified form in Fig. 6.18. In the CFB specification, the X_i, C_i variables are s-bits and the cipher outputs blocks of b-bits where $b > s$. In general, the s most significant bits of the output from the cipher is used in the exclusive-or operation. The cipher input accepts the ciphertext as the lower s-bits, and the upper (b-s) bits come from the lower bits of the previous input to the cipher. Details are available in Evans et al. (2001).

Output feedback mode is shown in Fig. 6.19. If the last plaintext block is u-bits where $u < b$, for b-bits cipher block length, then the most significant u-bits output from the cipher is used for the last block of the ciphertext. It is very interesting to note that if the initialization vector is known in advance, then all the cipher outputs can be precomputed. Thus, encryption only involves exclusive-or operations. This precomputation can also be performed for decryption and hence can be very efficient. Nevertheless to maintain security a different *IV* has to be used for each message, as discussed at the beginning of this section. For example, if the same *IV* is used for two messages, the security can be compromised.

The counter mode is shown in Fig. 6.20. Similar to the OFB mode, if the counter values, T_i, T_{i+1}, ... T_n in Fig. 6.20, are known in advance the values output by the cipher can be precomputed for efficiency. Unlike the OFB mode, there are counter

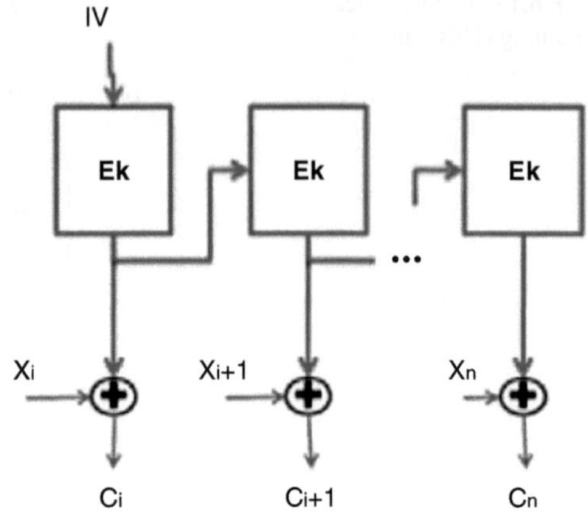

Fig. 6.19 OFB mode

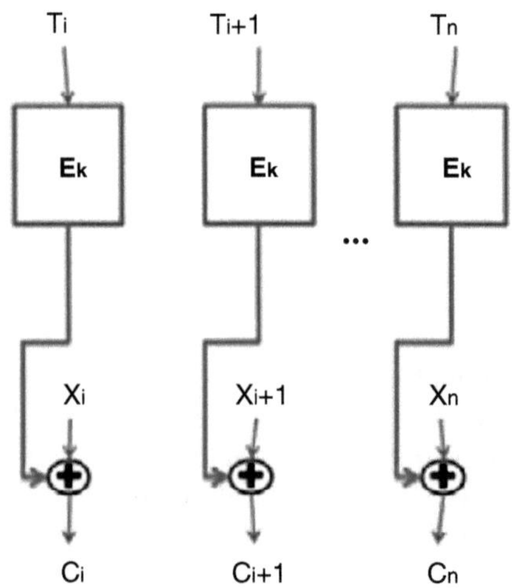

Fig. 6.20 CTR mode

value inputs to each cipher, which are independent of the output of the previous cipher. The counter values can be generated by using an incrementing function. The initial counter value must be chosen carefully, however, so that they are unique for each message.

6.5 Authenticated Modes for Encryption

Authenticated encryption provides not only ciphertext but also assurance of the authenticity of the confidential data and authentication of some data that may not be confidential (referred to as associated data). The notion of authenticated encryption arose largely due to the multitude of cases where designers had difficulty in securely combining two secure functions, encryption and authentication (to be discussed in the next chapter), as evident in the following quote.

> "...First was the realization that people had been doing rather poorly when they tried to glue together a traditional (privacy-only) encryption scheme and a message authentication code (MAC) ...; second was the emergence of a class of AE schemes ... that did *not* work by gluing together an encryption scheme and a MAC." (Bellare et al. 2003)

Hence, it is highly recommended to use a standardized authenticated encryption algorithm if data require both confidentiality and authentication.

The counter mode with cipher block chaining, or CCM, was the first developed mode, which offered authenticated encryption in 2004 (Dworkin 2004). In CCM the counter mode is used for the generation of the ciphertext whereas the cipher block chaining is used to provide the MAC functionality. The final block output from the CBC (may be truncated and) is used as the MAC.

The Galois/Counter Mode or GCM was standardized in 2007 (Dworkin 2007). It generally uses a cipher in counter mode for confidentiality along with a binary field multiplier, which multiplies data with a fixed hash subkey (referred to as a GHASH function) in order to compress the ciphertext into a single block, which is encrypted to produce the authentication tag. Figure 6.21 illustrates the ciphertext being produced using a general counter mode structure. The initialization vector is processed into a form ready for input to the counter mode structure. Each counter input is incremented using a specific method discussed in the standard from the previous counter value. The ciphertext ($C = (C_i, C_{i+1}, \ldots C_n)$) is processed and then fed into the GHASH function where the binary field multiplier ($*H$) is used. Figure 6.22 illustrates this general process. The final block output from the multiplier, S_m, is processed and sent through counter mode encryption to obtain the final truncated authentication tag, T. Further details can be found in (Dworkin 2005).

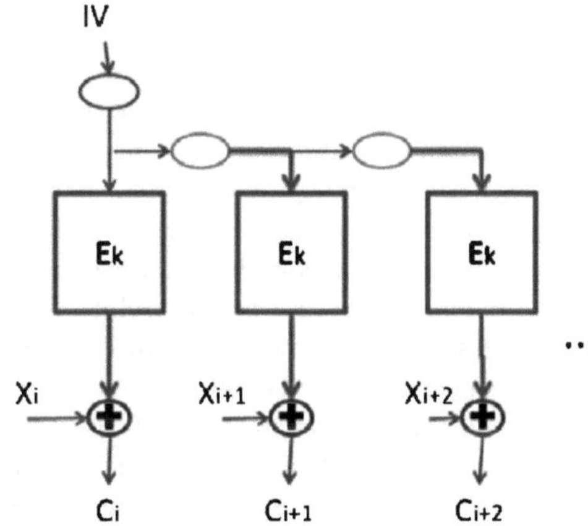

Fig. 6.21 Ciphertext produced using the counter mode

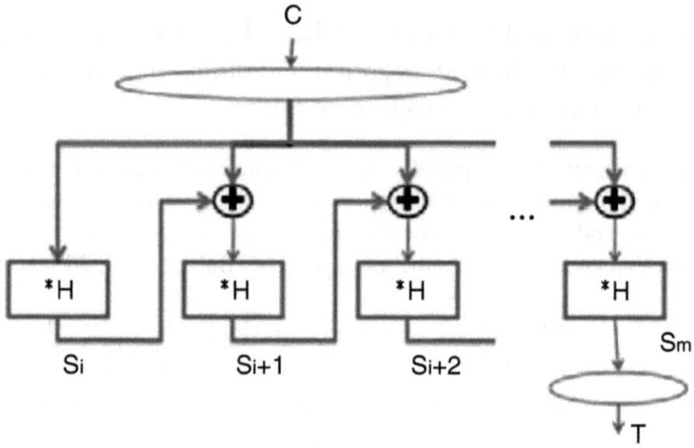

Fig. 6.22 The GHASH function

6.6 Embedded Systems Implications

This section will discuss some embedded implementations of the ciphers introduced in this chapter. Quantitative measures are used to provide some rough comparisons among various ciphers. However, detailed comparisons are not possible since implementations typically use different technologies, and architectures were designed with different objectives such as cost or performance. Measures as well as estimates taken from past research are listed in terms of area, energy, throughput, etc.

In general, AES is 1.5 times faster than DES per bit and 1.7 times more energy efficient. For example, energy consumption was measured for various ciphers (Potlapally et al. 2003) running on a Compaq iPAQ H3670. These results are presented in Tables 6.1 and 6.2 for various cipher modes using μJoules per byte (μJ/B). It is interesting to note that the energy calculated per block size is less than the key schedule energy except in the case of AES. However, typically the key schedule is performed only once and round keys are stored; hence, the more important figure is the cipher energy, where AES is most efficient. The various cipher modes for AES indicate that CFB mode has the highest energy. It is not surprising that ECB is the most efficient since it has no chaining or bit processing overheads.

Tables 6.3 and 6.4 provide further examples of some implementations of different ciphers using standard cell implementations and FPGA implementations. Metrics are reported from the library supported tools and FPGA tools. As expected the throughputs increase as the architecture increases its parallelism, thus using more gates. The GCM mode increases the number of gates (see rows 7 and 8) as one would expect since the Galois field multiplier is included. The lightweight ciphers have significantly lower areas (see PRESENT compared with DES) or equivalent areas but higher throughputs (see HIGHT roughly compared with AES-128). Table 6.4 illustrates a few examples of FPGA implementation of AES where increasing area leads to increased throughputs.

Table 6.1 Measured enegry costs of ciphers from Potlapally et al. (2003)

Cipher	DES	3DES (TDEA)	CAST	AES
Key setup (μJ)	27.53	87.04	37.63	7.87
Encrypt/Decrypt (μJ/B)	2.08	6.04	1.47	1.21

Table 6.2 Measured energy costs of AES modes from Potlapally et al. (2003)

Key size	Key setup (μJ)	ECB (μJ/B)	CBC (μJ/B)	CFB (μJ/B)	OFB (μJ/B)
128	7.83	1.21	1.62	1.91	1.62
192	7.87	1.42	2.08	2.30	1.83
256	9.92	1.64	2.29	2.31	2.05

Table 6.3 Standard cell implementations of some block ciphers

Cipher	Key size	Block size	Throughput (Kbps)	Tech (μ m)	Gate equiv
DES	56	64	44.4	0.18	2,309
HIGHT[a]	128	1	6400	0.25	3,048
PRESENT[b]	80	32	200	0.18	1,570
AES-128	128	128	12.4	0.35	3,400
AES[c]	–	–	311,000	0.13	5,400
AES[d]	–	–	2,600,000	0.13	21,000
AES-GCM[e]	–	–	2,560,000	0.13	34,500
AES-GCM[f]	–	–	34,700,000	0.18	498,000
AES-GCM[f]	–	–	62,600,000	0.18	979,000

[a] Hong et al. (2006)
[b] Bogdanov et al. (2007)
[c] Shirai and Mizuno (2007)
[d] Shirai and Mizuno (2007)
[e] Yang et al. (2005)
[f] Satoh et al. (2009)

Table 6.4 FPGA implementation of AES

Technology	Key size	Slices	RAMs	Tech	Gbps
Virtex3200E[a]	128	2,780	100	0.18	11.8
Virtex4-FX100[b]	128/192/256	3,800	114	0.09	17.9
XC2VP20–7[c]	128	5,170	84	0.13	21.5
XC2VP100(GCM)[d]	128	10,797	164	0.13	9.7

[a] Hodjat et al. (2004)
[b] Lemsitzer et al. (2007)
[c] Standaert et al. (2003)
[d] Wang and Ni (2004)

FPGA implementations of AES block ciphers and some stream ciphers are shown in Tables 6.4 and 6.5, respectively. Table 6.4 compares the area of the designs using slices, where a slice in the Xilinx technology refers to two 4-bit input lookup tables, two flipflops, and some other logic. In Table 6.5 gate equivalents are used for rough comparison purposes. As one can note from the table, RC4 is not particularly suited for hardware implementation, but instead is extremely efficient in software. However RC4 and DES are very old and generally no longer considered secure.

Table 6.5 Stream cipher comparisons

Cipher	Gate equivalent	Technology (μm)	Throughput (Mbps)
RC4[a]	~12,000	–	–
GRAIN[a]	1,300	–	–
GRAIN[b,c]	1,450	0.30	49
GRAIN[c,d]	1,450	0.13	282
GRAIN-128[e]	2,133–4,768	–	(1X–32X speedup)

[a] Robshaw (2007)
[b] Altera Max3000
[c] Hell et al. (2005)
[d] Altera Cyclone
[e] Estimates from Hell et al. (2005)

Table 6.6 AES-128 and ECC-233 technology comparison

	Power/throughput[a] (At 33 mMHz)	Tech (μm)	ASIC	RTAX Actel	Virtex-2[b]
AES-128 core[c]	9.6 mW/42 Mbps	0.25	14,500	8,900	5,100
ECC core[d]	56.8 mW/850 Kbps	0.25	27,600	19,300	12,900

[a] Throughput and power at 33 MHz using 0.25-μm technology ASIC
[b] xc2v6000–4
[c] Habinc (2006a)
[d] Habinc (2006b)

Table 6.6 provides more interesting data comparing an AES-128 core to an elliptic curve cryptographic core (ECC core). The ECC core provides encryption and decryption for an elliptic curve based on 233-bit key and point lengths. The increased power dissipation as well as area for the public key encryption vs. the symmetric key encryption. It is very interesting to note that the increase in power dissipation over AES-128 is almost six times higher whereas the area increase is only about two times. Note also that the throughputs for elliptic curve computations are significantly lower as well in the case of elliptic curves with 850 Kbps vs. 42 Mbps in AES-128. The RTAX and Virtex-2 numbers do not have throughputs or power numbers associated with them, but it is likely that their design or architectures are all very similar [see table in Gaisler (2008)].

In error-prone environments OFB or CTR modes of operation are best since one-bit error in ciphertext incurs only one-bit error in the plaintext. Countermode has the extra advantage of supporting parallelism and only requires an AES encryptor at both ends of the channel (no AES decryption module is required). Furthermore, if the initialization vector is known, then preprocessing is also possible. As mentioned earlier the more recent standards of CCM and GCM also work in counter mode.

The new set of lightweight block ciphers achieve low hardware cost by using very simple logical and arithmetic operations, which can easily and efficiently be implemented in hardware. Some do not even require S-boxes, but achieve nonlinearity by alternating simple operations. Additionally other lightweight ciphers do not require round key generation. Lightweight ciphers do not only provide useful confidentiality for many embedded systems, but they can also be used for strong identification as suggested in Paar (2008) using a challenge response protocol. For example the phone sends the battery a random value, r, and the battery returns $E_k(r) = y'$, which the phone also computes $E_k(r) = y$. The phone compares the result, y, with that received from the battery, y'. If the values do not match then the authorization fails (and it is rumored that the cellphone will then turn off its power management). Alternatively one could use a hash function instead of using a cipher. Hash and integrity functions will be discussed in the next chapter.

References

Adams C (1997) The CAST-128 encryption algorithm. RFC 2144, network working group, request for comment 2144

Anderson R, Biham E, Knudsen L (1998) Serpent: a flexible block cipher with maximum assurance, First AES candidate conference. http://www.cl.cam.ac.uk/~rja14/serpent.html

Barker WC (2008) Recommendation for the triple data encryption algorithm (TDEA) block cipher. NIST SP800–67

Bellare M, Rogawayy P, Wagner D (2003) A conventional authenticated-encryption mode 464 http://seclab.cs.ucdavis.edu/papers/eax.pdf

Bjorstad TE (2008) An introduction to new stream cipher designs. Presentation at 25th Chaos Communication Congress. http://events.cc.de/congress/2008/Fahrplan/events/2875.en.html

Bogdanov A et al (2007) PRESENT: a ultra-lightweight block cipher. CHES 2007 LNCS 4727:450–466

Dworkin M (2001) Recommendation for block cipher modes of operation – methods and techniques. National Institute of Standards and Technology, NIST special publication 800–38A

Dworkin M (2004) Recommendation for block cipher modes of operation – the CCM mode for authentication and confidentiality. National Institute of Standards and Technology, NIST special publication 800–38C

Dworkin M (2007) Recommendation for block cipher modes of operation – Galois/Counter Mode (GCM) and GMAC. National Institute of Standards and Technology, NIST special publication 800–38D

Evans DL, Bond PJ, Bement AL (2001) Recommendation for block cipher modes of operation. NIST special publication 800–38A

Feistel H (1973) Cryptography and computer privacy. Sci Am 228(5):15–23

FIPS-197 (2001) Specification for the advanced encryption standard (AES). Federal Information Processing Standards Publication 197

Fluhrer S, Mantin I, Shamir A (2002) Attacks on RC4 and WEP. CryptoBytes 5(2):26–34

Gaisler A (2008) AES/ECC. http://www.gaisler.com/cms/index.php?option=com_content&task=view&id=205&Itemid=152

Gladman B (2003) A specification for Rijndael, the AES algorithm. fp.gladman.plus.com/cryptographytechnology/rijndael/aes.spec.311.pdf:18–19

Habinc S (2006a) GRAES – advanced encryption standard (AES) IP core user's manual. Gaisler Research. http://gaisler.com/doc/grecc.pdf

Habinc S (2006b) GRECC – ECC IP core user's manual. Gaisler Research. http://gaisler.com/doc/grecc.pdf

Hell M et al (2005) Grain – a stream cipher for constrained environments. Intl J Wireless Mobile Comput 2(1):86–93

Hell M et al (2006) A stream cipher proposal: Grain-128. IEEE Intl Symp Info Theory 1614–1618

Hodjat A et al (2004) A 21.54 Gbps fully pipelined AES processor on FPGA. FCCM 2004

Hong D et al (2006) HIGHT: a new block cipher suitable for low-resource device. CHES 2006 LNCS 4249:46–59

Lemsitzer S et al (2007) Multi-gigabit GCM-AES architecture optimized for FPGAs. CHES 2007 LNCS 4727:227–238

Lim YI et al (2009) Implementation of HIGHT cryptic circuit for RFID tag. IEICE Electron Exp 6(4):180–186

Meyer CH, Tuchman WL (1972) Pseudo-random codes can be cracked. Electron Des 23:74–76

Meyer CH, Tuchman WL (1979) Design considerations for cryptography. Proc NCC 42:594–597

Paar C (2008) New directions in lightweight cryptographic primitives for RFID applications. Presentation at RFID CUSP Workshop

Potlapally R et al. (2003) Analyzing the energy consumption of security protocols. ISLPED, doi:10.1109/LPE.2003.1231830:30–35

Robshaw M (2007) Low cost cryptography. Presentation at http://uma.ensta.fr/conf/ifw-2007/talks/IFW2007-Robshaw.pdf

RSA website http://www.rsa.com

Satoh A et al (2009) High performance hardware architectures for galois counter mode. IEEE Trans Comput 58(7):917–930

Schneier B (1996) Applied cryptography. Wiley, New York

Shannon CE (1949) Communication theory of secrecy systems. Bell Syst Tech J 28(4):656–715

Shirai T, Mizuno A (2007) A compact and high speed cipher suitable for limited resource environment. 3rd ETSI security workshop presentation, Sophia-Antipolis, France

Standaert F-X et al (2003) Efficient implementation of Rijndael encryption in reconfigurable hardware. CHES 2003, LNCS 2779:334–350

Stinson DR (2006) Cryptography, 3rd edn. Chapman and Hall/CRC, Boca Raton, FL

Wang SS, Ni WS (2004) An efficient implementation of AES algorithm. ISCAS II:597–600

Wheeler DJ, Needham RM (1994) TEA a tiny encryption algorithm. Fast sofware encryption. LNCS 1008:363–366

Yang B et al (2005) High speed architecture for Galois/counter mode of operation (GCM), IACR tech report 146. eprint.iacr.org/2005/146.pdf

Chapter 7
Data Integrity and Message Authentication

The integrity of encrypted data was a cause for concern in the banking industry in the 1970s. They wanted to be sure that attackers were not able to modify data especially since the data, in this case, was financial. As a result many standards for integrity and authentication as well as authenticated encryption (Bellare et al. 2003) have been established (FIPS180-3 2008; FIPS198a 2002) and are being developed (SHA-3 2008). This chapter will introduce the basic properties and structure of data integrity functions including message authentication. In general, if confidentiality and assurance or authentication of confidentiality of data is important then authenticated encryption schemes (discussed in Sect. 6.5) should be used. Cryptographic hash functions and message authentication codes (MACs) are discussed in this chapter. Integrity trees will also be discussed as an application of these integrity functions. Some initial terminology is provided below as an aid to understanding concepts which will be developed in this chapter.

Data integrity: It provides assurance that the data has not been modified from its original state. Integrity is provided by hash functions. In general, these hashes are computed and then protected in some manner for verifying the data again at some time in the future. Hash functions combined with other properties (one-way, collision resistance) provide modification detection codes (MDCs) or message integrity codes (MICs).

Data-origin authentication: It allows one party to gain assurance that the data came from a party in possession of a secret associated by design with the party (Menezes et al. 1996). This is provided by digital signatures (where one often loosely refers to authenticating a user, see Chaps. 4 and 5). Additionally, a symmetric approach to data-origin authentication can be implicitly provided by MACs.

Data authentication: It provides assurance that the data is authenticated (e.g., came from the party in possession of a secret) and that the data has not been modified in transit (e.g., data integrity). This can be provided by MACs.

Transaction authentication: It provides data authentication along with the use of time variant parameters such as nonces, sequence numbers, time stamps, etc. Alternatively including a random number from a previous message in the next sequential

C.H. Gebotys, *Security in Embedded Devices*, Embedded Systems,
DOI 10.1007/978-1-4419-1530-6_7,

message also will provide replay attack detection, thus providing transaction authentication. Transaction authentication is typically used to thwart replay attacks where stale "replayed" messages are sent by the attacker.

Entity authentication: It provides assurance to a party of the identity of a second party involved in the protocol. Also the second party has participated or is active. Here there may not be a "message" unlike the above examples, and there is some timeliness guarantee. Examples include passwords, challenge–response with public key cryptography, etc. (see Chaps. 4 and 5 for more details).

Previously in Chap. 2, we defined integrity to mean that data has not been altered. It is important to differentiate the purpose of integrity from error checking approaches used in many designs. For example checksums or cyclic redundancy codes (CRC) are often used to detect (and even correct) errors in data. In contrast the data integrity and MACs are used to detect cryptographic active or malicious changes in the data. For example if CRC checksums are used, it would be easy for an attacker to modify data in transit and also modify the necessary CRC or checksum to be appended to this new data. This is unlike the MAC which is not easily forged.

Some examples are shown next in various protocols. When both confidentiality and authentication is needed, the safest method is to use authenticated encryption standards or an encrypt-then-authenticate protocol such as Protocol 7.4. In other cases, the protocols should be used carefully, see Menezes et al. (1996), and cryptographic publications should be checked for any recent attacks in this area.

Protocol 7.1. $Alice \rightarrow Bob: \; m, \; MAC_k(m)$

In Protocol 7.1, Bob receives the message in the clear but he will compute his own MAC using Alice and Bob's shared key k in order to verify that the message did come from someone who has possession of the key k.

Protocol 7.2. $Alice \rightarrow Bob: \; E_k(m, \; \mathrm{MDC}(m))$

In Protocol 7.2, Bob receives ciphertext from Alice which he will decrypt with their shared secret key k. The decrypted data includes not only the message but also the unkeyed hash of the message using the MDC function. Bob will compute MDC using the decrypted message and check that the output from the MDC is equivalent to the MDC which he obtained from the decryption. The inclusion of the MDC ensures that the ciphertext was not changed in transit.

Protocol 7.3. $Alice \rightarrow Bob: \; E_{k1}(m, \; MAC_{k2}(m))$

Protocol 7.3 has the advantage of using two keys, whereby if one key is compromised there is security provided by the second key. Furthermore the MAC should be independent from the encryption algorithm (Menezes et al. 1996).

Protocol 7.4. $Alice \rightarrow Bob: \; E_{k1}(m), \; \mathrm{MAC}_{k2}(E_{k1}(m))$

Protocol 7.4 is an encrypt-then-authenticate protocol. Note that there is no guarantee that the party creating the MAC knew the plaintext (Krawczyk 2001), nevertheless it is often used (Black 2004).

Protocol 7.5. *Alice* $\rightarrow$ *Bob* : $E_{k1}(m)$, $\mathrm{MAC}_{k2}(m)$

In Protocol 7.5, the MAC on the plaintext should not compromise the confidentiality of the plaintext. Also both keys used in the protocol should be independent of each other.

In some of these protocols, the use of two keys, one for encryption and another one for authentication, requires two passes over the data (which may be time consuming). The recommended method of combining encryption and authentication is to use authenticated encryption, such as GCM (Dworkin 2007) and CCM (Dworkin 2004), which was discussed in Chap. 6, or encrypt first followed by authentication of the ciphertext (Krawczyk 2001). In the protocols above we assumed that it was the message which should be authenticated. In fact, often other data sent with the message such as the header of a packet should be authenticated but not encrypted. This type of data is usually referred to as associated data and is dealt with in standard authenticated encryption modes (Dworkin 2007).

7.1 Properties of Hash and MAC

The main foundations of the hash and the MAC are the one-way function and the compression function. The one-way hash function and collision resistant hash function are often referred to as MDC. The following three properties will be used in this section to describe various functions:

1. *One-way property* (R_{O}) (also called preimage resistance): for any output it is computationally infeasible to find the input which generated that output
2. *Weak collision property* (R_{Wc}) (also called second-preimage resistance): it is computationally infeasible to find a second input, x_2, which hashes to the same y obtained from the first input, $y = f(x_1)$
3. *Strong collision property* (R_{Sc}) (also called collision resistance): it is computationally infeasible to find any two inputs which hashes to the same output

Figure 7.1a–e illustrates the one-way function, the hash function, the one-way hash function, the collision resistant hash function, and the keyed MAC (shown with key k at the side). Inside each box, or function, the relevant properties are shown, including C for compression and computation resistance shown as R_{Cr}.

The one-way function in Fig. 7.1a, is a function $y = f(x)$ with property R_{O}, or where it is easy to compute $f(x)$ but for all y it is computationally infeasible to find x such that $y = f(x)$. Note that the one-way function has no compression property. This function is easiest to explain since one can think of an egg where breaking it is easy but it is impossible to take a broken egg and recreate the original egg. However,

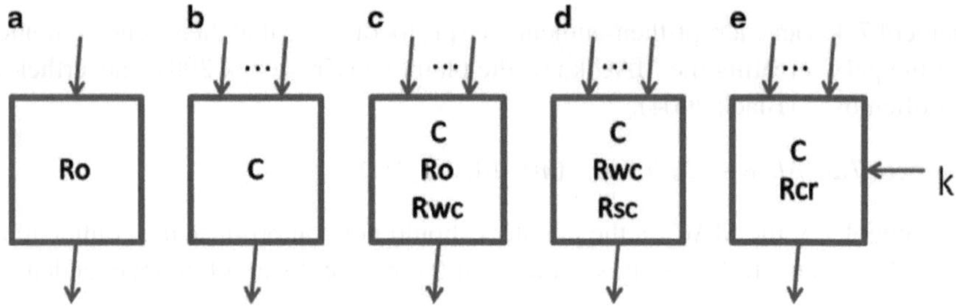

Fig. 7.1 One-way function in (**a**), hash function in (**b**), one-way hash function in (**c**), collision resistant hash function in (**d**), and MAC in (**e**)

it is also true that it is very difficult to mathematically prove that a given function is in fact a one-way function (Schneier 1996).

One-way functions can be built from any encryption function, $E_k(x)$, by defining the one-way function, $f(x)$, as shown in (7.1), for any known k (Menezes et al. 1996).

$$f(x) = E_k(x) \oplus x \tag{7.1}$$

Thus assuming k is unknown, it is computationally infeasible to obtain x given $f(x)$. Note that if k is known then the left hand side of equation (7.1) becomes $f(k, x)$ which is not a one-way function since we can use the decryption function to work from $f(k, x)$ back to x. It is important to note that encryption functions are typically not random functions by definition (since they are also invertible) and often exhibit behavior more like a random permutation (Menezes et al. 1997). Another example of a one-way function is multiplying two large prime numbers together. It is easy to multiply them together, but it is very hard to factor the product into the two large prime numbers. A one-way function with a trapdoor is another concept where one can only perform the function easily in the forward direction unless a user has the trapdoor in which case the function can be computed in either direction. An example of a one-way trapdoor function is the public key systems, discussed in Chaps. 4 and 5, where the private key is the trapdoor. In this case, the forward direction is encryption and signature verification, whereas the backward direction is decryption and signature generation.

The hash function, previously represented in Chaps. 3–5 as $H()$ and shown in Fig. 7.1b, is defined as a function providing compression, C, and which is easy to compute. It is used for data integrity specifically in digital signature protocols. For example, the hash of the message is signed instead of the message. In general it is a many to one mapping and therefore has the potential for collisions. A collision occurs when the hash function outputs the same data for two different inputs. However, in practice the hash function is designed to produce collisions with extremely low probability.

The MAC, also previously represented in Chaps. 3–5, as $\text{MAC}_k()$, is shown in Fig. 7.1e. It is also considered a hash function but in addition to accepting the message or data as input it also accepts a key as input. Thus it is a keyed hash function. It can be used to provide, in addition to integrity, symmetric data origin authentication and identification in symmetric key schemes. This function provides compression and is also designed to be easy to compute. However, unlike the hash function, the MAC also has computation resistance (shown as R_{Cr} in Fig. 7.1e). Specifically this means that it is computationally infeasible to compute for any input data, m, the MAC output, $\mathrm{MAC}_k(m)$. This is true even if one is given one or more pairs of $(m_i, \mathrm{MAC}_k(m_i))$. However, in general, the MAC can be thought of as a one-way key dependent hash function (Schneier 1996).

In addition to the general hash function and MAC function, there are other important unkeyed functions which are used in the security field. These are MDCs (or also sometimes called MICs). The MDC can be further subdivided into one-way hash functions, shown in Fig. 7.1c and collision resistant hash functions, shown in Fig. 7.1d.

A one-way hash function differs from a one-way function in that the one-way hash function has the compression property and the property that it is computationally infeasible to find a second input, x_2, which hashes to the same y obtained from the first input, x_1, $y = f(x_1)$ (which we refer to as W or the weak collision property). It is interesting to illustrate this property using the birthday analogy (Schneier, 1996). Consider a room filled with people. One might ask what are the odds that there is a pair in the room who have the same birthday. The one-way property is analogous to how many people must be in a room for the odds to be greater than half that a person has the same birthday as yours. The answer to this question is 253 people (since the probability that you share a birthday with n people in a room is $1 - ((365 - 1)/365)^n$, which is greater than 0.5 for $n = 23$). The second property known as the weak collision property or R_{Wc} (Schneier 1996), is analogous to how many people should be in a room for the chance to be greater than even for at least two of the people to share the same birthday? The answer to this is actually only 23. The answer is surprising but correct since there are, in fact, 253 pairs of people in the room.

A collision resistant hash function, shown in Fig. 7.1d, has the hash function property along with weak collision resistance property (R_{Wc}) and strong collision resistance (R_{Sc}). Unlike one-way hash functions it does not have the one-way property (R_{O}).

A summary of the various functions defined so far are provided below.

Hash on its own provides a compression function. It is used to provide a digital fingerprint, message digest, or imprint, which is then protected in some way for future data integrity checks. Additionally it has other uses in many protocols, such as digital signatures (see Protocol 5.5) where the hash of a message is signed instead of the longer message, and in encryption protocols (see Protocol 5.5) where the hash of a point on an elliptic curve is used to derive some keys, etc.

MAC message authentication codes are similar to hashes except they accept two inputs, the data and a key. They can be used to provide symmetric data origin authentication and data integrity. Examples include their use in challenge-response protocols, as well as key confirmation (see Protocol 5.3).

MDC or *MIC* provides a hash of data with additional properties. They can be classified into two types, one-way hashes and collision resistant hashes. One-*way hash* is a MDC where finding an input which hashes to the same hash-value is difficult. For example a one-way hash function can be used for key updating, $k_{\text{new}} = h(k_{\text{old}})$, to provide *backward security* (e.g., if the key is compromised, the previous keys cannot be determined due to the one-way property). Alternatively *forward security* can also be achieved with the one-way hash (e.g., by using a autokeying technique (Anderson 2001), hashing not only the key but all messages exchanged with that key, using $k_{\text{new}} = h(k_{\text{old}}, M_0, \ldots, M_n)$). *Collision resistant hash function* is a MDC where finding any two inputs which have the same hash value is difficult.

One should be very cautious when using integrity or MAC functions. It is best to use standardized algorithms whenever possible or refer to recent excellent books on this topic such as Menezes et al. (1996) and Stinson (2005). For example in 2005, SHA-1 was broken. In other words an attack was discovered which breaks one of its properties and notably takes less time than a brute force attack. SHA-1 is a one-way hash function with properties (1) one-wayness and (2) it is computationally infeasible to find a second input, which hashes to the same value as a given input. The second property was broken (Bellare 2005). Section 7.2 will discuss the structure of integrity and authentication functions.

7.2 The Structure of Integrity and Authentication Functions

One-way hash functions can be designed in an iterative fashion, as shown in Fig. 7.2. An example of a one-way hash function is shown in (7.2) which accepts input x_i and an initialization vector or H_0. The output of the hash function in (7.2) is some function of the last output, H_n, or $output = g(H_n)$

$$H_{i+1} = f(x_i, H_i), \quad 0 \le i \le n-1, \quad H_0 = IV \tag{7.2}$$

In this function, f provides the fixed size compression and it is used iteratively to further compress long data inputs.

One-way hash functions differ in practice by how they preprocess (padding and initialization vector), compress ($f()$), and perform the final output mapping, $g()$. Padding may involve adding as many "0" bits as needed or add a single "1" bit followed by as many "0" bits as needed. The later method of course is preferred in case the original data ends in one or more "0"s in which case the padding of "0"s will not identify the end of the original data, unless the recipients know beforehand how long the data will be. The initialization vector can be sent to the recipient along with the message if it is not known beforehand.

In general the security of hash functions is measured by the complexity of the most efficient known attack. However, it is important to understand that this only provides an upper bound on the security. Attacks attempting to reverse the hash (e.g., determine what data the hash has been made from) may randomly choose data to be hashed and check for equivalent hashes. Attacks on a MAC may include trying to determine the key or trying to forge a MAC of some data.

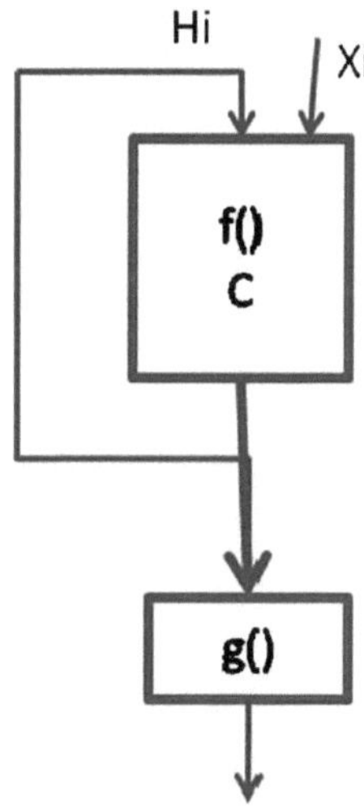

Fig. 7.2 General method for iterated hash function, where $f()$ performs compression

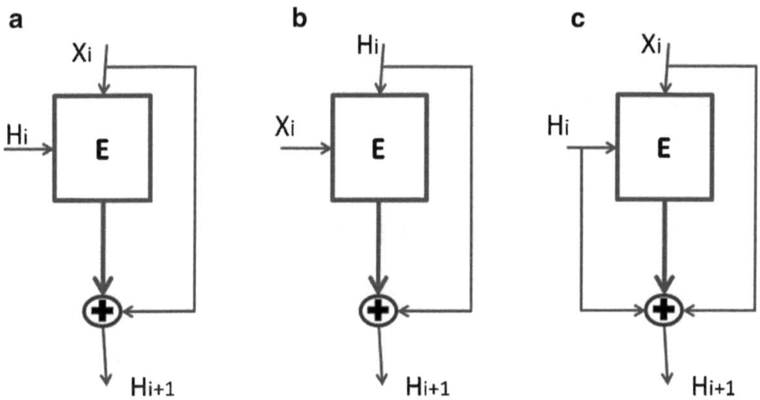

Fig. 7.3 Rate one MDCs using block ciphers

Building hash functions from encryption algorithm allows designers to reuse existing hardware or software already present in the device. These approaches are classified according to rate. A scheme which uses s block cipher operations to process each n-bits of message has a rate of $1/s$. In general we will refer to MDCs (since it includes both one-way hash functions and collision resistant hash functions). There are typically single (MDC-1) or double length (MDC-2) hashes, where the single outputs a hash length equivalent to the cipher output block length and the double outputs a hash length which is double the output block length of the cipher used. Figure 7.3 illustrates some rate one MDCs and Fig. 7.4 illustrates a double length or MDC-2 hash functions, both based on block ciphers. For example in Fig. 7.3a the bottom arrow is shown in (7.3).

$$H_{i+1} = E_{Hi}(X_i) \oplus X_i \tag{7.3}$$

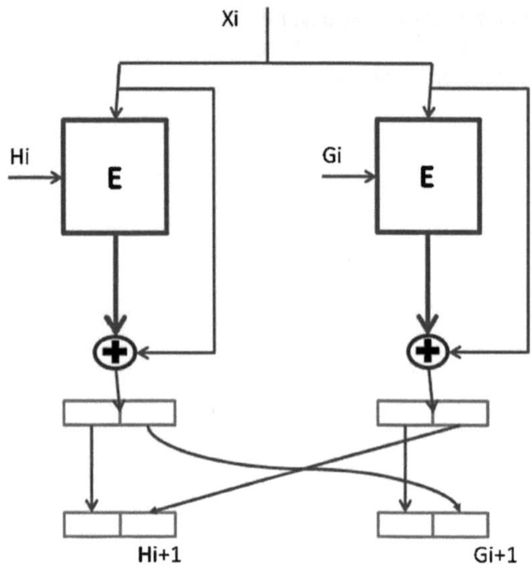

Fig. 7.4 MDC requiring two block cipher operations per block of plaintext input

In Fig. 7.3b the bottom arrow is shown in (7.4).

$$H_{i+1} = E_{Xi}(H_i) \oplus H_i \tag{7.4}$$

In Fig. 7.3c the computation is shown in (7.5).

$$H_{i+1} = E_{Hi}(X_i) \oplus H_i \oplus X_i \tag{7.5}$$

is performed. Also it is possible that in (a) and (c) the encryption key may be a function of the input shown such as $g(H_i)$ instead of H_i on left hand side in (a) and (c). Figure 7.4 shows a double length MDC defining H_{i+1} and G_{i+1}.

There are customized types of hash functions without MDC-1 or MDC-2 type of structure, such as MD-4, SHA-1 (designed from MD-4), SHA-2, etc. Current hash standards, such as SHA-2, are covered in (FIPS180-3 2008). In general the SHA-1,2 functions involve logical functions including exclusive-or, or, rotations, shifts, as well as many numerical constants, padding, and other functions performed together in an iterative algorithm. Additionally there are many proposals for new hashing standards which can be found at SHA-3 (2008).

Many MACs can be built from iterated hash functions, discussed in the previous section, where the output function ($g()$) or the input function to the compression function is a symmetric cipher (Schneier 1996). MACs built using block ciphers commonly use block-chaining mode, CBC mode (Menezes et al. 1996), a mode which is described in Chap. 6.3. Alternatively cipher feedback mode, CFB, can also be used (Schneier 1996). Both are shown in Fig. 7.5. CBC mode MACs are defined in (7.6) and CFB MACs are defined in (7.7). Again the MAC is taken as the last cipher block which is output.

Fig. 7.5 MACs built from block ciphers using CBC (**a**) or CFB (**b**)

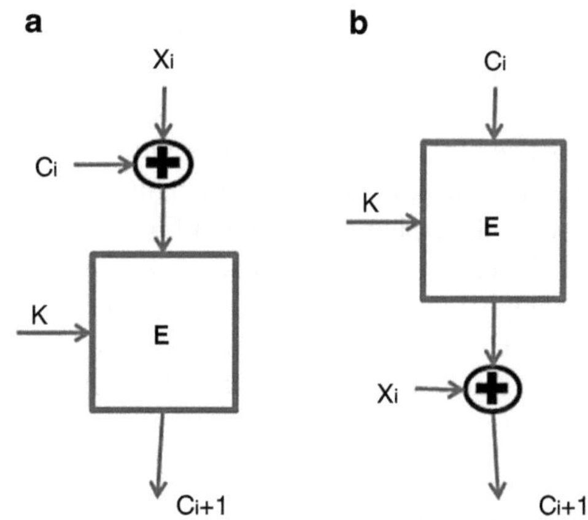

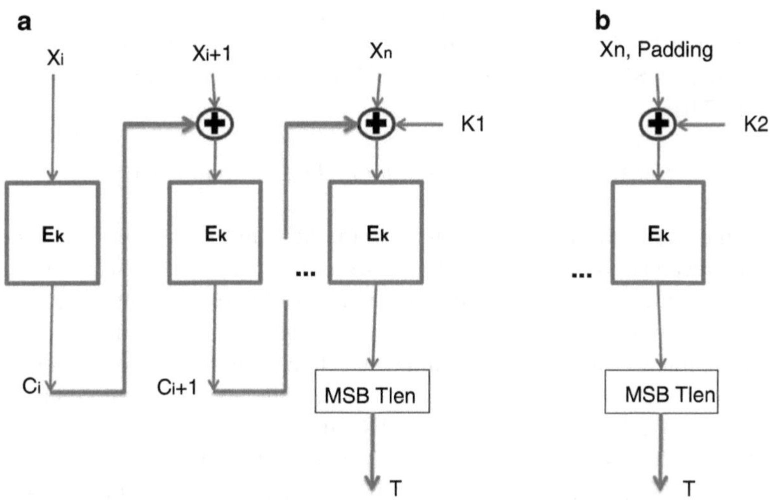

Fig. 7.6 CMAC cipher-based MAC standard

$$C_{i+1} = E_k(X_i \oplus C_i) \tag{7.6}$$

$$C_{i+1} = X_i \oplus E_k(C_i) \tag{7.7}$$

In 2005 the CMAC was standardized as SP800–38B (Dworkin 2005) which is a cipher-based MAC. Figure 7.6 illustrates the general specification. Two subkeys are generated, $K_1,\ K_2$, from the original MAC key K. The first subkey is generated from the left-shifted ciphertext of an input block of zeros, possibly exclusive-ored with a constant. The second subkey is generated from the first subkey through left shifting by "1" bit and again possible exclusive-oring with a constant. The message from which a MAC is to be generated uses the cipher-block chaining mode (see Sect. 6.4) processing the message until the last block is ready to be processed. If the message is a complete multiple of the block size input of the cipher, then the last message block is exclusive-ored with the first subkey as shown in Fig. 7.6a. The last message block is padded and exclusive-ored with the second subkey before being input to the cipher (as shown in Fig. 7.6b)). The output from this final execution of the cipher is then truncated according to the standard (the *Tlen* most significant bits are output). Figure 7.6a illustrates the first case and b illustrates the case of padding.

MACs can also be built from MDCs. In this way the key, k, is concatenated with the message, m, as input to the MDC. An example of a hash-based MAC is shown in (7.8).

$$MAC_k(m) = h(k, p_1, h(k, m, p_2)) \tag{7.8}$$

where p_1 and p_2 are the padding bits for k and m to reach full block length for compression (Menezes et al. 1996).

HMAC is a standard for keyed hashing providing message authentication, and documented in FIPS 198a (2002). It can be used with any iterative hash function, such as MD5 or SHA-1, SHA-2, etc. The HMAC function is generally defined as shown in (7.9).

$$HMAC(k, m) = [h((k \oplus opad), h((k \oplus ipad), m))]_t \tag{7.9}$$

In (7.9) the $h()$ is the hash function, k is the key, and *ipad, opad* are repeated constants specified by the standard. For example $ipad = (3636\ldots36)_{16}$ and $opad = (5C5C\ldots5C)_{16}$ in hexadecimal notation. These constants help to make collision finding much harder. The subscript t on the right-hand side indicates that the most significant t bytes of the quantity are extracted. For example, the key should be (padded if necessary to) B bytes long, whereas the text length should be n-bits where $0 \leq n \leq 2^B - 8B$. The output of HMAC takes the most significant t bytes of the final hash such that $L/2 \leq t \leq L$, where the hash function, $h()$, outputs L bytes. As an example, assume $t = L = 20$ bytes and the text is 9 bytes long and we are using HMAC with a 64 byte key. The length of $\{(k_{\oplus}ipad), m\}$ will be 73 bytes long and the hash of this quantity will be 20 bytes long. The concatenated length of $\{(k_{\oplus}opad),\ h((k \oplus ipad),\ m)\}$ will be 84 bytes. The final hash will be 20 bytes long.

7.2.1 The SHA-2 Function

Originally SHA was proposed as FIPS 180 SHS standard in 1993. A flaw was corrected and the standard was referred to as SHA-1 in 1994. SHA-2 refers to the FIPS PUB 180-3 standard published in 2008. This FIPS180-3 standard (FIPS180-3 2008) provides a description for all five secure hash standards, SHA-1, SHA-224, SHA-256, SHA-384, and SHA-512. In general SHA-x produces a x-bit message digest.

The SHA-256 algorithm can be divided into preprocessing and hash computation stages. The preprocessing stage first pads the input message in multiples of 512 bits, and then parses it into n 512-bit blocks, namely $M^{(1)}, M^{(2)}, \ldots, M^{(n)}$. In the algorithm below each 512-bit block is divided into sixteen 32-bit words in step 1. For example, message block i is represented as $M_t^{(i)}, 1 \leq t \leq 15$. The initial hash constant values, $H_0^{(0)}, \ldots, H_7^{(0)}$ are shown below.

$$H_0^{(0)} = \text{0x6a09e667},\ H_1^{(0)} = \text{0xbb67ae85}$$
$$H_2^{(0)} = \text{0x3c6ef372},\ H_3^{(0)} = \text{0xa54ff53a}$$
$$H_4^{(0)} = \text{0x510e527f},\ H_5^{(0)} = \text{0x9b05688c}$$
$$H_6^{(0)} = \text{0x1f83d9ab},\ H_7^{(0)} = \text{0x5be0cd19}$$

There are 64 message schedule words ($W_0, \ldots, W_{63}$), eight working variables (a, b, c, d, e, f, g, h), and seven hash values $H_0^{(i)}, \ldots, H_7^{(i)}$. In addition, 64 constant 32-bit words ($K_0, \ldots, K_{63}$) are used. Six logical functions are also used and are defined below, where *ROTRn*(x) (and *SHRn*(x)) are a rotation (and a shift) of x by n bits to the right.

$$Ch(x, y, z) = (x^\wedge y) \oplus (\bar{x}^\wedge y)$$

$$Maj(x, y, z) = (x^\wedge y) \oplus (x^\wedge z) \oplus (y \wedge z)$$

$$\sum_0 (x) = ROTR2(x) \oplus ROTR13(x) \oplus ROTR22(x)$$

$$\sum_1 (x) = ROTR6(x) \oplus ROTR11(x) \oplus ROTR25(x)$$

$$\sigma_0(x) = ROTR7(x) \oplus ROTR18(x) \quad \oplus SHR3(x)$$

$$\sigma_1(x) = ROTR17(x) \quad \oplus ROTR19(x) \oplus SHR10(x)$$

Four steps are performed on each message block $i,\ 1 \leq i \leq n$ as shown below:

1. Prepare the message schedule

$$W_t = \begin{cases} \sigma_1(W_{t-2}) + W_{t-7} + \sigma_0(W_{t-15}) + W_{t-16}, 16 \leq t \leq 63 \\ M_t^{(i)}, 0 \leq t \leq 15 \end{cases}$$

2. Initialize the eight working variables

$$a = H_0^{(i-1)}; b = H_1^{(i-1)}; c = H_2^{(i-1)};\ d = H_3^{(i-1)};$$
$$e = H_4^{(i-1)};\ f = H_5^{(i-1)}; g = H_6^{(i-1)}; h = H_7^{(i-1)}$$

3. For $t = 0$–63 do

$$T1 = h + \sum_1 (e) + Ch(e, f, g) + K_t + W_t$$

$$T2 = \sum_0 (a) + Maj(a, b, c)$$

$$h = g;\ g = f;\ f = e;\ e = d + T1;\ d = c;\ c = b;\ b = a;\ a = T1 + T2;$$

Compute the ith intermediate hash value $H^{(i)}$:

$$H_0^{(i)} = a + H_0^{(i-1)};\ H_1^{(i)} = b + H_1^{(i-1)};\ H_2^{(i)} = c + H_2^{(i-1)};\ H_3^{(i)} = d + H_3^{(i-1)};$$
$$H_4^{(i)} = d + H_4^{(i-1)};\ H_5^{(i)} = e + H_5^{(i-1)};\ H_6^{(i)} = f + H_6^{(i-1)};\ H_7^{(i)} = g + H_7^{(i-1)}$$

After processing all n blocks of message M, the final 256-bit message digest is obtained by concatenating the hash values as follows:

$$H_0^{(n)} \| H_1^{(n)} \| H_2^{(n)} \| H_3^{(n)} \| H_4^{(n)} \| H_5^{(n)} \| H_6^{(n)} \| H_7^{(n)}$$

An architecture for the SHA-2 (Juliato et al. 2009) is illustrated in Sect. 7.4, and it closely follows the above algorithm.

7.3 Integrity Trees

There are various attacks which may be launched between two devices, as discussed in Chap. 2. Consider attacks on the channel between a processor and external memory. These attacks include spoofing, splicing, and replay attacks and may be applied to data or memory blocks transferred on the channel. For example in the spoofing attack, a memory block is replaced with the attacker's memory block. In the splicing attack, the attacker permutes two memory blocks (e.g., memory block A is replaced with memory block B and vica versa). The replay attack involves copying the data D_T (or memory block) stored at address A at a time T and later at a time $T + N$ replaces the current data (or memory block), D_{T+N}, at address A with their copy of the old data, D_T, which was valid at time T. These types of attacks apply to many devices such as the channel between nonvolatile memory and a FPGA. Data integrity is one solution which can be used to prevent these types of attacks. In this section approaches to integrity will be discussed by referring to on chip storage as secure and external memory as untrusted storage. However, in general, readers should understand that trusted storage can be used in place of on-chip storage and untrusted storage in place of external memory.

Consider how one might employ data integrity to thwart these attacks. An iterated one-way hash function, as described in the previous section, could be used to compute the integrity of the entire external memory. This single hash value could then be stored in a trusted place such as stored safely on chip. However, when any data legitimately changes in the memory, the entire external memory has to be re-hashed in order to update the single hash value. One alternative is to use one hash value per memory block. The disadvantage to this approach is that there may be a large number of hash values (e.g., large number of memory blocks) requiring excessive storage on the chip, since if the hashes are stored in external memory they would not be resistant to a replay attack. The concept of integrity trees offers a middle ground solution to this problem. A binary tree configuration is assumed for illustration purposes, however, m-ary tree versions are also supported in most cases. Furthermore we refer to on-chip storage as a trusted place for storing the hash value, however, any trusted storage can be used.

The very first integrity tree was Merkles hash tree (Merkle 1989). Originally it was proposed for public key authentication, but later in Blum et al. (1994) it was applied as an integrity tree for protecting memory content. The method was referred to as tree authentication. A binary tree was used where the leaves contained the data and the nodes contained hashes of its children. Authenticating a leaf involved traversing the path from the leaf to the root. At each node of the tree, a hash function is performed. Let a hash function be represented by $H()$, which accepts 2 m-bit words, h_1 and h_2, and outputs one m-bit word, h_3. Equation (7.10) illustrates the hash function performed at each node of the tree.

$$h_3 = H(h_1, h_2) \tag{7.10}$$

For example, assume that the complete data consists of four *m-bit* words, specifically w_m^1, w_m^2, w_m^3, w_m^4. The tree will compute a total of three hashes, specifically: $H\left(w_m^1,\ w_m^2\right) = h_1$, $H\left(w_m^3,\ w_m^4\right) = h_2$, and $H(h_1,\ h_2) = h_3$. The value h_3 computed at the root of the tree would be stored securely on-chip. Hence the resultant hash computed at each node of the tree is used one level higher in the tree in a subsequent hash computation. An example of the binary tree structure is shown in Fig. 7.7a. One disadvantage of Merkle trees is that data writes (or data changes) involved several sequential hashes which could not be parallelized for efficiency. The rectangles in the figure represent the data stored in external memory (the original data and most computed hashes) whereas the data inside the rectangle shown with double lines is stored securely on-chip (the final hash). The ovals represent processing, where H is the hash function. The values $h12$ and $h34$ being hashed inside the double lined rectangle represent the computed hashes $h(d1,\ d2)$ and $h(d3,\ d4)$, respectively.

Another approach known as PAT (Hall and Jutla 2005), for parallelizable authentication tree, supports parallelizable operations in order to update the integrity tree during data writes. The binary tree version of PAT is shown in Fig. 7.7b. The authentication function is performed on three datums, the two data words and additionally a nonce. Thus the computation uses a nonce to move up to the next level in the tree, instead of using the hash output as in the Merkle tree. This allows parallelizable processing by making each tree level independent from each other. The computation illustrated in (7.11) is performed at each node of the tree. At the leaf nodes, $n,\ x_i,\ y_j$, represent a nonce and two datums. At intermediate nodes within

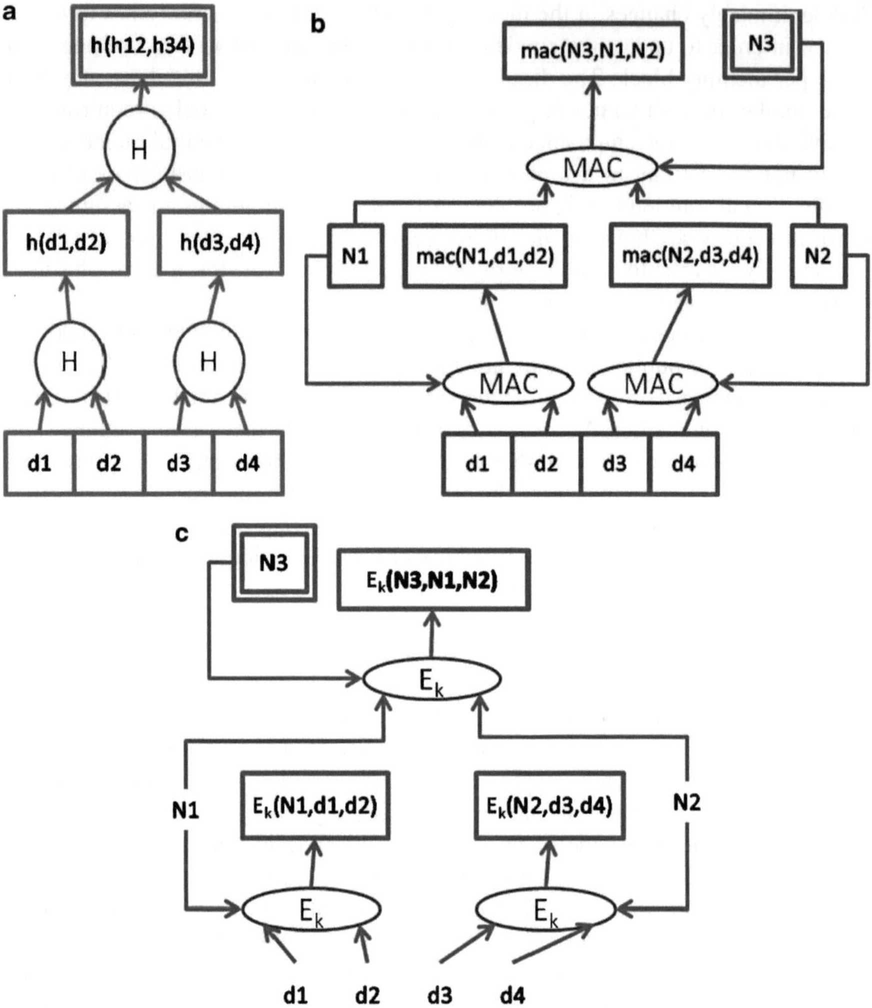

Fig. 7.7 Example of Merkles Hash tree (**a**), PAT-Tree (**b**), and TEC-Tree (**c**)

the tree, the values x_i, y_j, represent nonces from the left and right children of the current node in the tree. The MAC and nonce are stored at each node of the tree.

$$MAC_k(n, x_i, y_j) \tag{7.11}$$

The data writes are parallelizable since each tree node is generated from independently generated inputs. In PAT-tree approach all nonces and MACs are stored in external memory, except that the root nonce is securely kept on-chip while the root MAC is stored in external memory. The rectangles in Fig. 7.6 represent the data stored in external memory (the data, most nonces, and MACs) whereas the data inside the rectangle shown with double lines (nonce *N3*) is stored securely on-chip. A change to the data affects the MAC stored and new nonces are used to update the tree. To verify a data word, all MACs are computed all the way up the tree on-chip to the checking of the root MAC.

The TEC tree approach (Elbaz et al. 2007) is similar to PAT (with parallelizable read and writes) except it additionally provides data confidentiality. Encryption is performed at each node as shown in (7.12). The binary tree version of the TEC-tree is shown in Fig. 7.7c.

$$E_k(n, x_i, y_j) \tag{7.12}$$

Similarly to other techniques, at the leaf nodes of the tree the values x_i, y_j represent the data word and n represents the nonce (e.g., composed of the address of the data concatenated with a r-bit counter value incremented for each block of memory). At intermediate nodes the values x_i, y_j represent counter values from the left and right children in the tree along with a nonce n (e.g., composed of the address of the counter values and a counter value). Verifying integrity requires that a nonce retrieved in a decrypted node matches the corresponding nonce in the parent decrypted node. This checking process is applied to all nodes of a path from the leaf node to the root of the tree. When a data word is modified the relevant tree nodes are decrypted, nonces updated, and reencrypted.

Similar to PAT the TEC-trees support parallelizable data writes and authentication. In addition the TEC-tree approach provides confidentiality since data values are encrypted. Note that in PAT and Merkle trees, the data is stored in addition to the integrity values, whereas in TEC-tree the cipher data is stored as shown in Fig. 7.6. TEC and PAT have higher memory overheads than the Merkle tree. The common philosophy of all integrity trees is that (part of) the root of the tree must be tamper-resistant. PAT and TEC-tree must additionally provide secure storage for a key. Further details can be found in Elbaz et al. (2009) which contains a survey with many references.

7.4 Embedded Implications

Table 7.1 lists the energy measured for some HASH functions running on a Compaq iPAQ H3670 (Potlapally et al. 2003). This PDA device contains an Intel SA-1110 StrongARM processor running at 206 MHz. It also has 64 MB of RAM and 16 MB of FlashROM and is powered by a Li-Polymer battery with a 950 mAh rating. It is interesting to note that most hash functions have lower energy dissipation measurements than the ciphers, measured as $\mu J/B$. However, the energy dissipation of HMAC ($1.16\,\mu J/B$) is approaching the energy of AES ($1.21\,\mu J/B$).

The datapath design (Juliato et al. 2009) for SHA-256 is shown in Fig. 7.8. It is composed of three columns. The first column is the intermediate hash computation. The next column represents the compressor and finally the third column represents the message scheduler. It basically consists of 32-bit adders, shift-registers, logical

Table 7.1 Energy measurements of hash functions from Potlapally et al. (2003)

Algorithm	MD2	MD4	MD5	SHA	SHA1	HMAC
μJ/B	4.12	0.52	0.59	0.75	0.76	1.16

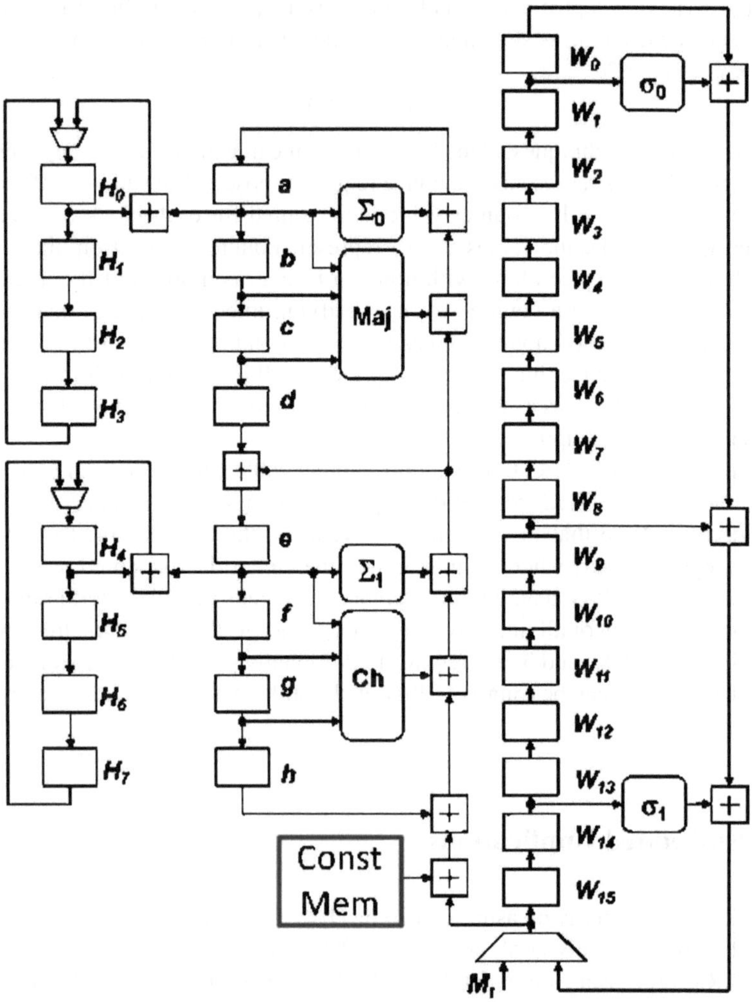

Fig. 7.8 Architecture for SHA-256 (Juliato et al. 2009)

operations, and a memory to store the algorithm's constants. The constants memory is implemented in the FPGA's RAM blocks, and utilizes seventy-two 32-bit words (2304 bits in total). The first step is the initialization of registers $H_0, \ldots, H_7$, which receive the initial hash values in the case of the first message block. Next, the message scheduler registers $W_0, \ldots, W_{15}$ are initialized by shifting in the first 16 words M_t of the message M. Simultaneously, the constants memory provides the initialization values for the working variables $(a, \ldots, h)$ as well as the initial hashes $(H_0, \ldots, H_7)$. These operations take 16 cycles. After that the compressor uses W_t to determine the new values of $a, \ldots, h$. As was shown in step 3 of Sect. 7.2.1, this is performed in 64 iterations and is controlled internally by a counter. In each of these iterations, registers $W_0, \ldots, W_{15}$ and $a, \ldots, h$ are shifted. At the end of the 64 iterations, the intermediate hash computation must be performed. This operation could be executed in one clock cycle, but it would require eight 32-bit adders for that. In order to save area, only two adders are utilized. This way, the computation of the intermediate hash is spread over the last four iterations $(t = 60, \ldots, 63)$ by computing two additions per clock cycle. For example, when $t = 60$, H_3 and H_7 are computed, when $t = 61$, H_2 and H_6 are computed, and so on. For multiblock messages, a new execution cycle initiates with 16 more words M_t being shifted into the module. Then, the same procedure described above is executed. For the last message block, eight read operations are performed to shift out the 256-bit message digest.

The FPGA implementations of various sizes of SHA are shown in Table 7.2 and HMAC is shown in Table 7.3. The estimated dynamic power per byte of message digest at 33 MHz varies from 1 to 1.3 or at their maximum frequencies from 2.2 to 3 mW per byte.

Lightweight hash functions would be extremely useful for RFID tags, satellite applications, and many other embedded systems. These can be made from lightweight ciphers using the techniques in Sect. 7.2 to construct a hash along with lightweight

Table 7.2 SHA FPGA[a] implementations (Juliato and Gebotys 2008)

	SHA-224	SHA256	SHA384	SHA512
Memory bits	2,304	2,304	5,632	5,632
Area (LEs)	1,559	1,591	3,275	3,341
Frequency (MHz)	76	79	61	60
Throughput (Mbps)	443	455	612	590
Dynamic power (mW)	78	81	147	143
Power at 33 MHz (mW)	36	34	83	81

[a] Altera Cyclone II EP2C35F672C6 FPGA

Table 7.3 FPGA[a] implementations of HMAC-SHA (Juliato and Gebotys 2009)

	HMAC-SHA224	HMAC-SHA256	HMAC-SHA384	HMAC-SHA512
Area (ALUTs)	2,262	2,347	4,240	4,601
Freq (MHz)	138	138	120	116
Reg/Mem (bits)	2K/2K	2K/2K	3.9K/5K	4K/5K
Throughput (Mbps) (1 blk messg)	–	–	–	–
1 blk new key	273	271	381	366
2 blk new key	182	181	254	244
Key reuse	547	543	763	733

[a] Altera stratix III EP3SE50F484C2

ciphers from Sects. 6.2 and 6.3. Alternatively newer hashes have been submitted to the Hash competition (SHA-3 2008), as a result of attacks on SHA-1 and other nonstandardized hash functions.

Chapter 8 will discuss side channel attacks. Although an attack on an encryption algorithm, AES, will be discussed, these types of attacks are also possible on hash functions.

References

Black J (2004) Authenticated encryption. In: Tilborg HCA van (ed) encyclopedia of cryptography and security, Springer (also http://www.cs.colorado.edu/~jrblack/papers/ae.pdf)

Bellare M (2005) Attacks on SHA-1, Initiative for Open AuTHentication, OATH, http://www.openauthentication.org

Bellare M, Rogawayy P, Wagner D (2003) A conventional authenticated-encryption mode http://seclab.cs.ucdavis.edu/papers/eax.pdf

Blum M, Evans W, Gemmell P, Kannan S, Naor M (1994) Checking the correctness of memories. Algorithmica 12:225–244

Dworkin M (2004) Recommendation for block cipher modes of operation – the CCM mode for Authentication and Confidentiality. National Institute of Standards and Technology, NIST special publication 800–38C

Dworkin M (2005) Recommendation for block cipher modes of operation: the CMAC mode for authentication, SP800–38B, Computer Security Division, Information Technology Laboratory, National Institute of Standards and Technology, Gaithersburg, MD

Dworkin M (2007) Recommendation for block cipher modes of operation – Galois/Counter Mode (GCM) and GMAC. National Institute of Standards and Technology, NIST special publication 800–38D

Elbaz R, Champagne D, Lee RB, Torres L, Sassatelli G, Guillemin (2007) TEC-Tree: a low cost and parallelizable tree for efficient defense against memory replay attacks. Proceedings of workshop on cryptographic hardware and embedded systems (CHES 2007), September 10–13, 2007, Vienna, Austria, pp 289–302

Elbaz R, Champagne D, Gebotys C, Lee RB, Potlapally N, Torres L (2009) Hardware mechanisms for memory authentication: a survey of existing techniques and engines. Trans Comp Sci IV LNCS 5430:1–22

FIPS180-3 (2008) Secure hash standard. Federal information processing standards publication, FIPS PUB 180-3, info tech lab, National Institute of standards and technology, Gaithersburg, MD 20899–8900

FIPS198a (2002) The keyed-hash message authentication code (HMAC). National Institute of Standards and Technology

Hall WE, Jutla CS (2005) Parallelizable authentication trees. SAC 2005, LNCS, Springer, Heidelberg 2006, 3897:95–109

Juliato M, Gebotys C (2008) SEU-Resistant SHA-256 Design for security in satellites. Proceedings of international workshop on signal processing for space communications, SPSC, doi10.1109/SPSC.2008.4686705: 1–7

Juliato M, Gebotys C (2009) High-performance HMAC processor based on the SHA-2 family of hash functions. CACR tech report http://www.cacr.math.uwaterloo.ca/tech_reports.html

Juliato M, Gebotys C, Elbaz R (2009) Efficient fault tolerant SHA-2 hash functions for space applications. Proceedings of IEEE aerospace conference doi:10.1109/AERO.2009.4839503

Krawczyk H (2001) The order of encryption and authentication for protecting communications (Or: How Secure is SSL)? International crytology conference on advances in cryptology 2001 LNCS 2139: 310–331

Menezes A, Van Oorschot P, Vanstone S (1996) Handbook of applied cryptography. CRC Press, Boca Raton, Fl

Merkle RC (1989) Protocols for public key cryptography. IEEE symposium on security and privacy, April 1980, Oakland, CA, 122–134

Potlapally NR et al. (2003) Analyzing the energy consumption of security protocols. ISLPED, doi:10.1109/LPE.2003.1231830:30–35

Schneier B (1996) Applied cryptography. Wiley, New York

SHA-3 (2008) Round one of the cryptographic hash algorithm. http://csrc.nist.gov/groups/ST/hash/sha-3/Round1/index.html

Stinson D (2005) Cryptography: theory and practice, 3rd edn. Chapman and CRC press, Boca Raton, Fl

Chapter 8
Side Channel Attacks on the Embedded System

I think we're about to start seeing it on smartphones... These attacks are not theoretical

(Mills 2009)

Design for security not only involves the design of security functions which meet price, performance, power, and other constraints, but also resistance to attacks. An attack may be successful in extracting the secret key solely from the side channel signals recorded from a device executing a cryptographic algorithm. In many cases the plaintext of an encryption algorithm additionally may be required by the attacker (known plaintext attack) or may have to be controlled by the attacker (chosen plaintext attack). In some cases the knowledge of the ciphertext (ciphertext attack) along with the side channel signals may be sufficient for an attack. This chapter will introduce the side channel attack and the following chapter will discuss resistance techniques.

8.1 The Side Channel

Side channel analysis (SCA) refers to the use of any information unintentionally leaked from a device while a cryptographic computation is performed. Unintentional information leakage could be in the form of timing information, acoustics, electromagnetic (EM) waves, power dissipation, etc. Figure 8.1 illustrates some of the side channels which have been investigated. The most well known are timing, EM, and power. However, there has been work on the acoustic side channel (Tromer 2004), optics (where light emitting diodes (LEDs) have indicated serial data being transmitted, etc), thermal, etc.

Although all electronic devices emanate electromagnetic waves, it was not known whether information in the waves emanating from a highly complex device could be used to determine the secret key of a cryptographic computation. Traditionally the secret key must always be protected in any embedded system, otherwise consequences could include identity theft, financial loss, etc. The very interesting history of the side channel is outlined in Chap. 2. The focus of this chapter is on practical techniques for analyzing the side channel. This chapter will illustrate that

C.H. Gebotys, *Security in Embedded Devices*, Embedded Systems,
DOI 10.1007/978-1-4419-1530-6_8,

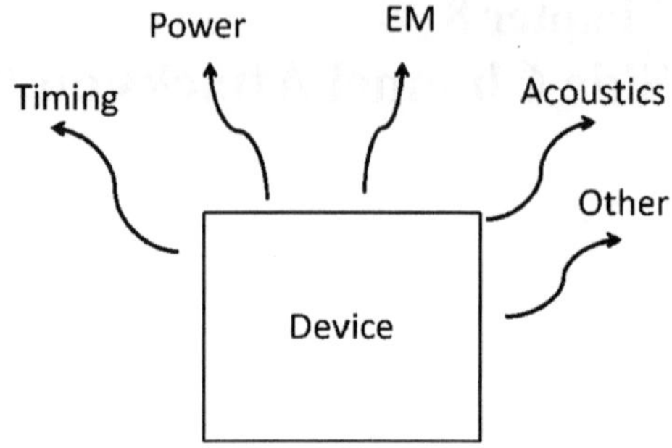

Fig. 8.1 Different side channels of a device

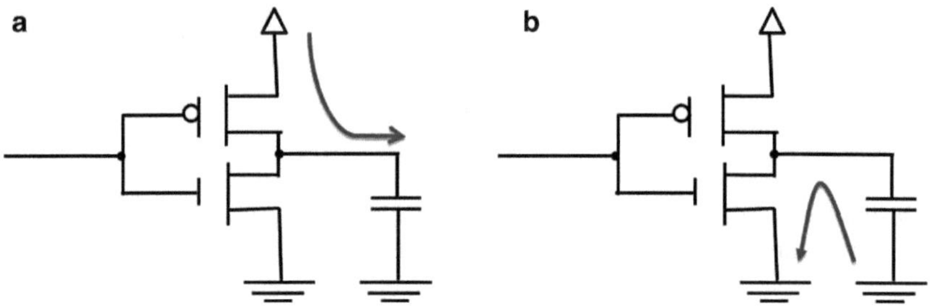

Fig. 8.2 Output switching $0 \rightarrow 1$, P_{01}, in (**a**) and switching $1 \rightarrow 0$, P_{10}, in (**b**)

EM side channel attacks on complex embedded systems are possible. We will use the term power traces in place of measured current traces. However in all experiments the current was measured with an inductive probe (and it is understood that a constant voltage source was used). Additionally we will use the term SCA to mean power or EM SCA.

8.1.1 Theory of the Side Channel

An understanding of the side channel starts with a brief discussion of the basic electronic components in any IC. This section will discuss the circuit behavior which is responsible for creating the current, power, and EM side channel.

Consider the simplest type of circuit, the inverter. A CMOS implementation of an inverter is shown in Fig. 8.2 (with the PMOS transistor on top and the NMOS transistor at the bottom). One can represent the instantaneous power dissipation for this circuit based upon its current state, i, and its next state, j, as P_{ij}. State is represented by the output voltage of the inverter. For example, P_{00} represents the condition where a zero value is output from the inverter (e.g., the input is a one value). The term P_{01} represents the condition where the output of the inverter makes a transition from a zero value to a one value. The input to the inverter has switched from a one to a zero value which will cause the output of the inverter to switch from a zero to a one, as shown in Fig. 8.2a. The switching at the output of an inverter from

1 to 0 is represented by P_{10} and is shown in Fig. 8.2b. The basis of the side channel attack is shown in (8.1). If we subtract the instantaneous power dissipation of the inverter when it is switching from zero to a one, from the instantaneous power dissipation of it switching from one to zero, we would get a difference, or power spike, which is greater relatively speaking than the power dissipated in the inverter when no switching is taking place.

$$|P_{01} - P_{10}| \gg |P_{00} - P_{11}| \tag{8.1}$$

This equation holds since we have the conditions shown in (8.2) and (8.3).

$$P_{01} \neq P_{10} \tag{8.2}$$

$$P_{00} \approx P_{11} \tag{8.3}$$

$$P_{11} + P_{01} \neq P_{00} + P_{10} \tag{8.4}$$

In (8.2), the instantaneous power dissipation for switching to one and for switching to zero at the output of the inverter is different. The power dissipations involved with no switching are similar relatively speaking, regardless of whether the inverter is holding a zero or a one value at its output, as illustrated in (8.3). These relationships can be reformulated as shown in (8.4) which is the basis of SCA. In particular the difference of the left- and right-hand side of (8.4) forms the basis of differential power analysis, or DPA. Generally in differential analysis instantaneous power dissipation traces are partitioned according to the value of a bit (as in this example, at the output of the inverter). Specifically the left-hand side of (8.4) represents the bit value of a "1" and the right-hand side illustrates a bit value of a "0."

Equation (8.5) provides a more detailed examination of the dynamic power dissipation which is relevant to SCA. For example, the side channel relevant component of dynamic power is switching power [first term of (8.5)] as well as short circuit power [second term of (8.5)]. The terms used in this equation α, f, V_{dd}, C_L, $\frac{\mu c_{\text{ox}} W}{12L}\tau$, V_{th} are defined as the activity factor (some value between 0 and 1), the clock frequency, the supply voltage, the capacitative load, parameters of the process, and the transistor threshold voltage.

$$P_{\text{dynamic}} = \alpha f V_{\text{dd}}^2 C_L + \frac{\mu c_{\text{ox}} W}{12L}(V_{\text{dd}} - 2V_{\text{th}})^3 \tau f \tag{8.5}$$

The two terms in (8.5) are directly relevant to SCA. In static CMOS circuits, the first component, switching power, is relevant to any activity in the circuit. Short-circuit current flows whenever there is a transition. Consider a transition of the input to the inverter from a 0 to a 1. The NMOS transistor turns on when the input voltage to the inverter, V_{in}, is greater than the threshold voltage, V_{tn}, or $V_{\text{in}} > V_{\text{tn}}$. The PMOS transistor does not turn off until the input voltage is greater than the PMOS threshold voltage, V_{tp}, subtracted from the supply voltage, V_{dd}, or $V_{\text{in}} > V_{\text{dd}} - V_{\text{tp}}$. Thus in between these two times, ${t_{\text{nmos}}}^{\text{on}} < t < {t_{\text{pmos}}}^{\text{off}}$, both transistors are "on" creating a short circuit from V_{dd} to ground. This current flow is referred to as the short circuit current. The transition from 1 to 0 at the input to the inverter also creates short circuit during the time t, where $t_{\text{pmos}}{}^{\text{on}} < t < t_{\text{nmos}}{}^{\text{off}}$. This is due to the PMOS transistor turning on before the NMOS transistor turns off. Additionally there are models used to estimate the amount of time spent in this short circuit region (Sylvester et al. 1999).

In summary if one could measure the instantaneous power dissipation of a circuit without noise, there would be a noticeable difference between a switching to a zero and a switching to a one. In practice many different measurements of instantaneous power are averaged in order to remove the impact of noise. The difference of these means (e.g. a mean trace where bit x is a '0' and a mean trace where bit x is a '1') would ideally represent $(P_{11} + P_{01}) - (P_{00} + P_{10})$. This is the basis of differential (power or EM) analysis (DPA or DEMA) which will be discussed in Sect. 8.3. The difference of these two means also removes the algorithmic noise when used in a differential attack on a cryptographic algorithm. The amplitude of the instantaneous power dissipation is not only proportional to the amount of switching (which is related to the number of bits which have switched or the hamming distance) but the size of the capacitance being switched, thus areas of the chip with larger capacitances (those with larger power dissipations) such as off chip transfers/etc will have larger side channel emanations. This is further discussed in Sect. 8.3 which further introduces differential analysis.

Cryptographic algorithms executing on an embedded processor or in a specific cryptocore will create current switching on the die. The current switching will include key dependent switching. Thus the power side channel can leak information concerning the key. The current drawn and switched by transistors also creates complex EM emanations through the interaction with inductances and capacitances. These interactions do not only include inductances and capacitances on the die, but also of the package, printed circuit board, and other components on the board, etc. Nevertheless, despite these complexities, analyses have shown that there is sufficient information within these EM emanations to leak information about processing with the key. Both simple EM analysis, SEMA, and differential EM analysis, DEMA, have been used to extract information about the key from some chips. Although near field EM signals have been used to compromise keys it may be possible that far-field EM signals may also compromise keys. Sect. 8.1.2 will illustrate the side channel experimental setup from a practical standpoint.

8.1.2 The Side Channel Attack in Practice

Typically EM or power analysis involves an attacker who captures the EM or power signals emanating from a device while it is executing a cryptographic process. The samples of EM or power signals acquired over a time interval at a suitable sample rate by the scope are often referred to as a trace (or sometimes a frame). Often the attacker collects many EM or power traces (e.g., one or more EM traces for each plaintext input to the cipher). The set of traces collected at one time from the

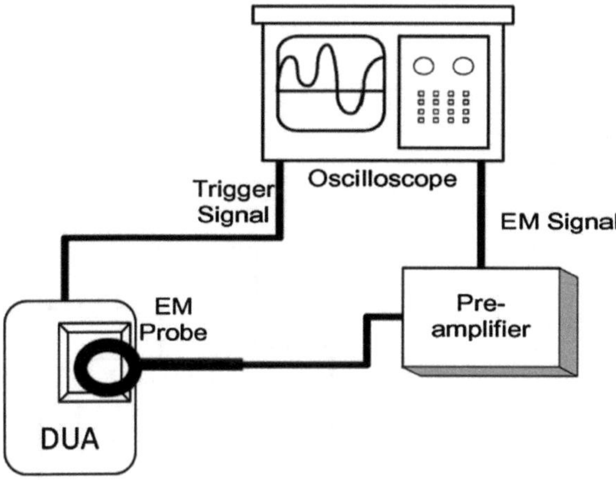

Fig. 8.3 Typical components of a side channel analysis for device under attack (DUA)

scope will be referred to as an acquisition. This section will discuss equipment for performing this capture as well as how the equipment is used to find side channels.

A typical experimental setup is shown in Fig. 8.3. The three important components to a side channel setup are the oscilloscope, the device under attack (DUA) including its trigger, and the probe. In this figure, the EM probe is connected to a preamplifier which is then connected to the scope. The EM probe and preamplifier are most relevant for complex embedded systems, where access to power lines is extremely difficult. However it has been suggested that the power pin of a USB port may provide a power side channel analysis for some embedded devices. The EM probe can be placed on top of the chip of interest in order to obtain side channel information. Although the EM probe with amplifier can be purchased or homemade, the device trigger often remains a challenge for real embedded systems.

Figure 8.4 shows another setup which is typically only possible using evaluation boards as opposed to a real complex embedded device. Here the side channel of a processor core within a specific chip is being analyzed. The components include the inductive probe for measuring the current drawn by the processor core in the chip. The PC and the multi-ICE support downloading of software programs into the processor in the evaluation board. The setup includes an EM probe as well so that both EM as well as power can be measured. From a side channel laboratory point of view, evaluation boards are ideal for conducting experiments on the side channel. Most evaluation boards are cheap in price and sometimes the processor being attacked in a real embedded system is available in an evaluation board. It may not be running the same full system as in the embedded device, but experiments on the evaluation board can often save time when analyzing the actual embedded device. For example, information on timing, side channel behavior, signal identification, etc can be very useful in order to help to characterize expectations of the embedded system.

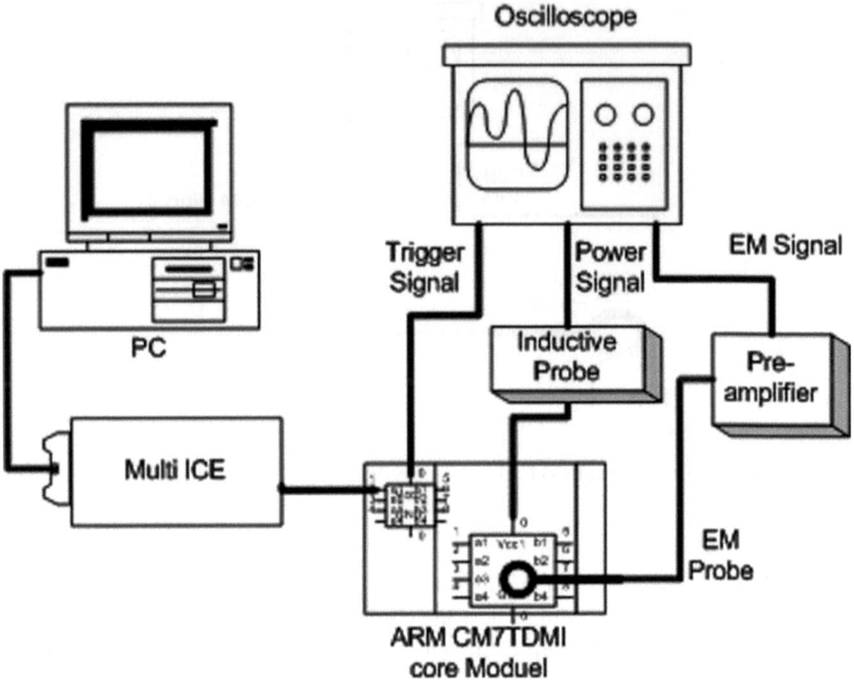

Fig. 8.4 Measuring both current and EM signals from a device under analysis

Evaluation boards sometimes utilize 0-Ω resistors connected to the power of the core processor in a chip, as was the case for the system shown in Fig. 8.5. A 0-Ω resistor looks like a resistor on the board, but it is just a wire inside. These can be removed and replaced with a resistor or inductive probe across the two terminals. In other cases, the power lines may be reachable and cut in order to support current measurements. If there are no power access points for the chip, the EM probes are easily used. EM probes can either be made or purchased. The trigger uses a current probe attached to a connector on board. The EM probe is shown held over the chip of interest, however in an attack it is strategically placed over the chip and fastened so that it will not move.

8.1.2.1 EM Probe

The EM probe is ideal for real embedded devices since it is the easiest to use. Few embedded devices conveniently support access to power lines. An exception to this are contact smartcards where power contacts are easily found with standardized interfaces. Nevertheless all possibilities for acquiring side channel data should be accessed, including battery connections. The EM probe is connected to an amplifier and then to the scope. The amplifier very importantly provides the boost of the

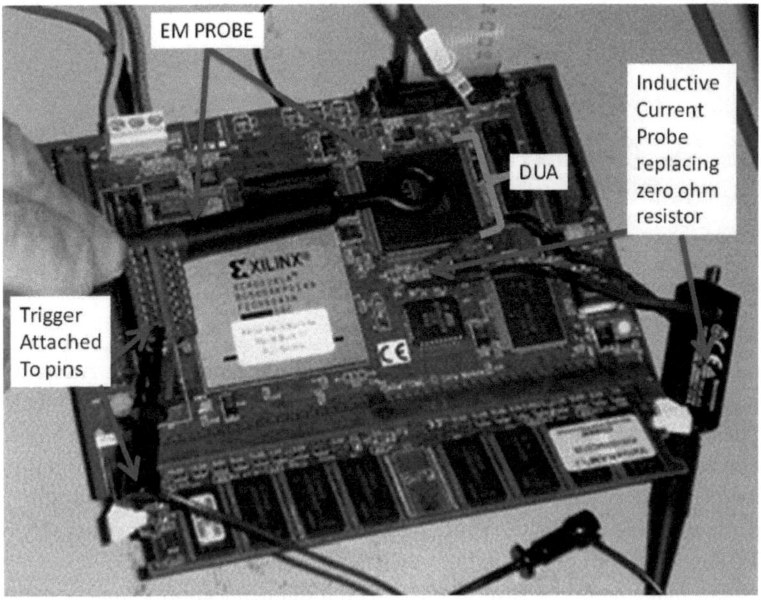

Fig. 8.5 Illustration of EM probe, trigger, DUA, and current probe

weaker signals which is important in SCA. The Electro-metrics EM-6992 probe (EM6992) is used with a preamplifier in all reported experiments. The typical gain of the amplifier is 22 dB. The probe has a 1 cm loop antenna which is sensitive to H-field frequencies from below 100 kHz to 1 GHz. The signal from the probe is connected with a 50 ohm coaxial cable and the scope was configured for matching termination.

An alternative to the EM probe, includes measuring the voltage across a resistor inserted in the path of the power to the device, although this is rarely feasible in most embedded systems. If the chip has an accessible power pin, then the pin wire would be cut. The resistor terminals would be connected to each cut end. Of course this is difficult if not impossible to do with most current chips which are currently embedded within a system. The voltage across the resistor can be measured with a differential probe. An inductive probe can also be used in place of a resistor, but like the resistor it has to be placed in the line of the power. The inductive probe directly connects to the scope. It may even be possible to measure the power of an embedded device by attaching an ADC to the power connector within a USB port as suggested by Adi Shamir (Shamir 2009).

The home-made variety of EM probes (Gerke and Kimmel 2002) are also quite good and offer comparable performance to the commercial probes. These are near field probes. Figure 8.6 illustrates a home-made EM near field probe. Another hand-made probe was created from a standard BNC cable. It had thin strands of aluminum core which were bent into a single elliptical loop (with the cable pointing along the major axis) about 7 mm by 4 mm. It was "flattened" out to create a ring about

Fig. 8.6 Some home-made EM probes

2 mm in thickness which was spiral wrapped in white electrical tape. Another probe was measured to have a peak frequency of approximately 104 MHz and was made of copper in a 1 cm loop configuration connected to a coax cable. The hand-made probe performed just as well as the commercial probe, however had slightly lower amplitudes. Other researchers have obtained the best results with an amplifier and a handmade solenoid probes made of coiled copper wire with outer diameters of 150 and 500 μm (Gandolfi et al. 2001).

You can make a coil of several turns with a diameter approximately equivalent to the chip under analysis. The coil is then connected to the coax cable. To avoid shorts, one should cover the probe wires and follow suggestions in (Carsten 1998). The coax cable's opposite end connects to the scope. The voltage output from the EM probe follows (8.6), where n is the number of turns of the probe, B is the incident magnetic field and A is the area enclosed by the loop.

$$V_m = n\omega BA\cos\beta \tag{8.6}$$

Commercial probes are often referred to as sniffer probes, and are typically used to find out which areas need to be modified in order to reduce the EM emanations for EM compatibility (EMC) or EM interference (EMI) reasons. Often a sniffer probe set can be purchased which has magnetic (H-probe) as well as electrical (E-probes) probes. The best probes for SCA are the loop H-probes. When the loop is sized to fit over the area of the chip, the measurements are correlated with the power grid of the chip (Mangard 2003). These EM probes measure the high frequency magnetic fields and thus currents in a circuit. Their range is typically 1–2 times the diameter of the loop (Gerke and Kimmel 2002).

In Fig. 8.7, a hand-made probe [denoted as *white probe*], (b), is compared to the commercial probe in (a) for acquiring EM traces for AES running on the experimental setup in Fig. 8.5. This plot has an x-axis ranging from 0 to 255 representing all 8-bit key guesses. The y-axis represents the height of the differential peak which will be discussed in Sect. 8.3. The highest peak (or y-value) in this plot, in theory, identifies the correct key. The commercial probe used in all experiments is the Electro-Metrics EM-6992. All the experiments were repeatable. In both cases the correct key was found. Typically the hand-made probe results are similar to the commercial probe results in most experiments.

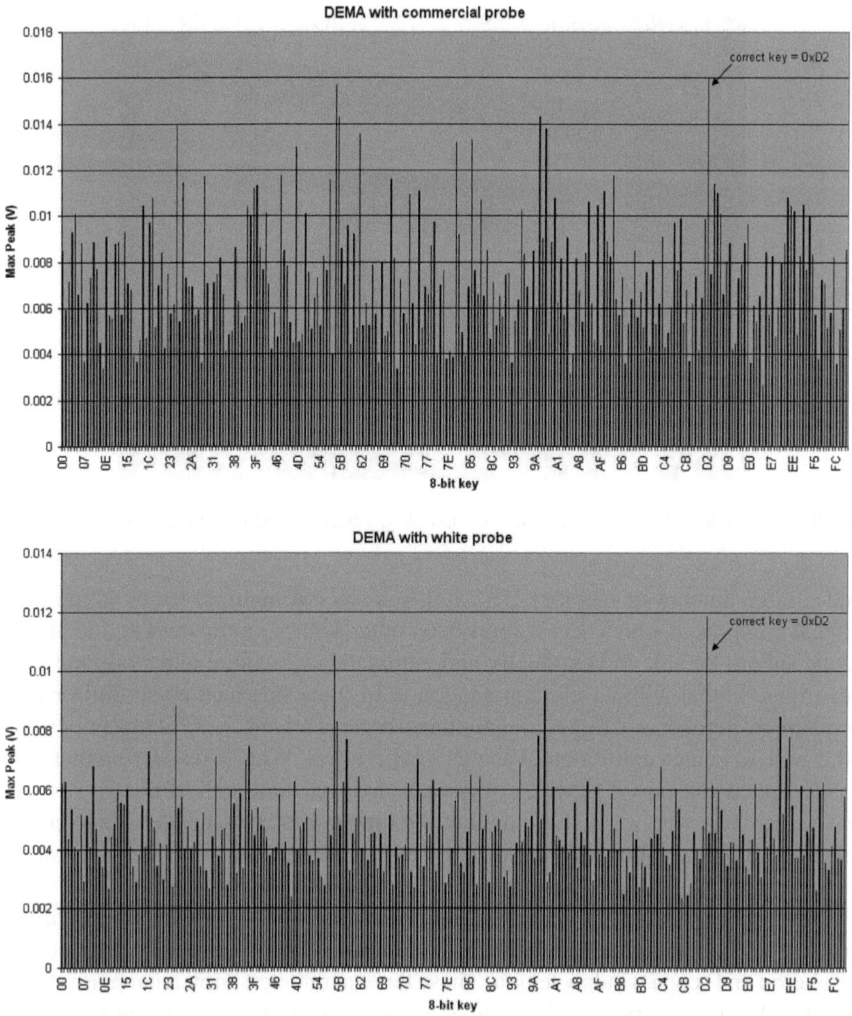

Fig. 8.7 Commercial vs. hand-made probe results in (**a**) and (**b**) respectively

8.1.2.2 The Oscilloscope

The oscilloscope is perhaps one of the most important parts of a SCA laboratory. The oscilloscope must have the ability to record multiple traces per acquisition. A trace is a series of samples of a (power or EM) signal acquired over a time interval (known as the scope window). Each trigger signal received tells the scope when to start sampling signals, though in general a user can easily program the scope to start any number of samples before or after the trigger. Traces need to be stored separately for each trigger received from the device. The set of traces, or acquisition, is later off-loaded for analysis. The oscilloscope must also have a sufficient sampling rate

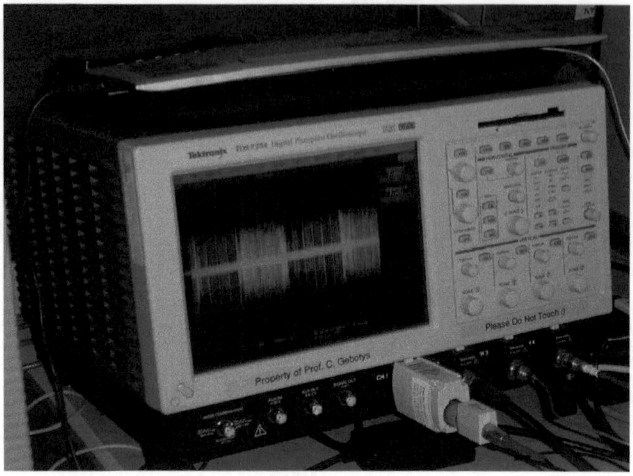

Fig. 8.8 Example of Scope with multitrace capability, Tektronix TDS7254 (Tektronix 2003)

and a large amount of memory. The ability to record multiple traces is the most crucial component which sets it apart from other scopes on the market. Ideally the scope should be able to continually and automatically acquire and store one trace per trigger signal without user interaction until the scope memory is full or until a specified number of trigger signals have been received. It is likely impossible otherwise to launch a differential analysis experiment. Why? Considering that most sophisticated analysis of the SCA of devices involves thousands of traces, there are few researchers who would have neither the time nor the patience to sit and collect one trace at a time. The scope in Fig. 8.8 can acquire multiple traces per acquisition, and this technique is called fast frames in Tektronix scopes.

The scope memory is typically the limiting factor in SCA. Since the scope memory size is fixed, an attacker is always trying to tradeoff the number of traces for resolution and scope window length. These three parameters form an important relationship. Let n_{traces} represent the number of traces in one acquisition that an attacker will acquire. Let S_{scope} represent the number of samples which can be stored in the scope. Let the sample rate be $f_{\text{sampleRate}}$ samples per second, which as a good rule of thumb is typically ten times the clock frequency of the device under analysis. If the scope window is t_{window} seconds wide, then (8.7) illustrates the scope memory relationship.

$$S_{\text{scope}} \geq n_{\text{traces}} \times t_{\text{window}} \times f_{\text{sampleRate}} \tag{8.7}$$

Let us assume that we first acquire one trace as shown in Fig. 8.9a containing $16x$ samples. Using the same resolution and memory, one could alternatively store 4 traces each of $4x$ samples as shown in Fig. 8.9b. For example assume that a single trace has 32 M samples. Now assume we will analyze a small section of this trace, in particular we wish to have 32,000 traces. Using the same sample rate, we would only be allowed to use a window of 1,000 samples. In theory if we had the maximum amount of memory in our scope, we could run the experiment several times,

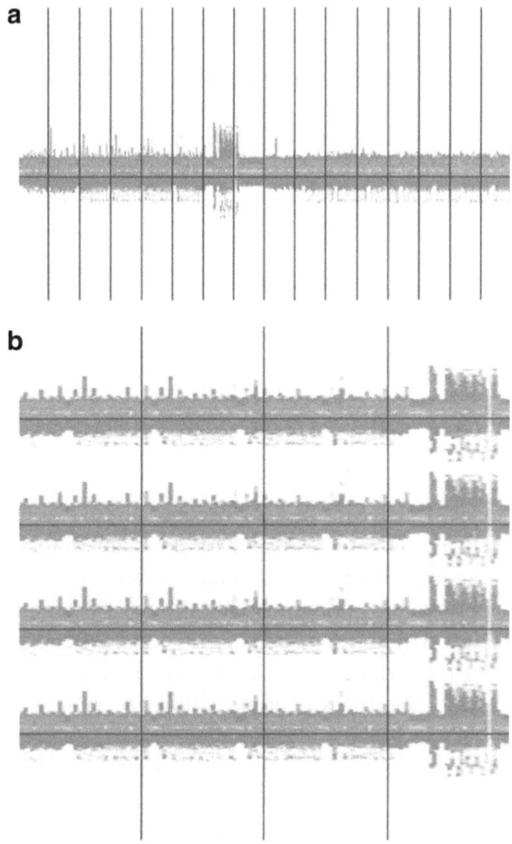

Fig. 8.9 Scope memory illustration in (**a**) single high resolution trace acquisition and (**b**) multiple trace acquisitions with smaller scope windows

each time recording the maximum number of samples, and processing all acquisitions later for our analysis. However, this becomes very tedious if we require more than, say, ten acquisitions. It is ideal for a SCA scope to have memory supporting 32–64 M samples or more. It is interesting to note that most scopes have a PC within them, thus it may be possible to run the analysis of traces on the scopes PC as well. Based upon our experience, it was worth the added time to offload the traces to a faster PC with larger amounts of data to run the analysis.

In SCA, the more traces you can acquire the higher likelihood that a successful attack is possible. This is even true for attacking devices with built-in countermeasures. Hence in theory, if an attacker had sufficient funds, it may be possible to use numerous scopes to record enormous amounts of data for analysis. Performing the subsequent analysis of these traces would also require numerous computers.

8.1.2.3 Device and Trigger

The trigger must be generated typically from the device under analysis (as described in Sect. 8.5) or some aspect synchronized with the device (as described in Sect. 8.7 and Chap. 9). A rising edge or more sophisticated trigger signal definition is possible

```
Loop for i=1 to T
{
data = key ^ plaintext(i)
...
Set trigger signal high
...
% attack point
Load S-box(data)
..
Reset trigger signal
...
Wait # of cycles
...}end
```

Fig. 8.10 Example of trigger in software

as supported by the oscilloscope. Normally trigger signals can be obtained from evaluation boards using one of the many output signals of various connectors. Finding a trigger signal from a real embedded device is often a challenge. The attacker has to be able to activate the trigger, by generating the signal on the specific wire which will be connected to the scope. Often software methods are used to generate the trigger right before the program under analysis is initiated. Figure 8.10 illustrates pseudocode for triggering within an cryptographic algorithm such as the general cipher in Fig. 8.13. Depending upon the latency between the trigger being set in software and the trigger signal generated and sent to the scope, it may be necessary to shift the scope window in order to find the attack point or set the trigger signal earlier (as shown in Fig. 8.10). If the trigger is extremely fast, and can be inserted in between processor instructions, then it may be placed directly before the instruction under attack. However this is not possible in real embedded systems thus triggers often are set before the entire cryptographic application is executed. There are often connectors on embedded devices which may be used to generate triggers. For example it may be possible to send particular signals to connectors of the device in order to generate the trigger signal such as through USB, HDMI, GPIO, etc. On a PDA device, a possible trigger could be the wires at each end of the LED light or control through other ports or interfaces. The LED trigger is used for experiments reported in Sect. 8.5. Contact cards may use serial data lines of the standard contact interface to trigger. On a contactless card, the RF signal itself could be used as a trigger, though not as reliable as a wired trigger, as discussed in Sect. 8.7. More details on setting up attacks with triggers will be provided in Sect. 8.1.3.

8.1.3 Setting up a SCA

This section will discuss the steps to acquiring and verifying a side channel. The term side channel activity will be used in place of EM or power activity. It will refer to the variation of EM or power acquired by the scope.

Before SCA is attempted, the embedded system should be simplified as much as possible. If it is a portable communications device such as a PDA, then the radio should be turned off. If there are applications that can be turned off, then they should be turned off, etc. The idea is to try to isolate the algorithm under attack in the embedded system. If it is not possible to perform any isolation, then the attack will be more complex and typically take many more traces.

Given an embedded device, the side channel can be initially investigated using a receiver which demodulates at the frequency of the clock of the device running the security application. If it is not known, a guess can be made and normally the receiver still does a good job. The receiver allows the attacker to listen to the side channel. This helps to locate which area of the device is causing the most emanations from your algorithm under attack. By running the algorithm in an infinite loop, the receiver antennae can be moved around various parts of the device. The louder signals can be used to locate regions of interest in the device. Additionally running the algorithm for half a minute or so will verify that the sounds you are hearing are specific to the algorithm of interest. There are receivers which can be used to specifically demonstrate the side channel, such as the ICOM IC-R10 (http://www.icomamerica.com). This section will outline the necessary requirements of equipment used for establishing a side channel experimental setup.

Once the regions of interest are located, the EM probe and scope can be used to further identify which chip or area of the chip is actively emanating the side channel waves. Positioning of the probe is important. It is often at first best to place the scope into a mode where traces are not stored, but continually displayed as they are acquired. Thus the EM signal magnitudes can be observed while positioning the probe. The probe should be fastened to preserve its position and orientation where maximum amplitude was attained. Even slight changes to the probe position can make a large impact on signal amplitudes and shapes captured in traces.

It is important to verify that you are observing the side channel of interest. In particular it is easy to see side channel patterns on the scope and be overconfident that they relate to your algorithm, only to find that they have nothing to do with your algorithm. Typically in embedded systems there are many other processes running which are not readily identified. Embedded systems are so complex that typically hundreds of scientists and engineers contribute to various parts of the system. Thus when you notice some EM events or timed events, it is not easy to find answers to what that event may be. In other words you need to verify that what you are seeing is actually the execution of your algorithm of interest. We will consider two scenarios for this. The first is an ideal scenario (e.g., evaluation board) where you have control over the system. The second scenario is a real embedded device where system activity cannot be controlled and possibly the attacker has no knowledge of the system at all.

In an ideal scenario, the algorithm of interest should be sandwiched with very low side channel activity. For example in Fig. 8.10 the region in between the setting and resetting of the trigger, can consist of several hundred inlined NOP instructions placed before and after the attack instruction. Thus ideally the scope window should show low side channel activity, then the side channel activity of the attack point followed again by low activity. If this is possible, one could verify the side channel by replacing the attack instruction with a larger number of high activity instructions. In these cases if the side channel activity with high amplitude captured by the scope also widens, respectively, you could be confident that what you are seeing is in fact the side channel of those instructions. This approach worked well with the ARM7TDMI evaluation board. However results may vary with other systems. For example, it is interesting to discuss "low" side channel activity. An interesting experiment is to place a "NOP" instruction within a loop so that the processor executes hundreds of NOPs before the algorithm under attack. Although one might expect low side channel activity, the loop is actually switching away, as it counts down through the loop iterations. Thus instead of low activity, one actually sees high switching activity. In a real embedded system, unlike the ARM7TDMI system, the inlined NOP instructions created high activity on the scope. The high activity was a result of repeated NOP instruction fetches into cache.

Sometimes specifics of the embedded system can be used to locate the approximate start and end times of the algorithm under attack. Since ciphers typically execute in rounds, these rounds are often seen as repeating patterns in the EM activity. Memory accesses or jumps to the start of loop, typically create visible activity per round. The next section will introduce and demonstrate simple analysis, followed by differential analysis and correlation analysis.

8.2 What is Simple Analysis

Simple power or EM analysis refers to the analysis of one trace from the DUA. Typically it is assumed that detailed analysis of another device, or characterization device, was performed beforehand.

Hamming weights are illustrative of a simple analysis. In theory, the higher the hamming weight, the more bits have switched from 0 to 1. Thus the side channel (power or EM) amplitude should be related to the hamming weight of the data. However, in real embedded systems, assuming the data is placed on a bus, the switching may not be directly related to the hamming weight of the data. For example the bus may not contain all zeros before the data is placed onto the bus. It may be precharged or it may have held the previous data which was placed onto the bus. Also, the height of the power measurement may be related to previous computations performed not directly reflecting the current data on the bus. Figure 8.11a illustrates each hamming weight and corresponding mean trace (generated from averaging 1500 power traces of loading data with a fixed hamming weight using an ARM7TDMI (Chen 2004)). Each of 1500 power traces were obtained from a load instruction surrounded by NOP instructions. The value of the data loaded in each of 1500 traces varied however its hamming weight was constant as shown. These 1500 traces had to be preprocessed such that the peak of the previous load instruction (which loaded zeros into a register) was at the same amplitude in all traces. After this amplitude alignment, the traces with the same hamming weights were averaged

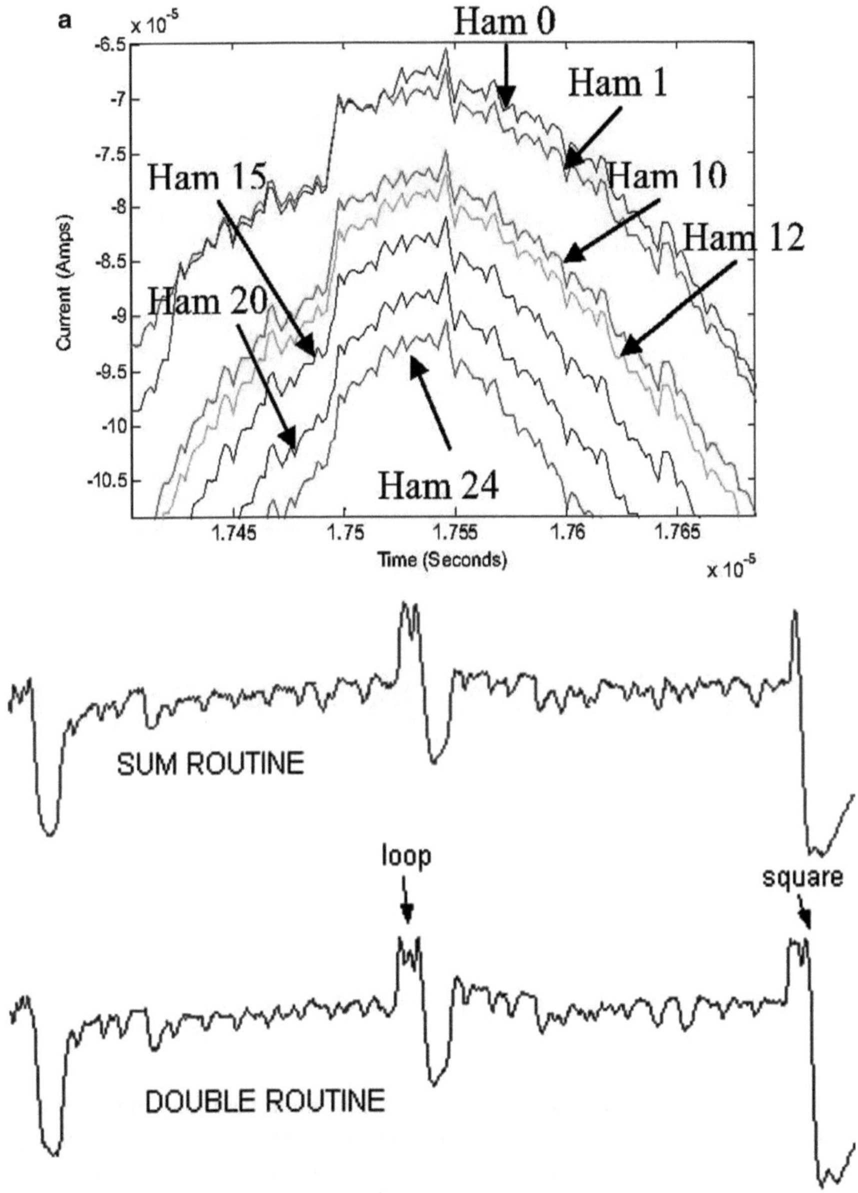

Fig. 8.11 Hamming weights vs. current of a ARM7 in (**a**) and SPA illustrating differences in top and bottom traces of a DSP in (**b**)

and the results are shown in Fig. 8.11a. However it was also observed that the noise was too high for one to be able to deduce the hamming weight from a single trace.

SPA is typically more useful for determining algorithmic structure as opposed to hamming weights. Recall from Chap. 5 that during an elliptic curve point

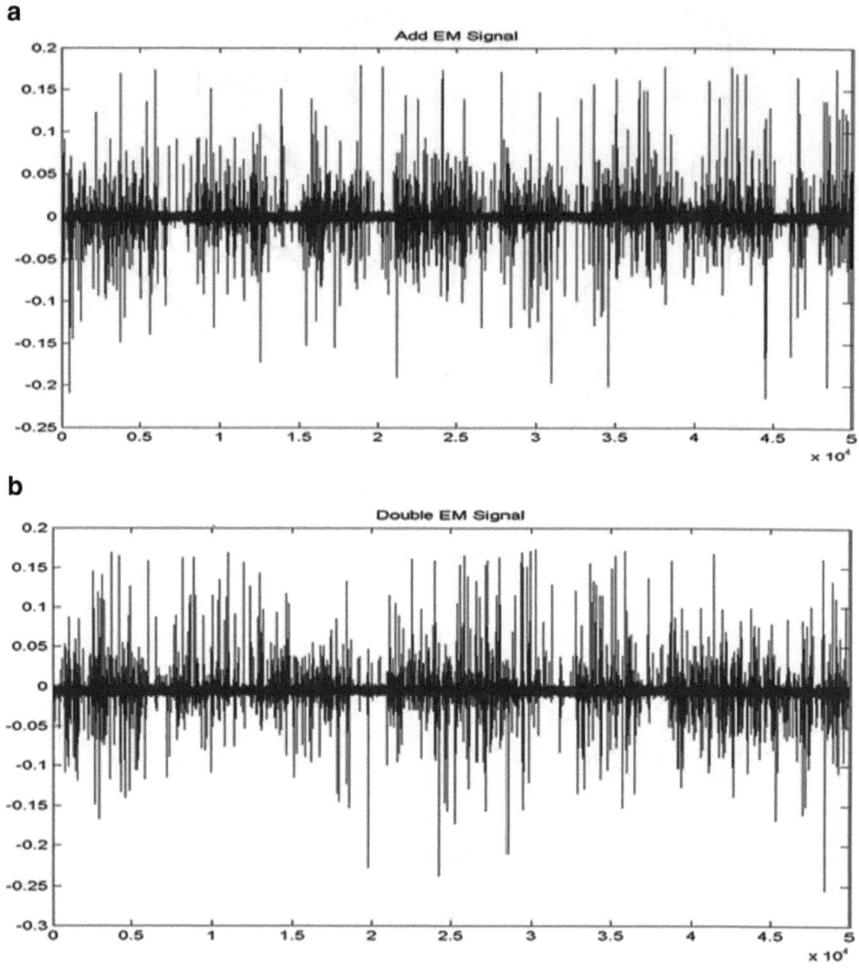

Fig. 8.12 Part of EM trace for ECC point addition in (**a**) and ECC point doubling in (**b**) on a ARM7TDMI

multiplication, the key determines the sequence of point additions and point doublings. Thus if an attacker can differentiate a point double from a point sum through the side channel, then the key could easily be determined. Figure 8.11b illustrates the power consumption of an elliptic curve single point addition (at the top) and single point doubling (at the bottom) executing on a DSP processor (Gebotys 2004). The differences between the two traces are identified. Using these differences, an attacker could capture the side channel trace and extract the secret key. This is another example of SPA which successfully extracts the secret key because key dependent operations are performed. The traces in Fig. 8.11b were obtained from the SC140 DSP processor Star*core 2000 where high levels of parallelism were utilized to optimize performance. As a comparison, Fig. 8.12 illustrates the same point addition and doubling on an ARM7TDMI. As one can see simple analysis appears to be more difficult since both traces look like random noise. However, a closer look reveals differences and in fact more sophisticated analysis of these traces using neural networks was able to identify EC point additions and doublings (Ho 2004) in many cases.

8.3 Differential Analysis

An introduction to differential analysis is presented in this section. First-order differential analysis will be discussed in this section, whereas higher order analysis will be covered in Chap. 9. Consider Fig. 8.13 which illustrates a common part of many cryptographic algorithms. In fact this is very common in block ciphers executed on 32-bit processors since table look-ups offer a much faster implementation than most bit level computations as discussed in Chap. 6. Typically the input data or plaintext and key are involved in some computation (shown as an exclusive-or) whose result is indexed into a table look-up (or S-box table). Assume for now that the plaintext and key are exclusive-ored together and then indexed into a table, as in the table method of the AES advanced encryption standard (Gladman 2003) or fast implementation of DES (see Sect. 6.2).

Assume the attacker knows the plaintext values and also knows the values in the S-box table (according to some well-published encryption algorithm). The attacker wishes to determine the key value. There are two ideal attacks shown. One is at the output of the exclusive-or operation and the other is at the output of the S-box as shown in Fig. 8.13. However, from an attack point of view, it is generally not practical to attack the output of a logical operation like an exclusive-or in a 32-bit embedded processor (see experimental evidence in (Gebotys 2006)). However it may be possible if the chip has been decapsulated and a very small EM probe is placed very close to the silicon. Alternatively it may be possible in some older 8-bit processors which store their results to memory frequently (since off-chip memory accesses produce strong side channel signals) or as part of their instruction set architecture. On the other hand, the output of the exclusive-or operation is correlated with the input to the S-box and thus this would be possibly correlated with an attack of the address used to access S-box memory. Even though specific memory addresses used by the processor do not have the same value as the S-box

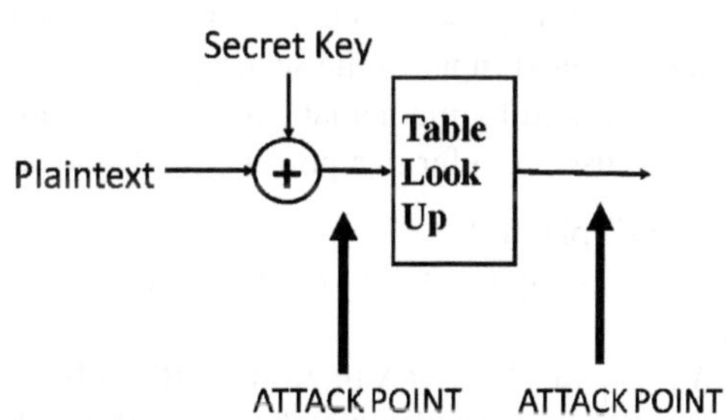

Fig. 8.13 Two-side channel attack points on a cipher

address (which range from 0 to 255), they may be correlated to enable a potential attack. However this attack is more difficult on medium sized (32-bit) processors, especially when the address width (8-bit input to S-box) is less than the data width accessed from memory (32-bit output from S-box). If the S-box is stored off-chip there is greater likelihood that the attack will be successful. This is due to the fact that operations which consume or dissipate large amounts of power dissipation, such as external access, will generally have noticeable side channels (since the capacitances of the chip to chip interconnects being switched are significant) and, therefore have greater chance of attack. On-chip memory and on-chip bus switching will be more difficult to attack compared to off-chip data transfers. The side channel signal strength depends largely upon the underlying embedded system. For example, some older 8-bit processors (within older smartcards) utilize memory operands and store their data values to memory due to limited register storage. In these systems the processor-memory system utilized dual chips (processor chip and memory chip). Hence exclusive-or operations were easily attacked (due to the off-chip memory accesses) and remained a focus for much side channel research. It is interesting to note that 8-bit processors are still used in smartcards today, however processors and memory are completely integrated into one chip, and this together with many implemented countermeasures make these newer cards more secure. Typically on-chip register or logical and arithmetic operations which have lower power dissipation are not attackable in complex embedded processors unless the chip is decapsulated and smaller probes are placed closer to the silicon. These on-chip logic/register operations involve smaller capacitances being switched on the chip which are likely closer to the noise level. Furthermore this switching is likely also insignificant relative to other parallel activity within the processor and embedded system which varies over time. Thus differential analysis which tries to expose this logic/register switching is extremely difficult in these complex chips without getting the probe close to the silicon area of interest. This is unlike other embedded platforms where the only activity on the chip may be the switching of the cryptocore/etc. Further details on chosen plaintext and ciphertext attacks are discussed in Sect. 9.2.2.

The procedure and mathematical functions are defined below for a differential attack at the input and output of the S-box. For either attack (at the input or output of the S-box), we first record the EM or current drawn from the device (or processor) while the device is executing the cryptographic algorithm. For example, let n traces be represented by $\mathrm{C}_0, \mathrm{C}_1, \ldots \mathrm{C}_{255}, \mathrm{C}_{256}, \ldots \mathrm{C}_{n-1}$ which are generated from n plaintexts $\mathrm{p}_0, \mathrm{p}_1, \ldots \mathrm{p}_{255}, \mathrm{p}_{256}, \ldots \mathrm{p}_{n-1}$. Note that $p_i = i$ for $i = 0, \ldots, 255$, and $\mathrm{p}_{256} = 0$, $\mathrm{p}_{257} = 1$, etc ..., or $p_i = i \, mod \, 256$. Many traces are acquired for each plaintext input to the security algorithm. Some terminology will be presented first followed by mathematical functions and procedures. Again side channel trace will be used to refer to a power or EM side channel trace.

$p_i = i^{\text{th}}$ plaintext
$C_i(t) =$ sample at time t of the ith side channel trace corresponding to the ith plaintext
$\{X\}_j = the\ j^{\text{th}}\ bit\ of\ X$ (e.g., $X = (01011011)$, *the* 0th *bit is* 1, *the* 1th *bit is* 1, ...)
$K = key\ guess$ (*which may be values* 0 *to* 255 *for an* 8-*bit key*)

$$A_i^K = \begin{cases} p_i \oplus K, & \text{at input of S-box} \\ S(p_i \oplus K), & \text{at output of S-box} \end{cases}$$

$D^{j,K}(t)$ = *the differential trace based upon partitioning with the* j^{th} *bit of attack point* A_i^K *for key guess K*

$D_S^{j,K}(t)$ = *the significant differential trace based upon partitioning with the* j^{th} *bit of attack point* A_i^K *for key guess K*

n_b = *number of traces whose plaintext* p_i *obeys* $\{A_i^K\}_j = b$ (*note that it is also a function of j,K, however for simplicity we do not show the superscripts j,K*) where $b = \{0, 1\}$

The Differential Trace

$$\mu_b^{j,K}(t) = \frac{1}{n_b} \sum_{i|\{A_i^K\}_j=b} C_i(t),$$

$$D^{j,K}(t) = \mu_0^{j,K}(t) - \mu_1^{j,K}(t) \tag{8.8}$$

The Significant Differential Trace

$$D_S^{j,K}(t) = \begin{cases} \mathrm{D}^{\mathrm{j,K}}(t) - 2s, & \text{if } \mathrm{D}^{\mathrm{j,K}}(t) > 0 \\ \mathrm{D}^{\mathrm{j,K}}(t) + 2s, & \text{if } \mathrm{D}^{\mathrm{j,K}}(t) < 0 \end{cases}$$

where

$$\sigma_b^{j,K}(t) = \sqrt{\frac{1}{n_b} \sum_{i|\{A_i^K\}_j=b} (C_i(t) - \mu_b^{j,K}(t))^2}$$

$$\mathrm{s} = \sqrt{(1/n_0)(\sigma_0^{j,K}(t))^2 + (1/n_1)(\sigma_1^{j,K}(t))^2} \tag{8.9}$$

The differential trace and significant differential trace equations are defined in (8.8) and (8.9) with respect to the two attack points shown in Fig. 8.13. In general, means are taken to remove the noise, and the difference of means are taken to remove the algorithmic noise, so finally data dependent variation is left in the differential traces. The term s is the standard deviation of the difference of means. Hence the significant differential trace is represented by all parts of the differential trace which are outside of plus or minus two standard deviations of the difference of means. These parts of the trace are statistically significant. These can be generalized for other ciphers where an attack point is defined as a general function of a key guess (K), a partitioning bit (j) and a plaintext value (i). We will now present the procedure for launching a side channel attack at the input of the S-box. It is assumed that the traces containing the attack point have already been acquired.

Procedure 8.1. Attack at Input of S-box

Input: $C_i(t)$, p_i
Output: k_j, $j = 0, \ldots, 7$ (secret key value)

1. Set $K = (00000000)$
2. For $j = 0$ to 7
   (a) Compute $D_S^{j,K}$ (t) or $D^{j,K}(t)$
   (b) If the plaintext load spike is in the same direction as the S-box load spike then $k_j = 0$ else $k_j = 1$

In the procedure 8.1, a zero byte for K is used in line 1. We do not guess any keys in this attack. Instead we examine each differential trace in line 2b. which is generated for each bit of the plaintext. The differential trace should have two spikes. The first spike (from left to right in direction of increasing time) should correspond to the plaintext load and the second spike should correspond to the S-box load. Line 2b. in procedure 8.1 requires the attacker to determine the direction of the second spike (up or down) relative to the first spike. If the spike is in an opposite direction we would expect the key bit had flipped this bit, $k_j = 1$.

For example a differential trace is shown in Fig. 8.14, where $j = bit\ 0$, or the least significant bit of plaintext was used to partition the traces. The differential traces shown in Fig. 8.14 are focused on the load of the exclusive-or output. This single load produced one positive spike and one negative spike with many oscillations in between. Although the plaintext spike is not shown, it was very similar to Fig. 8.14a. The key bit values for Fig. 8.14a–d are 0110 as indicated on the plots. Part of the differential traces are shown with plus or minus two standard deviations in Fig. 8.14a–d. In Fig. 8.14b, c differential traces rises high first and in the second part drops. This can be contrasted to the differential trace in Fig. 8.14a and d, where the trace drops first and rises in the second part. This difference is due to the key bit. Hence the direction of the difference in this attack can be used to determine each key bit value. The results in Fig. 8.14 utilize power measurements on the ARM7TDMI processor running at 40. Plus and minus two standard deviations are plotted along with the differential traces.

The procedure for the attack at the output of the S-box is shown in procedure 8.2. Unlike the previous attack at the input to the S-box where each bit is used to partition, at the output of the S-box we only need one bit which performs the partitioning. Often the least significant bit is used in the attack, but every device should be analyzed to determine which bit is best. The procedure uses (8.10).

$$H_K = \max_t \left| D_S^{j,k}(t) \right| \tag{8.10}$$

Procedure 8.2. Attack at Output of S-box

Input: $C_i(t)$, p_i
Output: k (8-bit secret key value)

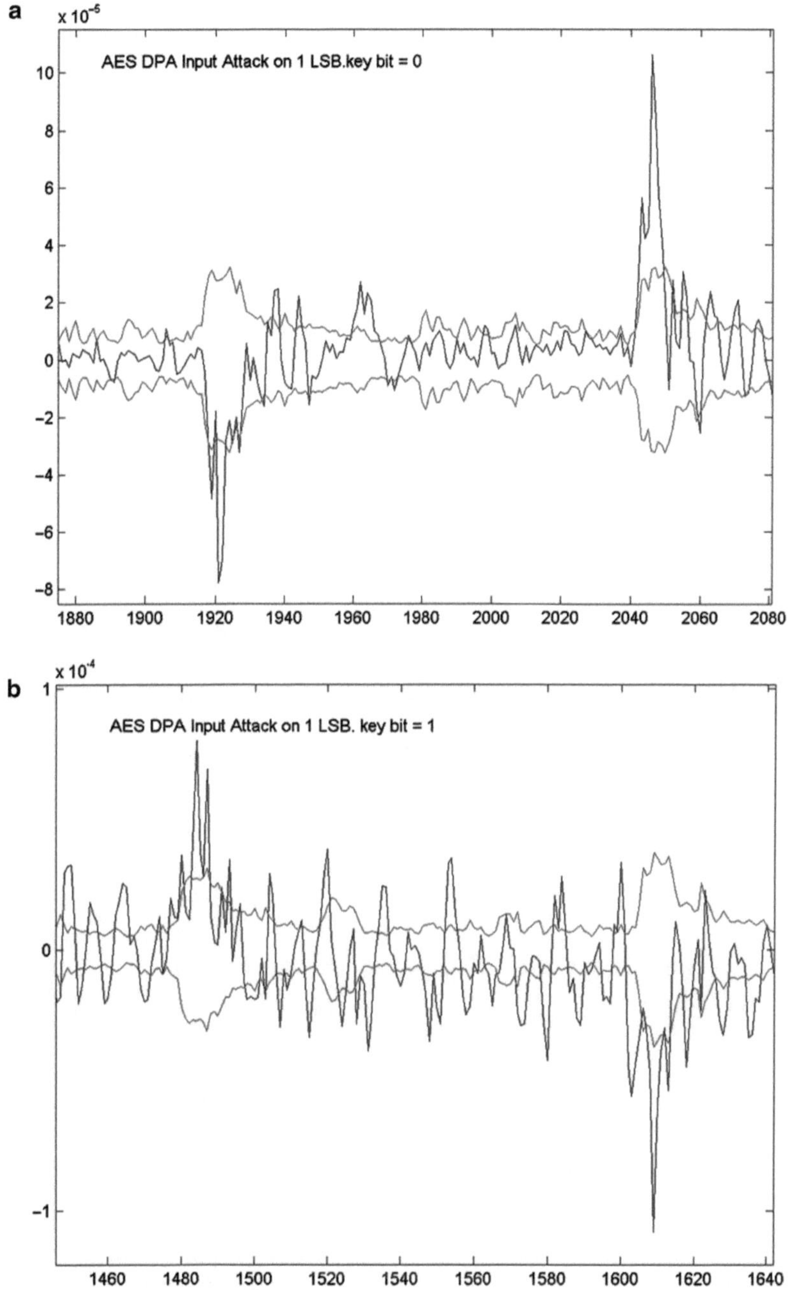

Fig. 8.14 Differential traces at input to S-box (a–d) indicating key bits 0, 1, 1, 0 respectively

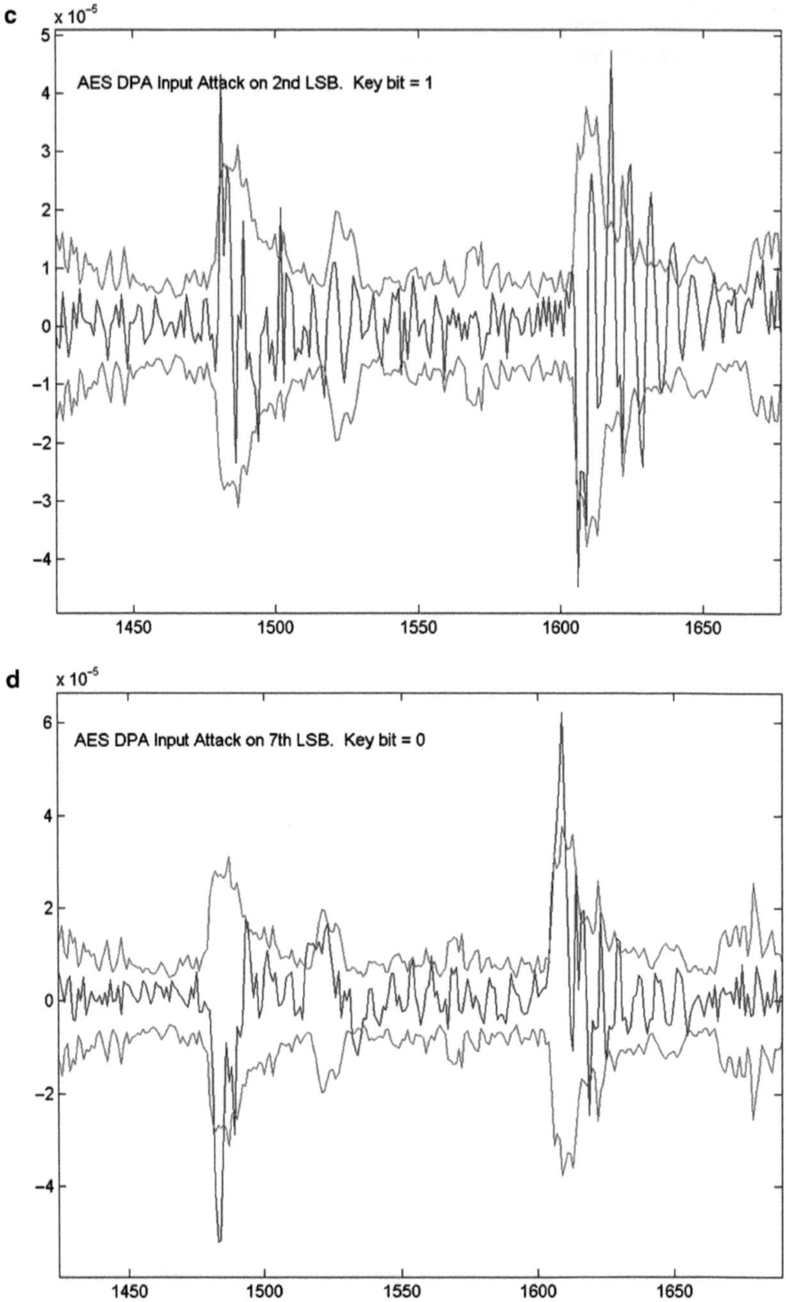

Fig. 8.14 (continued)

1. Select a partitioning bit, $j, 0 \leq j \leq 7$
2. For $K = 0$ to 255
   (a) Compute $D_S^{j,K}$ (t)
   (b) Calculate H_K (see 8.10)
3. $k = \{K | H_K > H_i \; for \; all \; i \neq K\}$

The procedure 8.2 provides one key byte, hence this routine is repeated for all 16 bytes of plaintext (recall from (6.8) in Chap. 6, there are 16 S-box accesses) in order to obtain the entire 128-bit key (since the first round key is equivalent to the 128-bit AES key). The maximum absolute value of the significant differential trace is computed as defined in (8.10) and stored for this key guess. The value in (8.10) will be referred to as the height of the differential, H_K. A plot of H_K vs. K is referred to as an all keys plot. Finally the maximum height of the differential over all keys, determines the correct secret key. Often researchers will launch differential analysis using the differential trace, however we have found the significant differential trace to be quite useful providing successful attacks with fewer number of traces.

An enlarged differential trace, partitioned with a correct key guess in AES, is shown in Fig. 8.15c along with the corresponding two mean traces from which it was generated. These traces were obtained from measured power of an ARM7TDMI processor using the setup in Fig. 8.5. Plus or minus two standard deviations are also plotted above and below the zero axis. The parts of the differential trace which emerge above and below the standard deviations indicate significant differences. The peak of the differential trace identifies the output of the S-box.

The height of another differential trace measured above the two standard deviations is highlighted in Fig. 8.15a. Figure 8.15b shows an all keys plot where the maximum differential heights, each calculated using (8.10) for a key guess, are plotted. The largest differential height identifies the correct 8-bit key from all possible 256 keys.

The correct key height is quite higher than any other key heights in Fig. 8.15b. This is indicative of a strong attack, where the attacker can be very confident in the key obtained. In other experiments, there may be several keys with similar heights, thus the attacker must typically acquire more traces in order to find the correct key. When all keys have approximately the same height, it is clear that results are random and the attack has failed or more traces are required.

Next we will show how noise affects the progress of the DEMA in terms of number of traces required for a successful attack. The ARM7TDMI evaluation board is used for this experiment.

Figure 8.16a presents the plot of DEMA height (solid line) vs. number of EM traces for the attack for a correct key guess at the output of the S-boxes in AES. In all plots where the x-axis denotes the number of traces, the numbers reflect the number of traces in one of two partitions in the differential analysis. Hence the total number of acquired traces in these experiments is actually double the number shown on the x-axis. The dashed line indicates two times the standard deviation of the difference of means (indicating significance when the DEMA height exceeds

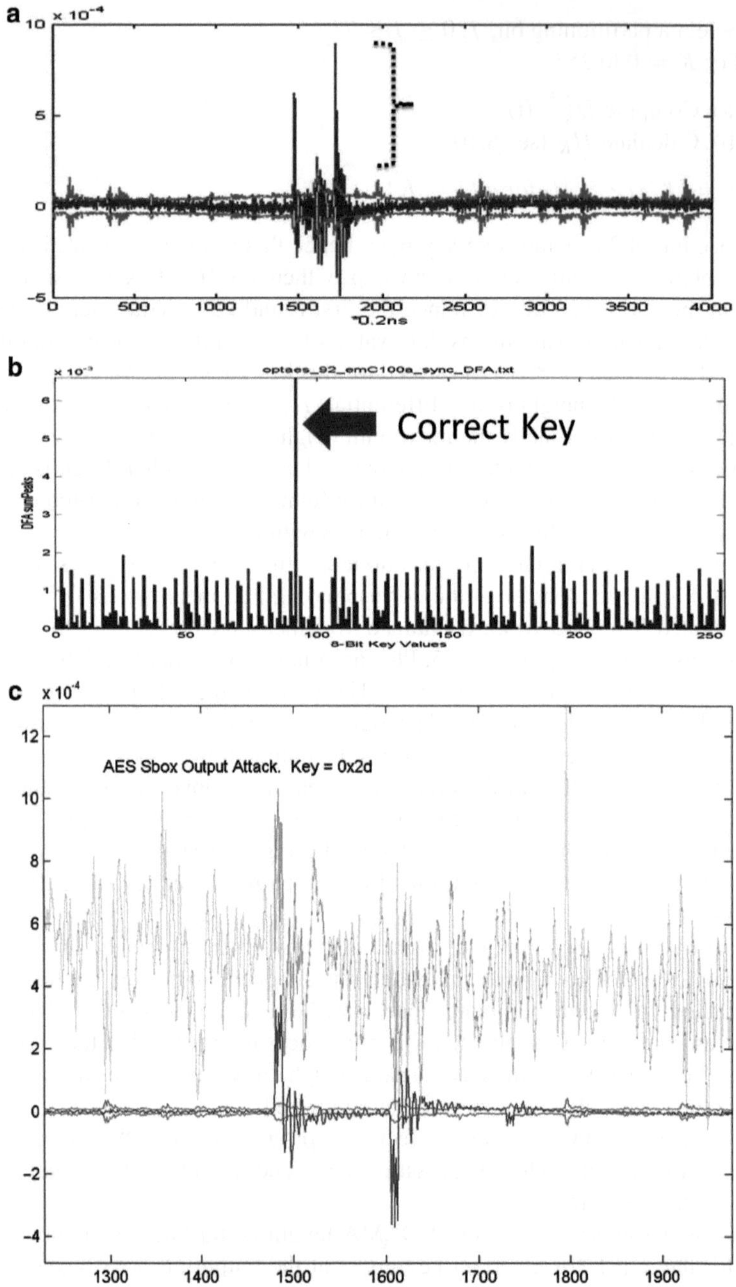

Fig. 8.15 Example of the height of the differential trace from (8.10) in (**a**) and all keys plot in (**b**) and the two mean traces producing a differential trace in (**c**)

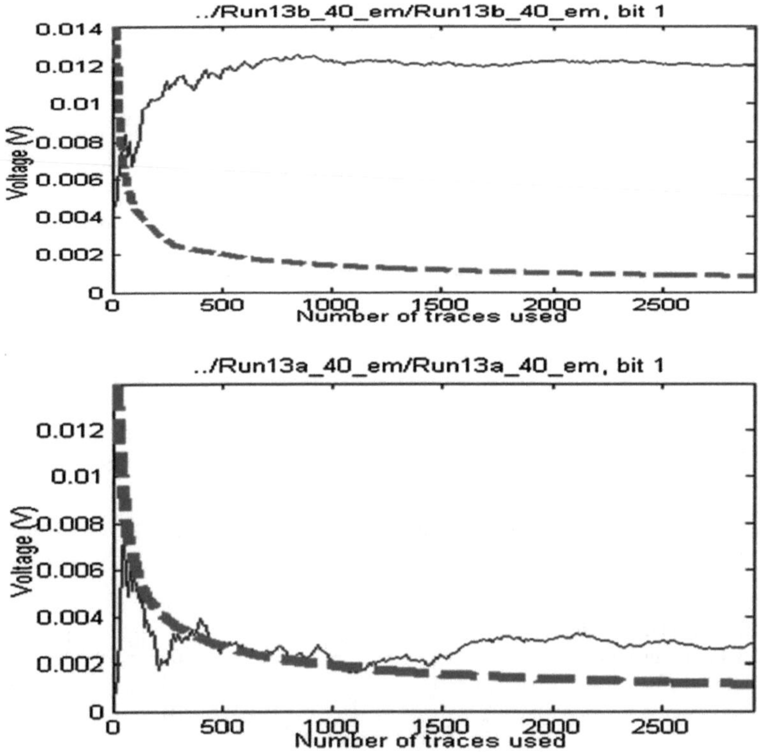

Fig. 8.16 Maximum height of differential trace vs. number of traces with 2 standard deviations shown as a dashed line

the dashed line). Two example plots are shown in Fig. 8.16a, b. In both cases bit 1, the least significant bit, was used to partition EM traces and produce the first-order DEMAs. Two different acquisitions with the same set of data were used to generate the two figures; hence the EM noise is the main factor in the difference in the EM analysis results. In all cases, the two standard deviation drops off as the number of traces increases and the DEMA height levels off. It is interesting to note that in Fig. 8.16a only 100 EM traces (in each partition, totaling 200 EM traces) are required for a significant DEMA peak, whereas in Fig. 8.16b over 1,000 EM traces (in each partition) are required for the same data, acquired at different times. The DEMAs, for bit 1 through bit 4, are shown in Figs. 8.17 and 8.18 which correspond to the data acquired in Fig. 8.16a and b, respectively. The faster converging data has stronger DEMAs as evident by the blue DEMA exceeding the red two standard deviations in Fig. 8.17. However, in Fig. 8.18 the DEMAs are not as strong, hence supporting the slower convergence time shown in Fig. 8.16b for the same set of data.

It is interesting to note that for some acquisitions, the smallest number of EM traces required in order to obtain the correct key guess (for attacks at the output of the S-box) using the ARM7TDMI evaluation board was 64 traces. Multi-bit

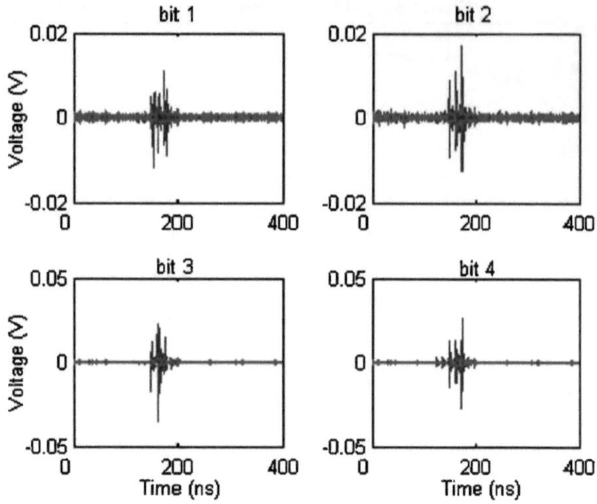

Fig. 8.17 Some DEMAs for bits 1,2,3,4 of acquisition shown in Fig. 9.13a

Fig. 8.18 Some DEMAs for bits 1,2,3,4 of acquisition shown in Fig. 9.13b

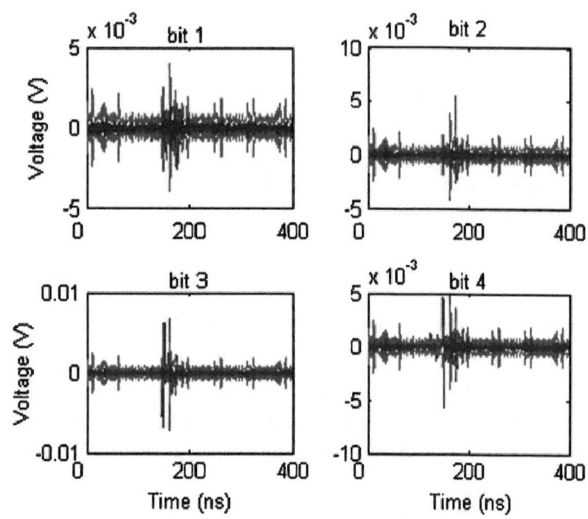

differential analysis can also be launched by summing the differential heights for a key guess over all different partitioning bits. Section 8.4 will discuss another analysis technique, known as correlation analysis, which we have found to be especially powerful for more recent technologies (with higher clock frequencies and lower supply voltages).

8.4 Correlation Analysis

Correlation analysis is a very powerful technique that relies on correlating the power or EM measurements with the hamming weight or distance of the predicted data. The formula is provided in (8.11).

Correlation Analysis

$$
\begin{aligned}
n &= \text{the number of traces} \\
p_i &= i^{\text{th}} \text{ plaintext} \\
K &= \text{the key guess} \\
W_i^K &= HW(S(p_i \oplus K)) \\
&= \text{the hamming weight or distance of the predicted S-box output for the } i^{\text{th}} \\
&\qquad \text{plaintext based on key guess } K \\
C_i(t) &= \text{sample at time } t \text{ of the } i^{\text{th}} \text{ trace generated using plaintext } i \\
\rho^K(t) &= \text{the correlation factor at time } t \text{ based on the key guess } K \\
&= \frac{\operatorname{cov}(C, W)}{\sigma_C(t)\sigma_W^K} = \frac{E_{C,W}^K(t) - \mu_C(t)\mu_W^K}{\sigma_C(t)\sigma_W^K}
\end{aligned} \tag{8.11}
$$

where

$$
\begin{aligned}
E_{C,W}^K(t) &= \frac{1}{n}\sum_i C_i(t) W_i^K \\
\mu_C(t) &= \frac{1}{n}\sum_i C_i(t),\ \mu_W^K = \frac{1}{n}\sum_i W_i^K \\
\sigma_C(t) &= \sqrt{\frac{1}{n}\sum_i (C_i(t) - \mu_C(t))^2},\ \sigma_W^K = \sqrt{\frac{1}{n}\sum_i \left(W_i^K - \mu_W^K\right)^2}
\end{aligned}
$$

$$
H_K = \max_t\{|\rho^K(t)|\} \tag{8.12}
$$

For example, an attack at the output of the S-box involves guessing a key (K) and then predicting the values at the output of the S-box for each plaintext. The hamming weight of the predicted value is used to correlate with each sample point in the specific trace. Each trace will generate a correlation trace, such as the one shown in the top plot of Fig. 8.19. The top plot is the correlation factor vs. sample, see equation (8.11). It shows a high correlation just past 1 μs. The corresponding EM trace has a glitch at this location, which turned out to be the attack point. The shaded boxes are placed over parts of the EM trace to protect the vendor, since this complex portable embedded system utilized advanced technologies.

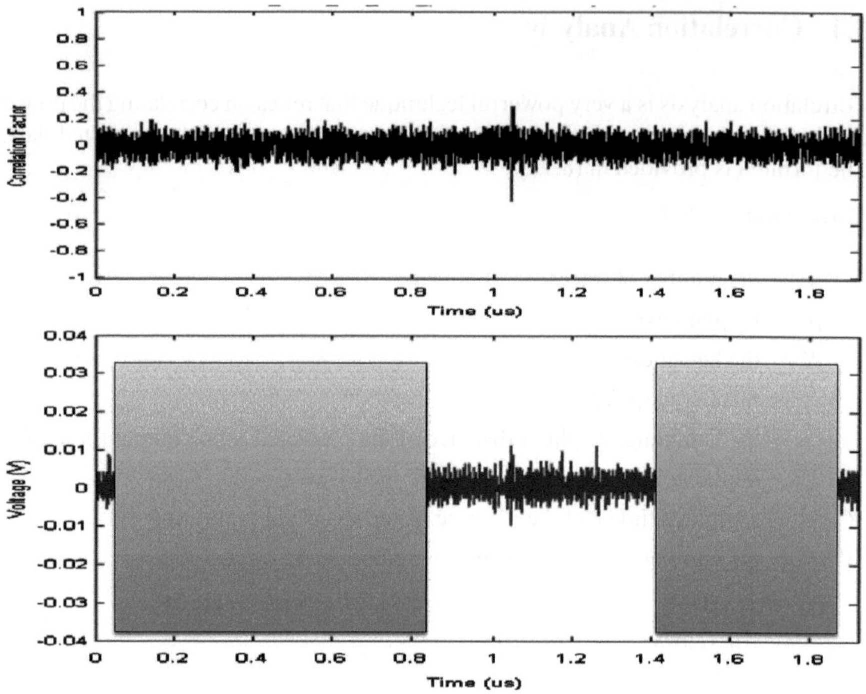

Fig. 8.19 Correlation plot for a correct key guess at top and trace at bottom

8.5 Differential Frequency Analysis

The frequency-based differential analysis (Gebotys and White 2008a; Gebotys et al. 2005) is an extension of differential analysis. In general, instead of computing the differential signals in the time domain (as in (Kocher et al. 1999) and in almost all previous research), the computation is performed in the frequency domain (see differential frequency analysis (DFA), algorithm below). Analyzing signals captured in the frequency domain solves the problem of misalignment (or time-shifts) in traces since fast fourier transform (FFT) analysis is time-shift invariant. Frequency analysis is important for attacking real embedded systems where temporal misalignment (or time-shifting) of traces (typically caused by the triggering signals, the operating system, or dynamic events) is a big concern. Additionally, frequency analysis may reveal loops and other repeating structures in an algorithm that is not possible with time domain analysis.

The following terminology is used to describe the differential frequency analysis formulas in equations (8.13) and (8.14), specifically: $i \in \{0, \ldots, n-1\}$ is the trace number; $b \in \{0, 1\}$ is the set number; C_i^b is the EM signal of set b and trace i; $t \in \{0, \ldots, m-1\}$ is time; f is the frequency index, $f \in \{0, \ldots, m/2-1\}$.

Differential Frequency Analysis

$$
\begin{aligned}
C_i(t) &= \text{trace corresponding to plaintext } p_i, i = 0, 1, \ldots, (n-1) \\
K &= \text{key guess} \\
C_i^b(t) &= \text{trace } C_i(t) \text{ where } \{S(p_i \oplus K)\}_j = b \\
FFT() &= \text{fast fourier transform} \\
P_i^b(f) &= \frac{\left|FFT(C_i^b(t))\right|^2}{n} \\
D_j^K(f) &= \frac{1}{n/2}\sum_i P_i^0(f) - \frac{1}{n/2}\sum_i P_i^1(f) \qquad (8.13) \\
H_j^K &= \sum_f D_j^K(f) \qquad (8.14)
\end{aligned}
$$

Some differential traces for a specific key guess, K, computed using equation (8.13) are shown in Fig. 8.20a and b. The all keys plot is obtained by summing the difference in amplitudes over all frequencies using equation (8.14). Again one could subtract two standard deviations of the difference of means from the differential signals in the frequency domain defined in (8.13). The analysis methodology involves first using DFA on different parts of a trace to locate possible areas of attack. Next the attacker can focus in on smaller areas of interest within the traces which show interest.

8.6 Experiments on PDAs

Attacks on the PDA or cellphone may arise not only from device theft or loss but also during everyday use where unintentional electromagnetic (EM) waves radiated from the wireless device during cryptographic computations may leak confidential data to a nearby attacker. For example, a lunch-time attack (previously described in Chap. 2) may be successful in obtaining the secret keys stored in confidential memory in a wireless device. This attack may be possible through loss or theft of the device, or alternatively through temporary access to the device by monitoring the EM waves emanating from the device while performing cryptographic computations. In the latter case the attack may be able to extract the encryption keys, making future wireless communications insecure.

Figure 8.21 illustrates a methodology for EM attack on a Java-based embedded system. It is assumed that the attacker has access to a device, called the characterization device (used in Fig. 8.21a), which is similar to the device under attack (used in Fig. 8.21b). Also the attacker can execute and modify the same encryption algorithm implemented in Java on the characterization device but only execute it on the device under attack. This section will examine a methodology for attacking a real embedded device. Unlike 8-bit microcontrollers or the 32-bit ARM7TDMI, a

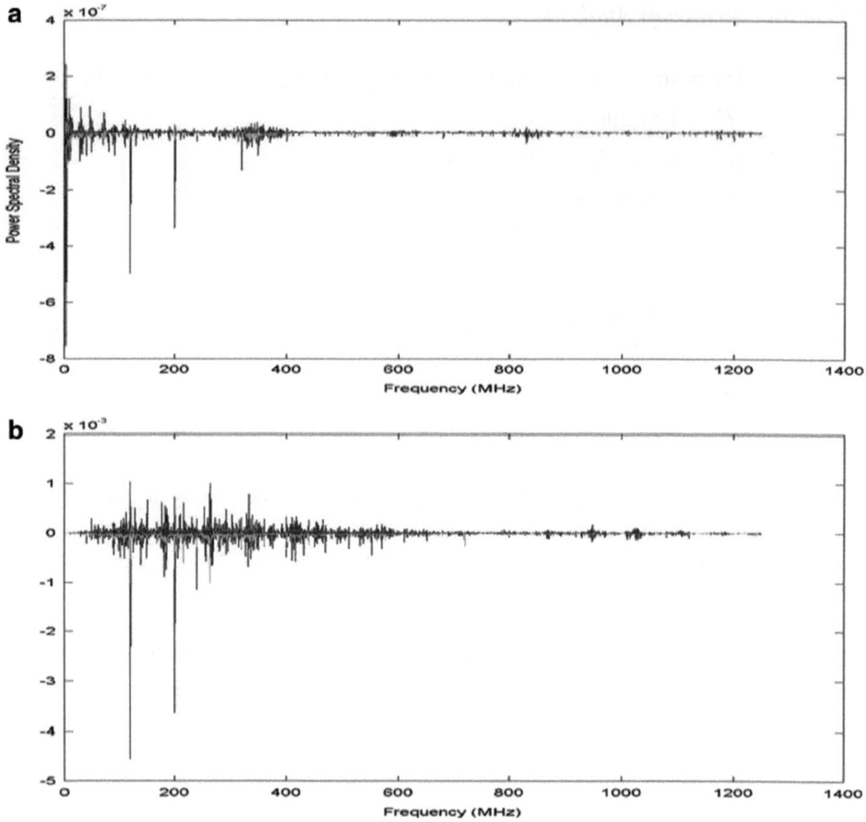

Fig. 8.20 Differential trace where x-axis represents frequency

complex real embedded device may have misaligned traces due to system interrupts, poor triggers, etc. Hence this section will detail a pattern recognition approach for aligning traces followed by use of DFA (described in Sect. 8.6) which can extract keys without requiring perfect alignment (unlike previous differential analysis in the time domain). A Java-based implementation of AES running on a PDA was attacked. The SEMA details are provided as well since these were crucial to narrowing down where the attack point may be. Since Java was very slow performance-wise it was difficult to obtain a sufficient number of traces with required resolution over the expected first round region.

The first step is to acquire the EM signals from the characterization device while it is executing the cryptographic algorithm (called an EM trace). A SEMA should identify the rounds of the encryption algorithm in step 2 of Fig. 8.21a. Next the attacker must be able to find round one and the specific region of attack (i.e., where the Sbox outputs are accessed for the AES encryption algorithm). This may require modification of the Java algorithm (such as truncation of the code followed by

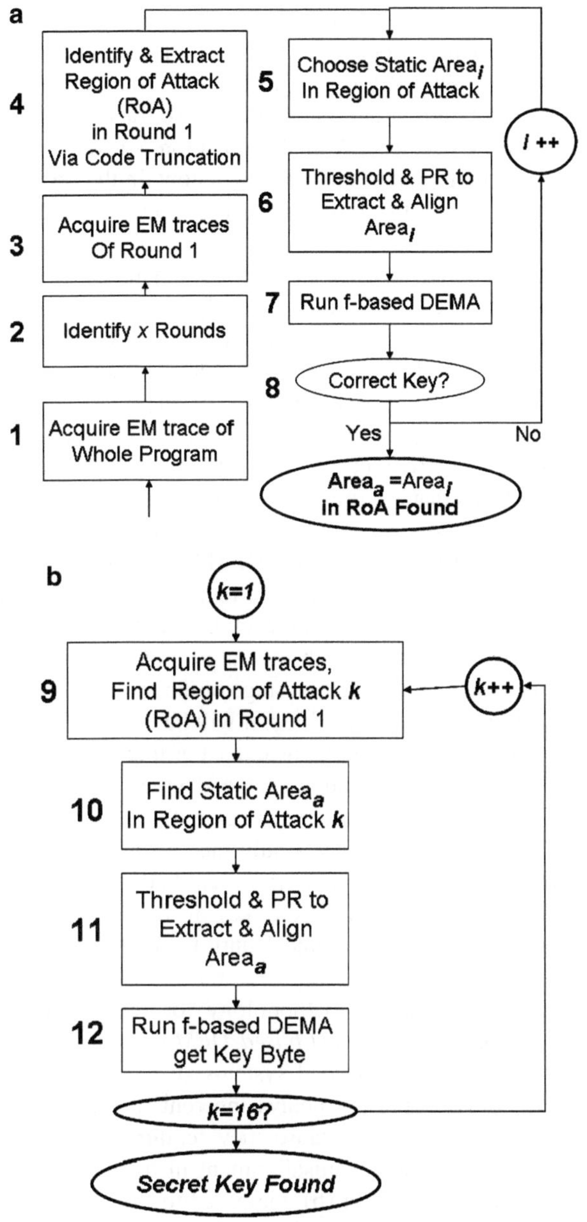

Fig. 8.21 Methodology for characterizing EM (**a**) for attack (**b**)

SEMA) for verification (see step 4 of Fig. 8.21a). The attacker must characterize the trace-to-trace misalignments associated with the acquisition of traces. For example, the attacker will acquire several traces of EM signals that center on the region of attack and further explore a random selection of these traces to determine the time shifts or delays that have occurred. A final acquisition is taken with proper resolution and samples such that most of the traces will contain the region of attack (i.e., the region of attack width plus two times the average trace-to-trace delay is typically the trace width).

In step 4, the attacker develops a pattern recognition algorithm so that the region of attack is extracted from each EM trace. This algorithm, involves EM magnitude thresholding, identifying regions of activity, and finally developing a set of rules to extract the region of attack. In step 5, the attacker examines various areas within the extracted region of attack in each EM trace. If an area is dynamic, or present in some traces but not others, then it is not a candidate for analysis since it likely represents some dynamic activity in the PDA, which is not consistent or associated with the Java code. Areas that are present in all EM traces (static areas, $Area_i$) are candidates for further analysis. In step 6, these static areas are examined, through extraction and subsequent frequency-based DEMA in step 7. If the correct key is found, in step 8, then the area of interest in the region of attack has been found, and is referred to as $Area_a$. A successful frequency-based DEMA on the specific area extracted from the region of attack with several sets of acquired EM traces using different keys verifies the characterization stage of the methodology.

Once the attack area, $Area_a$, is identified, it will be found in subsequent EM acquisitions using the developed thresholding and patterning algorithms. In steps 9–12 of Fig. 8.21b this area, $Area_a$, is found in each of 16 regions of attack ($k = 1$–16). Step 9 includes acquiring EM traces for each key byte attack. This requires SEMA to find and extract the region of attack (multitrace acquisition), subsequent identification and extraction of the area, $Area_a$, identified in Fig. 8.21a, and finally frequency-based DEMA to obtain the secret key of the device. This process is repeated for each key byte (k) within round 1 until the complete 128-bit encryption key is found.

An integral part of the attack methodology for real embedded systems is frequency-based differential analysis (*f-based DEMA* in Fig. 8.21 and defined in Sect. 8.5). This is crucial for embedded systems since it does not require exact alignment of traces, thus supporting delays inherent in the system. DEMA and correlation analysis do not work because they require perfect trace alignment. Figure 8.22 illustrates the inherent misalignment in traces from a complex embedded system such as this PDA in (b) compared to the processor present in an evaluation board with no misalignment in (a). The averaged trace is shown as the darker color. In Fig. 8.22b the averaged signal is sitting at the zero axis due to large amounts of misalignment whereas in (a) the average signal resembles one trace.

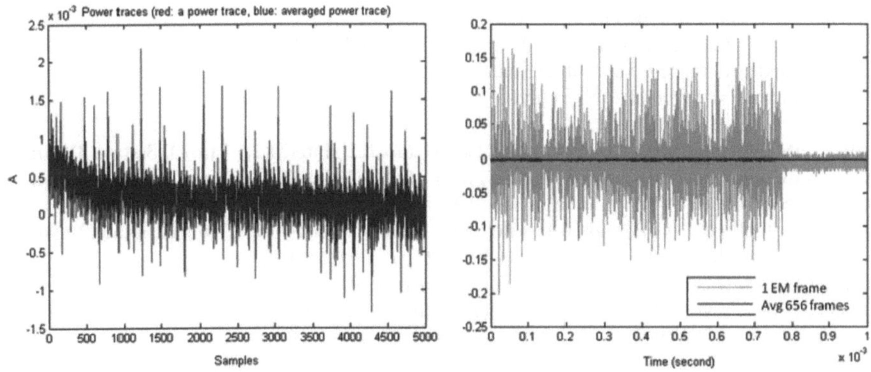

Fig. 8.22 Well-aligned traces with similar mean trace on the left vs. misaligned traces on the right with zero mean plot

8.6.1 EM Results

This section focuses on attacking the output of the S-box using EM traces acquired from a PDA executing AES on a PDA. Figure 8.3 illustrates the experimental setup used to acquire EM signals from the PDA device, the device under attack. A high sample rate oscilloscope, a 1-cm loop EM probe, wide band preamplifier, and a PDA (which was opened to expose the packaged chip over which the probe was placed) were used to acquire EM traces. In all cases, 512 EM traces were captured in each scope acquisition. A trigger signal was generated from the PDA using the Java code to turn the LED on and off. The voltage across the terminals of the LED was used to trigger the scope. The wireless radio of the PDA was turned off during acquisitions.

The AES encryption algorithm [implemented using the table-based method of (Gladman 2003)] was written in Java, compiled into bytecode, and loaded onto the PDA device. Figure 8.23a) illustrates part of the Java code of the AES cipher (*optAES*, see also Sect. 6.2.1). The plaintext is input to the cipher (*in*), and the output is the ciphertext (*out*). The cipher uses the table method (recall from Chap. 6) where four S-box tables (*Te*0 through *Te*3), each providing 32-bit outputs, are utilized. The beginning of the loop code involves the S-box table accesses. The 32-bit outputs from the S-boxes are exclusive-or'd together to create $t(0)$ through $t(3)$. The next part of the code involves state creation, where $t(0)$ through $t(3)$ are shifted and transformed into 16 state bytes, *state*()(). The last part of the loop code exclusive-or's the state with bytes of the next round key, *rkey*, using a routine called "*AddRoundKey*." This resulting new state is then used to address into the S-box tables of the next loop iteration. Each round is represented by a loop iteration, except the last round which follows the loop.

The complete AES encryption is executed in a loop on the PDA device for a finite number of iterations using different plaintext input. Specifically only the most significant byte of the 128-bit input plaintext is changed in each loop iteration and the other bytes are held fixed (or constant). The value of the most significant byte

a) public void optAES(byte() in, byte() out) {

```
   wCount = 0; Copy.copy(state, in); t(0) = 0; t(1) = 0; t(2) = 0; t(3) = 0;
   AddRoundKey(state); // xor with expanded key
   for (int round = 1; round < Nr; round++)
   {
t(0) = tab.Te0(state(0)(0)) ^ tab.Te1(state(1)(1)) ^ tab.Te2(state(2)(2)) ^ tab.Te3(state(3)(3));
t(1) = tab.Te0(state(1)(0)) ^ tab.Te1(state(2)(1)) ^ tab.Te2(state(3)(2)) ^ tab.Te3(state(0)(3));
t(2) = tab.Te0(state(2)(0)) ^ tab.Te1(state(3)(1)) ^ tab.Te2(state(0)(2)) ^ tab.Te3(state(1)(3));
t(3) = tab.Te0(state(3)(0)) ^ tab.Te1(state(0)(1)) ^ tab.Te2(state(1)(2)) ^ tab.Te3(state(2)(3));
   state(0)(0) = (byte) (t(0) >> 24); state(1)(0) = (byte) (t(0) >> 16);
   state(2)(0) = (byte) (t(0) >> 8);  state(3)(0) = (byte) (t(0));
   state(0)(1) = (byte) (t(3) >> 16); state(1)(1) = (byte) (t(3) >> 8);
   state(2)(1) = (byte) (t(3));       state(3)(1) = (byte) (t(3) >> 24);
   state(0)(2) = (byte) (t(2) >> 8);  state(1)(2) = (byte) (t(2));
   state(2)(2) = (byte) (t(2) >> 24); state(3)(2) = (byte) (t(2) >> 16);
   state(0)(3) = (byte) (t(1));       state(1)(3) = (byte) (t(1) >> 24);
   state(2)(3) = (byte) (t(1) >> 16); state(3)(3) = (byte) (t(1) >> 8);
   AddRoundKey(state); }// xor with expanded key

   t(0) = (tab.Te4(state(2)(3)) & 0xff0000L) ^ (tab.Te4(state(0)(0)) & 0xff000000L) ^
(tab.Te4(state(0)(2)) & 0xff00L) ^ (tab.Te4(state(2)(1)) & 0xffL);
   t(1) = (tab.Te4(state(3)(2)) & 0xff0000L) ^ (tab.Te4(state(1)(3)) & 0xff000000L) ^
(tab.Te4(state(1)(1)) & 0xff00L) ^ (tab.Te4(state(3)(0)) & 0xffL);
   t(2) = (tab.Te4(state(0)(1)) & 0xff0000L) ^ (tab.Te4(state(2)(2)) & 0xff000000L) ^
(tab.Te4(state(2)(0)) & 0xff00L) ^ (tab.Te4(state(0)(3)) & 0xffL);
   t(3) = (tab.Te4(state(1)(0)) & 0xff0000L) ^ (tab.Te4(state(3)(1)) & 0xff000000L) ^
(tab.Te4(state(3)(3)) & 0xff00L) ^ (tab.Te4(state(1)(2)) & 0xffL);
   state(0)(0) = (byte) (t(0) >> 24); state(1)(0) = (byte) (t(1) >> 8);
   state(2)(0) = (byte) (t(2) >> 24); state(3)(0) = (byte) (t(3) >> 8);
   state(0)(1) = (byte) (t(2) >> 16); state(1)(1) = (byte) (t(3));
   state(2)(1) = (byte) (t(0) >> 16); state(3)(1) = (byte) (t(1));
   state(0)(2) = (byte) (t(0) >> 8); state(1)(2) = (byte) (t(1) >> 24);
   state(2)(2) = (byte) (t(2) >> 8); state(3)(2) = (byte) (t(3) >> 24);
   state(0)(3) = (byte) (t(2)); state(1)(3) = (byte) (t(3) >> 16);
   state(2)(3) = (byte) (t(0)); state(3)(3) = (byte) (t(1) >> 16);
   AddRoundKey(state); Copy.copy(out, state); } // xor with expanded key
private void AddRoundKey(byte()() state) // AddRoundKey: xor a portion of expanded
key with state
   { for (int c = 0; c < Nb; c++)
      for (int r = 0; r < 4; r++)
         state(r)(c) = (byte)(state(r)(c) ^ rkey(rkeyCount++)); }
```

b) public void optAEStrunc(byte() in, byte() out)

```
   {wCount=0; Copy.copy(state,in);t(0)=0;t(1)=0;t(2)=0;t(3)=0;
   AddRoundKey(state);
   for (int round = 1; round < 2; round++)
   { t(0) = tab.Te0(state(0)(0));
      state(0)(0) = (byte) (t(0) >> 24);
         ...
      state(3)(3) = (byte) (t(1) >> 8);
         AddRoundKey(state); }}
```

Fig. 8.23 Partial code in (**a**) and truncated code in (**b**) of plaintext starts at 0 and continues sequentially through to a value of 255 and then repeats from 0. Only the most significant byte of the plaintext changes in order to allow only the first S-box table, *Te*0, in round 1, to change. All other round one S-box table outputs remain constant, hence the noise created by these table accesses is minimized. This greatly simplifies the attack since it can attack the output of *Te*0 as well as $t(0)$ and the input to Sbox tables in round 2. This approach helped to maximize the probability of a successful attack (since it was not known exactly where the *Te*0 load or $t(0)$ store/load was located in the EM traces). The plaintexts are then modified and new EM acquisitions are used in order to obtain each key byte [as shown in step 9 of Figure 8.21b)].

The characterization of the device, followed by the attack on the PDA device, will be described next. The experimental results follow the methodology presented in Fig. 8.21. The information gained through the SEMA, the truncated AES code analysis, and the frequency-based DEMA is then utilized in a full attack on the complete AES code running on the PDA.

8.6.1.1 SEMA of AES on the PDA

Recall that the simple EM analysis, SEMA, involves analyzing a single trace of EM signals emanated from the chip within the PDA device while the Java cryptographic algorithm is executed. A single trace of EM data is acquired with a window set wide enough (about 10% wider than the duration of the algorithm) to view the entire cryptographic algorithm. This was approximate since the delay from the software setting of the trigger to the actual trigger signal received by the scope was not known exactly due to LED circuitry and wiring delays.

Scope plots of the acquired EM signals with labels of the ten rounds of AES executing on the PDA are shown in Fig. 8.24a). Each of the ten rounds can be seen, thus illustrating a SEMA on the device. Since each of the rounds generally utilizes the same instructions (with the exception of the last round), the EM signals are expected to show similar activity in each round. Hence from analysis of the EM trace, it is clear that each round is illustrated with EM signals grouped together with high amplitude separated by low EM amplitude. In each round the high EM signals are expected to be associated with the S-box table accesses and the lower EM activity (shown as separating the rounds) is expected to represent the state creation and add round key operations described in the previous section. This observation is made since it is well known that memory operations dissipate more power than logic type operations, and hence their expected EM emanations would be larger. The final area of large EM signals shown at the end of AES likely involves writing the ciphertext back into memory. The period of low EM activity after the cipher completes, involves the Java thread being placed into sleep mode. In this SEMA experiment a thread was created to call the AES encryption algorithm. However, after AES's execution has completed, the thread is programmed to sleep. Therefore, there are minimum EM signals as shown in the graph when the thread is in the sleep mode. Figure 8.24e illustrates the scope capture of EM signals from PDA running a 192-bit AES, where 12 rounds are evident.

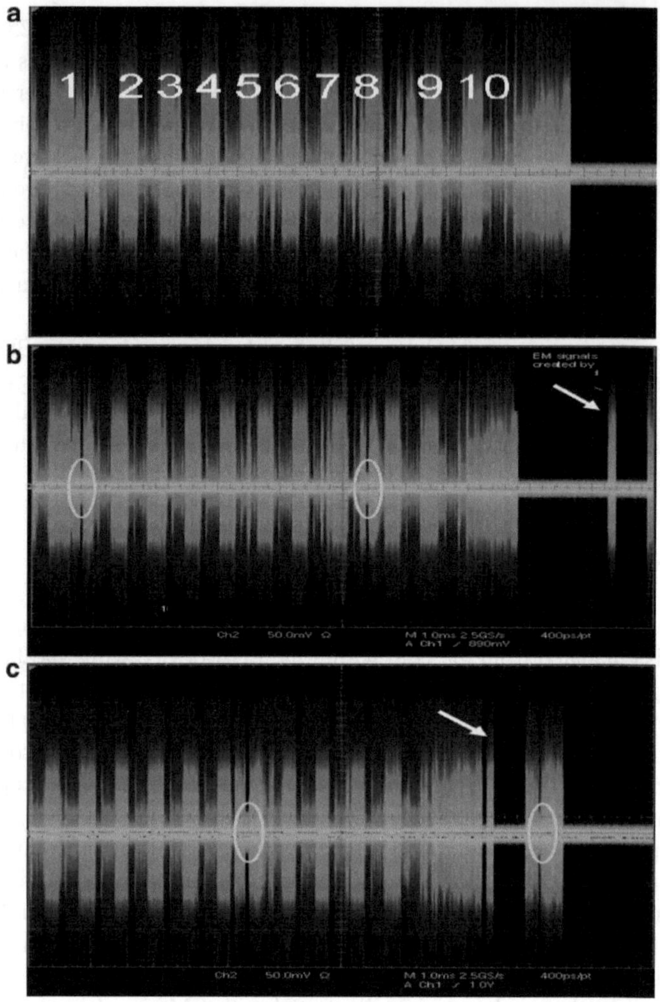

Fig. 8.24 SEMA attack on a PDA device running Java-AES with 10 rounds scope plot in (**a**–**d**), 12 rounds in (**e**)

There are also other interfering EM signals which are present in these scope plots. For example at the end of the scope plot of Fig. 8.24b there is a short burst region of high EM activity, illustrated by the white arrows. This is likely created by a timer interrupt which occurs to check if a thread needs to be activated. This signal also appears ins Fig. 8.24c, d. Other interfering signals which seem to interrupt the AES code are shown with ovals in Fig. 8.24b–d. These appear to be timed interrupts since they occur at regular intervals. These interrupts seem to be shifted in some acquisitions, but in most cases seem to occur at the same times. For example in Fig. 8.24c, d the interrupts occur in the same positions within the AES code. However, in Fig. 8.24b the interrupts occur at different locations in the AES

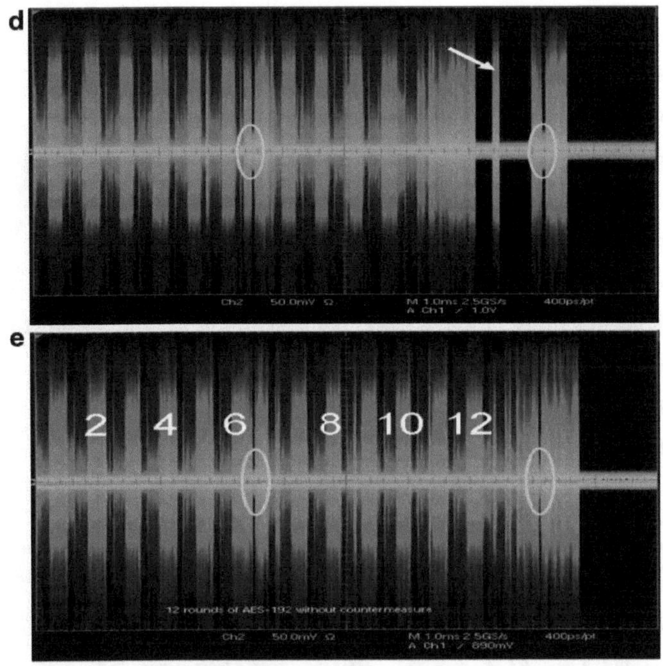

Fig. 8.24 (continued)

code. Within one acquisition the locations of the timed interrupt are consistent. From further analysis it appears that the interrupts do not significantly interfere with the overall AES execution. For example, in Fig. 8.24b two interrupt events directly stop the AES execution, whereas in Fig. 8.24c, d only one event interrupts AES. The difference between the time when AES completes in these two cases also appears to be equivalent to approximately one-timed interrupt.

Part of the AES execution is shown in Fig. 8.25a in a plot of trace 4 of the EM acquisition for key $= 92$ decimal. Four regions that represent the four table accesses, (which we will refer to as Group 1–4) for creation of $t(0)$, are labeled with arrows. A different EM trace is shown in Fig. 8.25b, which illustrates the severe misalignment (or time shifting) of the traces. Time shifts of up to 40 μs were generally observed in the EM acquisitions.

It is interesting to note that the first trace, see Fig. 8.26, of most EM acquisitions involve EM signals that are unrecognizable. This likely occurs since the first trace may correspond to the translation of byte code into native code and the subsequent execution of the resulting native code. Additionally, these extra computations likely shift the EM signals of interest outside of the current acquisition window of the scope. However, the remaining executions of the loop of AES corresponding to traces 2–512 only perform execution of native code.

Figure 8.27 shows the EM samples from the high activity region or *Te*0 access. In Fig. 8.27, each identifiable area ($Area_i$) of the extracted *Te*0 access region trace

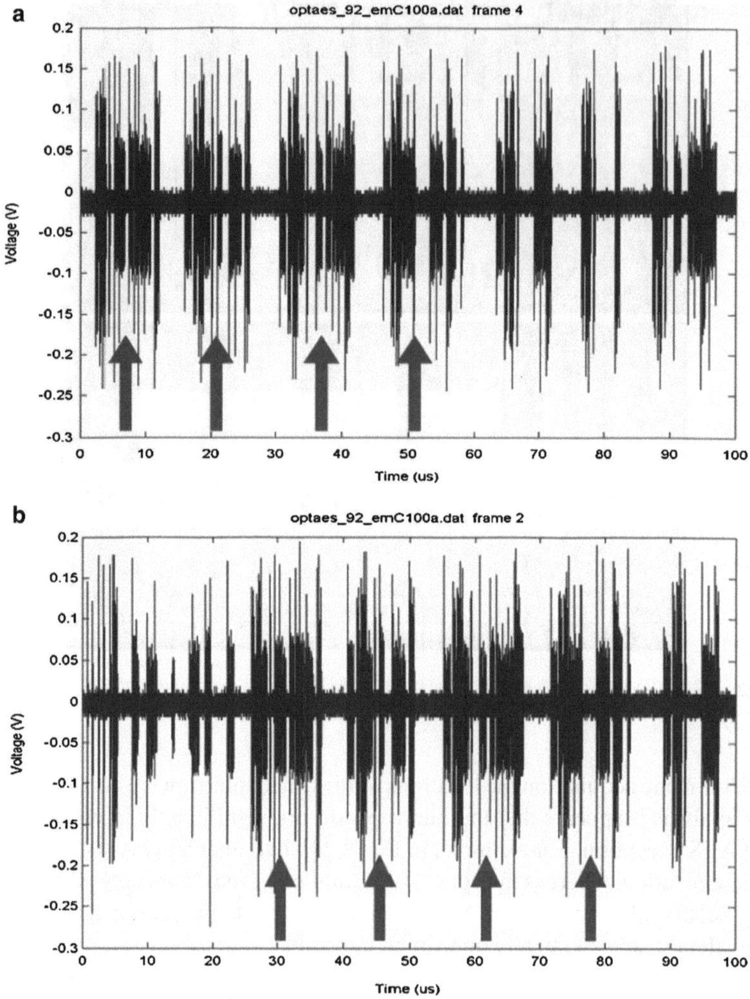

Fig. 8.25 EM acquisition trace 4 (**a**) and trace 2 (**b**) capturing 265–365 μs showing 4 groups with *arrows*

plot is labeled with a letter from A to N ($i = A\ to\ N$). However, in some general table accesses of the same or different trace, not all areas were present. Sections that were missing were either replaced with very low activity close to 0 μs wide or with very low activity for $1/2$ of the sections' typical width.

Figure 8.28a shows 15 μs of Group 1 zoomed in and section labels are assigned by comparison to the characteristic shapes and section labels A–N of Fig. 8.27. Similarly, Fig. 8.28b–d illustrates a zoomed in view of the 15 μs of Group 2–4 with section labels. Section N, which corresponds to the *Te*0 or $t(0)$ or possibly any *Te** table output load, only occurs in Group 4. The section N characteristic shape does not occur in Group 1–3.

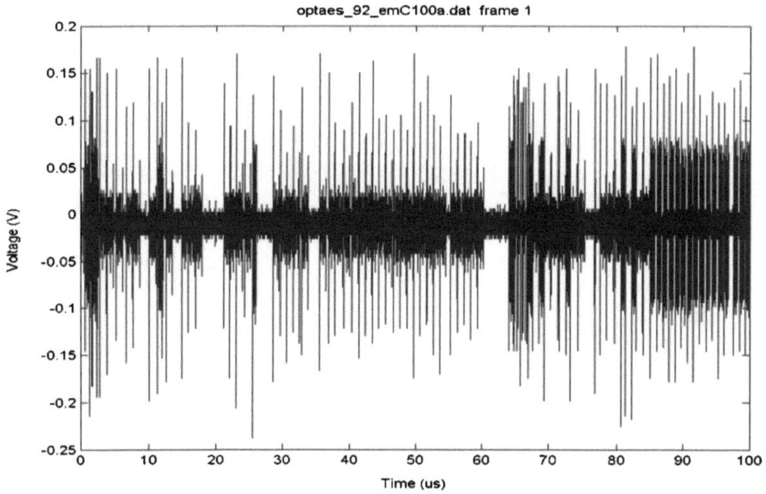

Fig. 8.26 EM acquisition "abnormal" trace 1

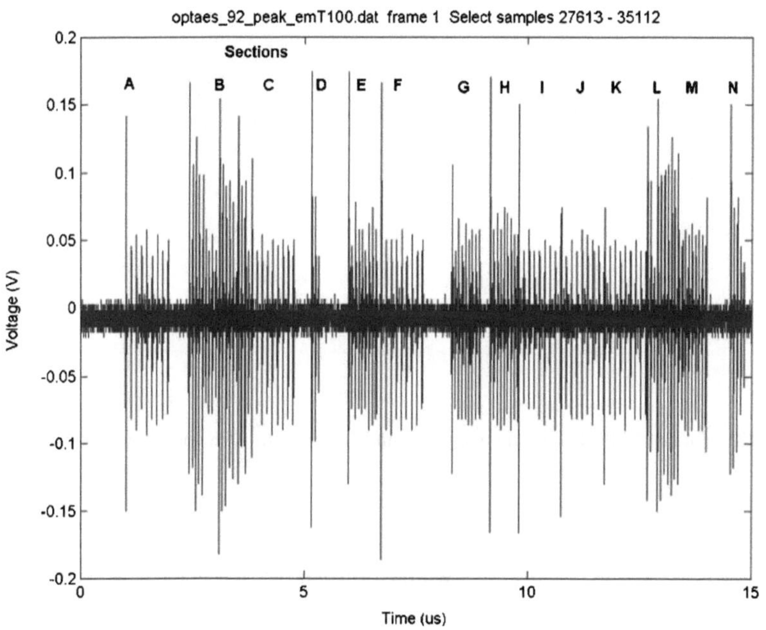

Fig. 8.27 EM acquisition trace 1, extracting 15 μs *Te*0 region with labels A to N

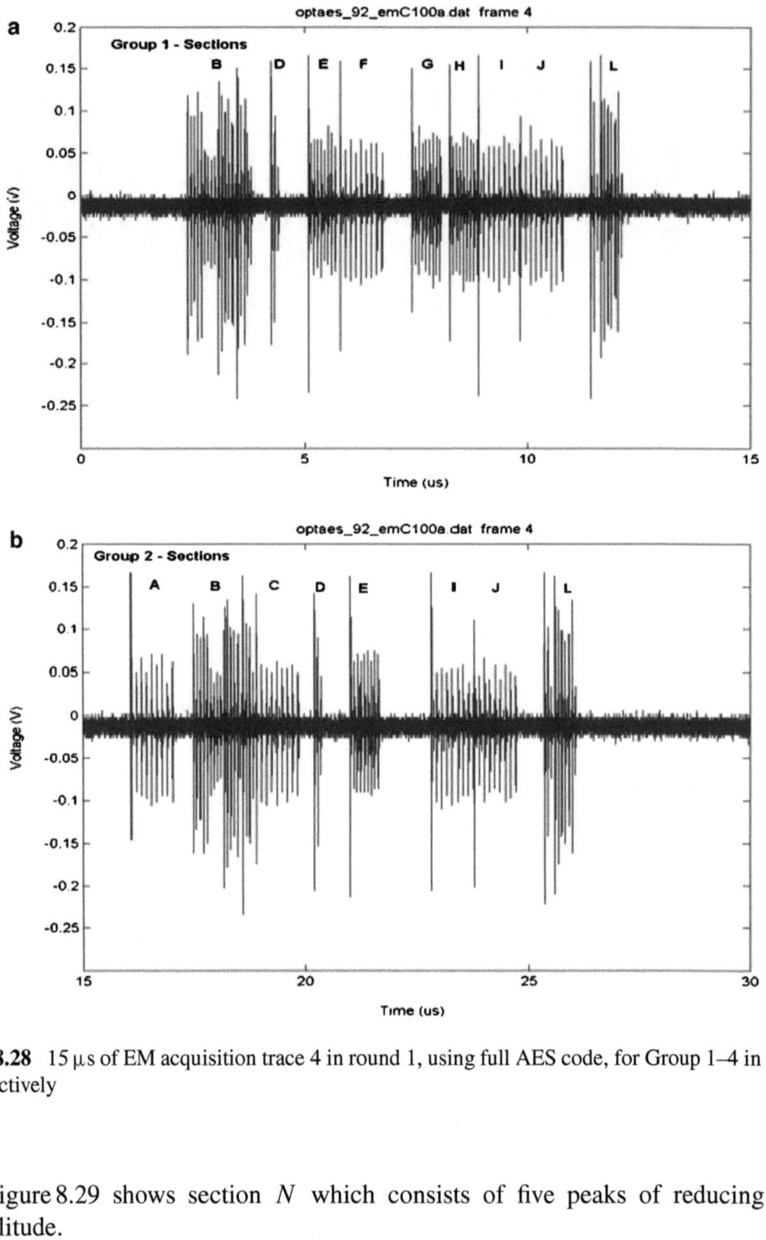

Fig. 8.28 15 μs of EM acquisition trace 4 in round 1, using full AES code, for Group 1–4 in (**a**–**d**) respectively

Figure 8.29 shows section *N* which consists of five peaks of reducing EM amplitude.

Figure 8.30 illustrates the SEMA of the remaining code, including the state code assignments (rectangles) in Fig. 8.30a and the "*AddRoundKey*" code lines in Fig. 8.30b (small circles). The larger ovals (labels as G#, where # goes from 1 to 16) represent the table accesses previously discussed. In Fig. 8.30b since the "*AddRoundKey*" occurs at the end of each round, one can see the start of round 2 with group 1 and group 2 identified.

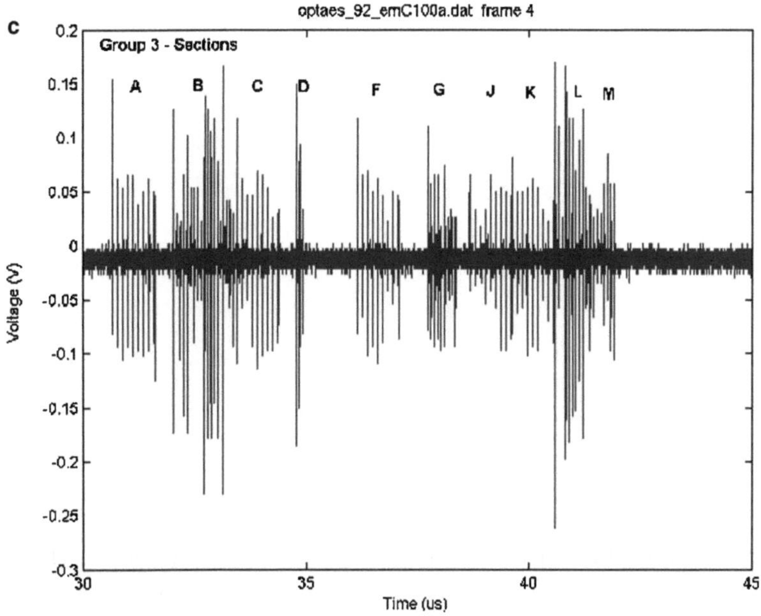

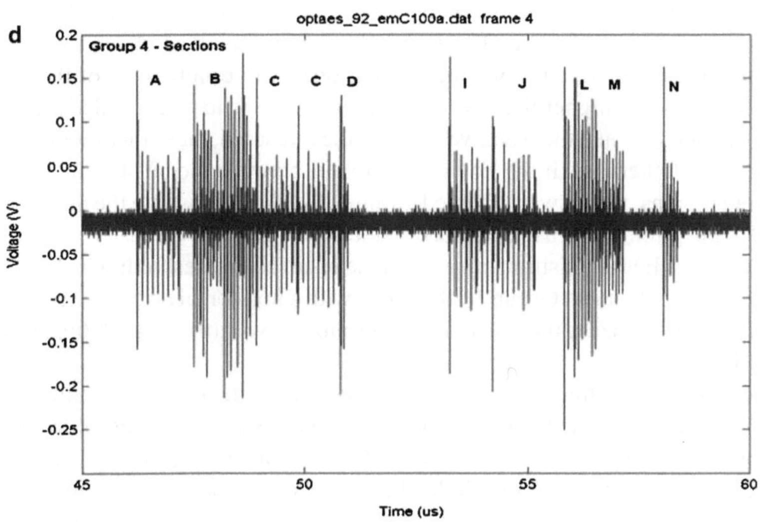

Fig. 8.28 (continued)

8.6.1.2 Truncated code Analysis of the PDA

Once the SEMA is performed, and round 1 has been identified, the region of attack needs to be automatically extracted from each EM trace within the scope acquisitions in step 4 of the methodology. A pattern recognition algorithm was developed to extract and align the region of attack from each EM trace. The pattern recognition

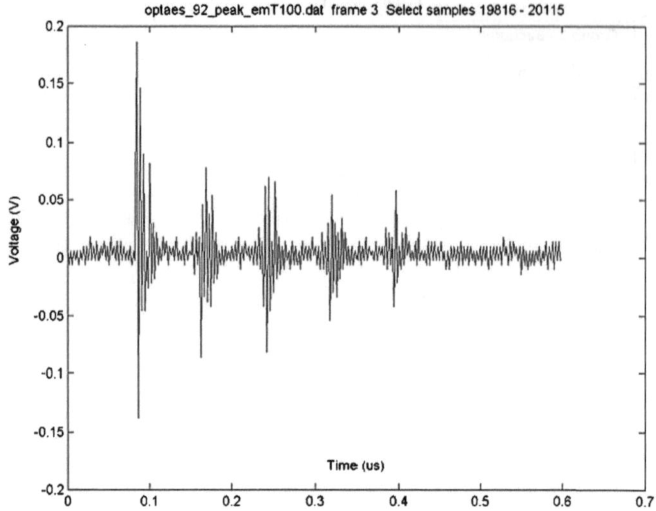

Fig. 8.29 0.6 μs of EM acquisition trace 3 using truncated AES, showing section *N* of *Te*0 region

algorithm consisted of a thresholding step followed by a rule based extraction performed on each EM trace. The thresholding step involved changing each EM trace into a trace of voltages of value zero or one. A threshold level of 0.04 V was set since it was just higher than the noise levels. If the voltage exceeded the threshold or was smaller than the negative value of the threshold, then it was replaced by a value of one. Otherwise, if the EM sample fell in-between the positive and negative threshold regions, then it was assigned a value of zero. To illustrate the thresholding, an EM acquisition of the truncated code called optAEStrunc was used. Figure 8.31a shows a plot of EM acquisition trace 3 and the result after thresholding in Fig. 8.31b, with the region of attack, Group 1, illustrated with a horizontal arrow. Since the truncated code was utilized, there is only one group shown [t(0) = tab.Te0(state(0)(0); in Fig. 8.23b)].

After the thresholding is performed, the region of attack or specifically the Te0 access region is extracted. In order to extract the region, the thresholded trace is transformed into regions of high activity (dark regions) and regions of low activity (long continuous strings of thresholded zeros). Within the region of interest (Te0 region) which, in this case, has high activity, there are mixed samples of zeros and ones. The longest number of continuous zeros within this region is used to set a lower bound on the tolerance parameter. A few EM traces, typically 10, are examined to get this typical value. The minimum number of continuous zeros before and after the region of interest sets an upper bound on the tolerance parameter. The tolerance parameter is arbitrarily set to be some value in-between these two bounds. The EM trace of zero and ones is transformed into a series of regions of high activity and low activity using the tolerance parameter. Hence each continuous set of thresholded zeros in the EM trace which is greater than the tolerance parameter (set to 1,000 in this case) is kept as a region of zeros, otherwise it is merged into an

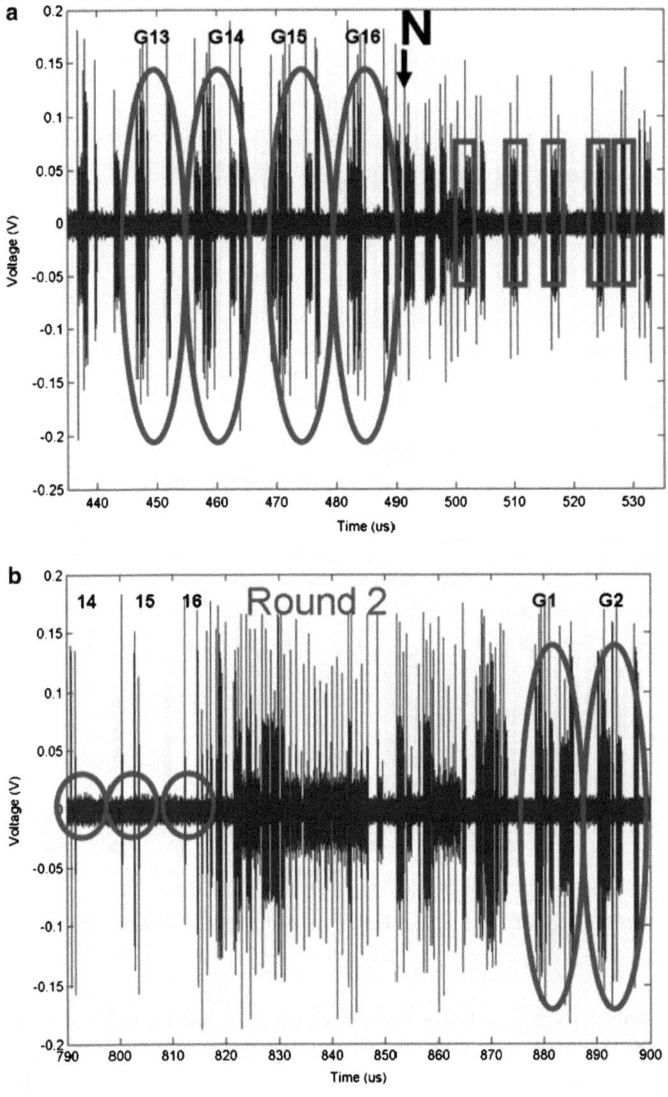

Fig. 8.30 EM activity of last $t(4)$ and state assignment code in (**a**) and the last *AddRoundKey* code lines (small circles), followed by beginning of round 2 in (**b**)

area of high activity. The small vertical arrows in Fig. 8.31b identify these zero or low activity regions. It is important to note that the pattern recognition problem is now reduced from dealing with 50,000 V per trace to dealing with ten regions of low activity.

Pattern recognition tests are next developed to extract the relevant high activity region. In this case, the region of interest is identified by using the width of the low activity region immediately before it and the width of the Te0 access area of interest.

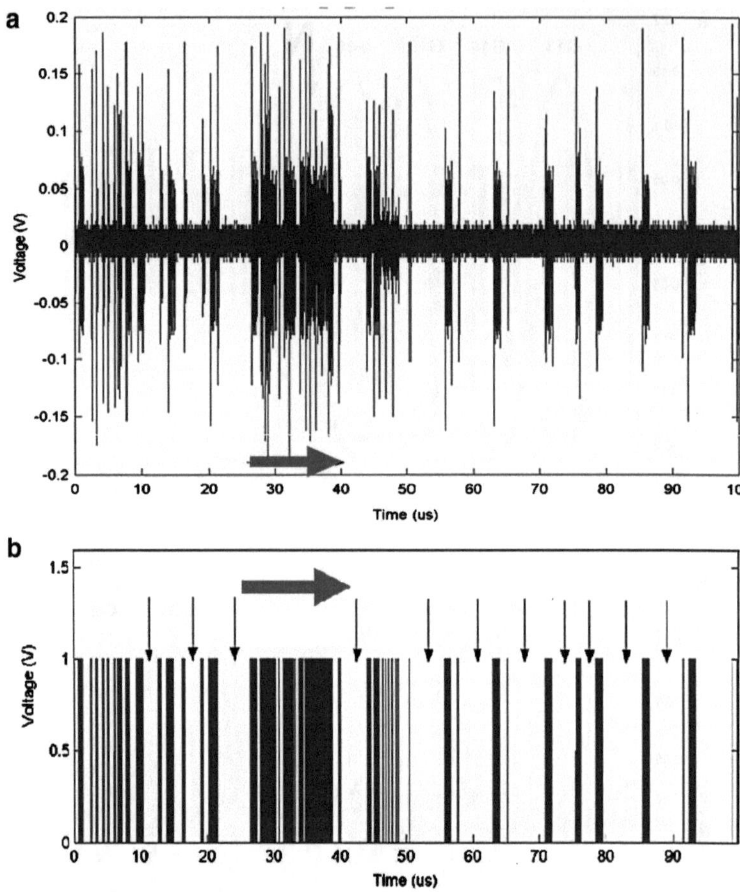

Fig. 8.31 Trace 3 of acquired EM signals before (**a**) and after thresholding (**b**)

In this example, any low activity region which is at least 2,000 samples wide and at most 3,000 samples wide with a high activity region next to it with a width of at least 5,000, is chosen. In Table 8.1 only index 3 satisfies this criterion. If more than one region satisfies the criteria, the first region is chosen. Table 8.1 illustrates the number of samples between high activity regions. Each row of Table 8.1 illustrates a region of low activity with *start* and *end sample numbers*. The *width* of the low-activity regions are also shown (*width* = *end sample #* − *start sample #*). The final column in each row provides the *width* of the adjacent high-activity region. Hence the next *start sample* = *previous end sample* + *high activity width.*

Step 5 of the methodology examines the various areas ($Area_i$) within each extracted region of attack. However, not all areas are present in each trace that has been acquired. For example, Fig. 8.27 showed the trace with all areas A through N labeled. However, the trace shown in Fig. 8.28d indicates that areas E,F,G,H and other areas are not present. Table 8.2 indicates which areas (or *sections*) labeled

Table 8.1 Long strings of thresholded zeros for EM acquisition Trace 3 of Fig. 8.31

Index	Starting sample number	Ending sample number	Low activity width (samples)	High-activity width (samples)
1	5,265	6,434	1,169	1,937
2	8,371	9,558	1,187	1,217
3	10,775	13,242	2,467	6,773
4	20,015	22,020	2,005	3,222
5	25,242	27,866	2,624	1,069
6	28,935	31,578	2,643	1,061
7	32,639	35,472	2,833	515
8	35,987	37,742	1,755	327
9	38,069	39,270	1,201	513
10	39,783	42,751	2,968	514
11	43,265	45,770	2,505	907

Table 8.2 EM analysis of extracted 15 μs of Te0 access region

Trace number	Sections													
	A	B	C	D	E	F	G	H	I	J	K	L	M	N
01	..	..	..	..	..	..	..	..	..	..	..	..	..	..
02	..	..	..	..	..	..	..	..	..	..	nl	..	nl	..
03	..	..	..	..	..	..	..	..	..	..	..	..	nl	..
04	..	..	ms	..	..	..	..	..	..	..	nl	..	nl	..
05	..	..	ms	..	..	..	..	..	..	..	..	..	nl	..
06	..	..	ms	..	..	..	..	..	..	ms	ms	..	nl	..
07	..	..	ms	..	..	..	..	..	..	..	ms	..	nl	..
08	..	..	..	..	..	..	..	..	..	ms	ms	..	nl	..
09	..	..	ms	..	..	..	..	..	..	..	..	..	nl	..
10	..	..	ms	..	..	..	..	..	..	ms	ms	..	nl	..
11	..	..	ms	..	..	..	..	..	..	..	..	..	nl	..
12	..	..	ms	..	..	..	..	..	..	ms	ms	..	nl	..
13	..	..	..	..	..	..	..	..	..	..	..	..	nl	..
14	..	..	ms	..	..	..	..	..	..	ms	ms	..	nl	..
15	..	..	ms	..	..	..	..	..	..	..	ms	..	nl	..
16	..	..	ms	..	..	..	..	..	..	ms	Ms	..	nl	..
17	..	..	..	..	..	..	..	..	..	..	..	..	nl	..
18	..	..	ms	..	..	..	..	..	..	..	Nl	..	nl	..
19	..	..	ms	..	..	..	..	..	..	..	..	..	nl	..
20	..	..	ms	..	..	..	..	..	..	..	Nl	..	nl	..
21	..	..	ms	..	..	..	..	..	..	..	..	..	nl	..
22	..	..	..	..	..	..	..	..	..	ms	Ms	..	nl	..
23	..	..	ms	..	..	..	..	..	..	..	Ms	..	nl	..
24	..	..	..	..	..	..	..	..	..	..	Nl	..	nl	..
25	..	..	ms	..	..	..	..	..	..	..	..	..	nl	..
26	..	..	ms	..	..	..	..	..	..	ms	Ms	..	nl	..
27	..	..	ms	..	..	..	..	..	..	ms	Ms	..	nl	..
28	..	..	ms	..	..	..	..	..	..	..	Nl	..	nl	..
29	..	..	ms	..	..	..	..	..	..	..	..	..	nl	..
30	..	..	ms	..	..	..	..	..	..	ms	Ms	..	nl	..
31	..	..	ms	..	..	..	..	..	..	..	Ms	..	nl	..
32	..	..	..	..	..	..	..	..	..	..	Nl	..	nl	..

A through *N* (as columns) are present (labeled as ". .") or missing (labeled as *nl* or *ms*) in the extracted region of attack of several traces (see *Trace Number*). It was observed that areas that were not present were either replaced with a very short period (close to 0 μs wide) of low EM activity (shown as *ms*, for missing, in Table 8.2) or were replaced with very low activity for 1/2 of the original section's typical duration (*nl*, for null activity). The first row of Table 8.2 shows that all sections A to N are present in trace 1, as was illustrated in Fig. 8.27. The second row indicates that trace 2 had sections K and M both replaced with null activity. For example trace 6 had missing sections C, J, K, and had section M replaced with null activity. A total of 32 traces were analyzed in Table 8.2 and the results indicate some variability.

Section *N* was chosen as the first area to analyze since it was present in each trace and appeared toward the end of the trace (hence was candidate for representing the table value being accessed). Section *N* was extracted, in step 6 of the methodology in Fig. 8.21a, by taking samples immediately before the end of the region of attack and 100 samples following the end of the region of attack as well. This processing created only 300 extracted samples per trace, as compared to the 50,000 V samples in the original traces. Figure 8.32 shows the extracted section *N*, which consists of five peaks of reducing EM amplitude. Analysis of other sections that appeared in all EM traces (such as section *L*) was also performed, but did not reveal the correct key.

The frequency-based DEMA, step 7 of the methodology, was run on the extracted section *N*. It successfully found the correct key as shown in Fig. 8.33.,New EM traces were acquired with several different keys to confirm the methodology. It is interesting to note that time-based DEMA analysis was not successful in finding the correct key. Further investigation showed that although the extracted section *N*

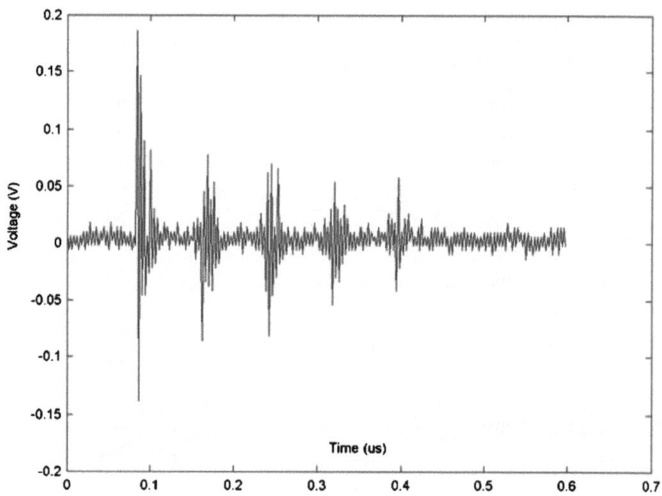

Fig. 8.32 Trace 3 of EM acquisition showing (0.6 μs section *N*) $Area_k$ extracted from region of attack

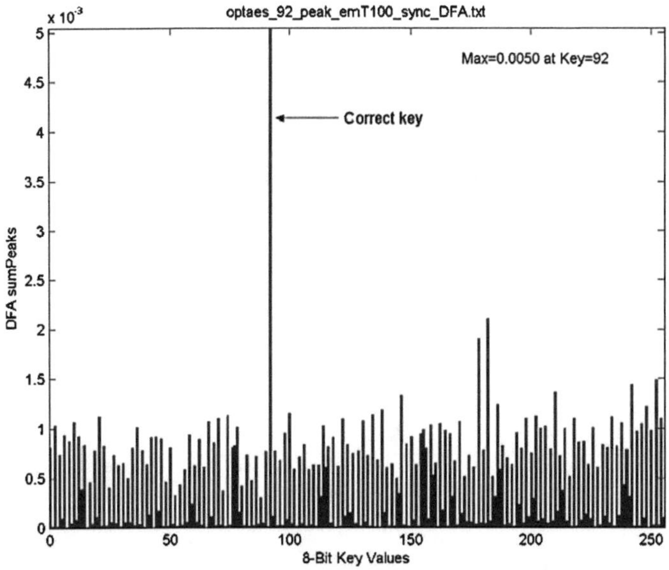

Fig. 8.33 All keys search DFA results, extracting 0.6 μs section N of *Te*0 region for PDA truncated optimized AES encryption, with correct key = 92 decimal

samples were almost aligned, there were still a few nanoseconds of remaining time shifts. These small time shifts were enough to thwart the DEMA.

In summary, based upon analysis of the truncated code, section N was determined to represent the attack area. The next section will utilize the full AES encryption code to illustrate a complete attack as outlined in Fig. 8.22b.

8.6.1.3 Attack of the Device Using Full AES

Analysis of the truncated code was helpful in determining that section N represented the attack point. This section describes the attack of the full (not truncated) AES encryption Java code. Thresholding and pattern recognition are both used again to extract the appropriate regions for analysis.

Figure 8.34a shows a plot of trace 4 of the EM acquisition (using key 92 decimal). Four regions that look similar to the one *Te*0 access region for the PDA truncated AES are labeled with arrows pointing up and will be referred to as Group 1–4 from left to right. These four regions of attack corresponded to the four S-box table accesses. From the SEMA performed, it was observed that there were four groups, each consisting of four S-box table accesses. Each group corresponded to processing of the table accesses, specifically *Te*0, *Te*1, *Te*2, *Te*3. Section N consistently only appeared in every fourth group. For example, Fig. 8.28a illustrates Group 1 (the first arrow on the left in Fig. 8.34a) of the first candidate region of attack, where there is no section N. Section N, which only appeared in Group 4, was used again as the pattern recognition target for isolation and extraction.

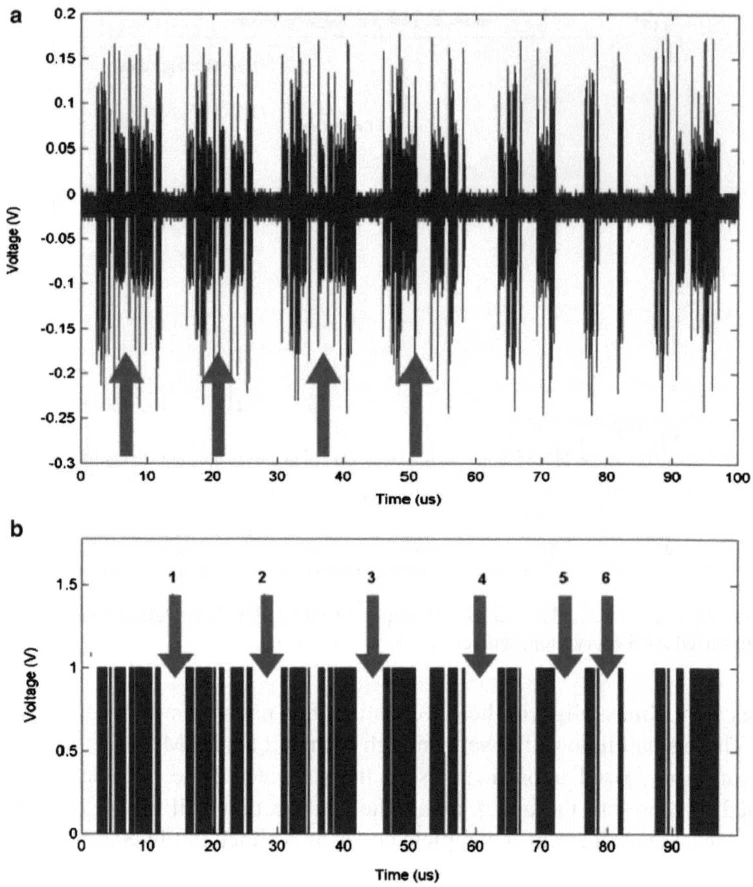

Fig. 8.34 EM acquisition trace 4 (**a**) before thresholding and (**b**) after thresholding

The next step, step 11, in the attack methodology was to perform thresholding and pattern recognition in order to extract section N. The thresholding step was the same as described in the previous section. Figure 8.34b illustrates the thresholded result corresponding to the original EM acquisition trace 4, shown in Fig. 8.34a. For creating the regions of high and low activities, the tolerance parameter is set to be greater than any gaps (strings of zeros) within the Group 1–4 regions but less than the width of the strings of zeros preceding and following the Group 1–4 regions. Again the pattern recognition problem is reduced from dealing with 50,000 V samples per trace to dealing with five regions of high activity.

The arrows in Fig. 8.34b identify the long strings of thresholded zeros. Again the pattern recognition was designed to search for a width (between 2,000 and 3,000 samples in this case) and separation pattern (separation of greater than 5,000 samples) which identifies the thresholded zeros region preceding the Group 4 region.

Using the EM acquisition section N was extracted from each trace. The extraction took 200 samples (0.4 μs) from the end of the Group 4 region (section N) and 100 samples (0.2 μs) following the end of the Group 4 region. There were now only 300 extracted samples per trace. The pattern recognition and extraction of Group 4 section N was repeatable over all traces which were extracted. Figure 8.35a, b illustrate two extracted section N's. The number of traces, where pattern recognition failed, was at most 2 for each of the eight EM acquisitions using four different keys. Mostly this failure is a result of the first acquired EM trace which is sometimes "abnormal," not having the usual pattern of groups and sections.

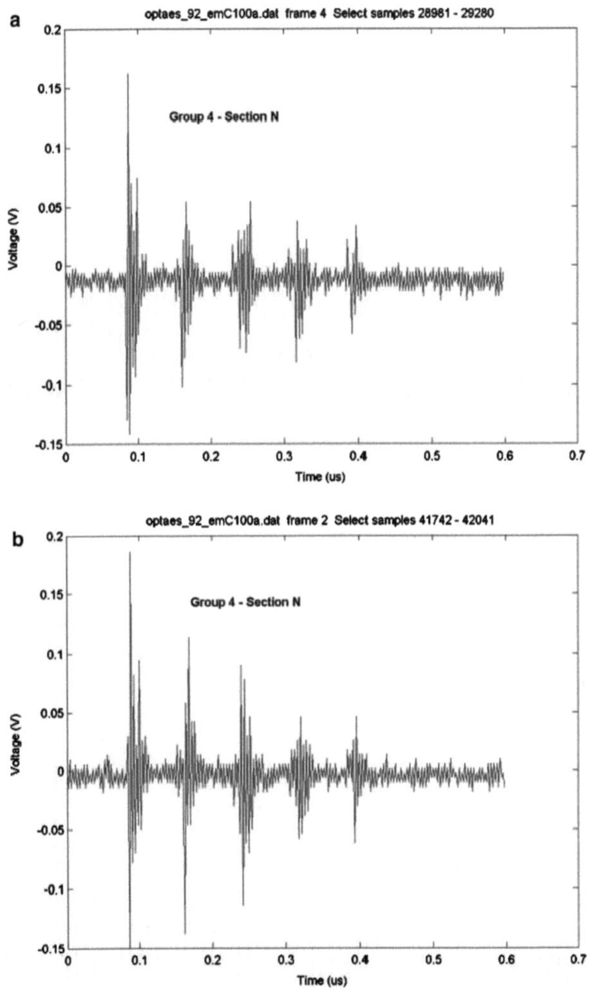

Fig. 8.35 EM acquisition trace 4 and trace 2 for AES in (**a–b**) respectively, extracting 0.6 μs section N of Group 4 region

The frequency-based DEMA (see equations 8.13 and 8.14) was run on the extracted Group 4 section N for each trace. It successfully found all correct key bytes for each of the eight EM acquisitions using four different keys. The next highest magnitude from the frequency-based DEMA results was only 28–44% of the correct key magnitude. Figure 8.36 shows detailed plots of the all keys search

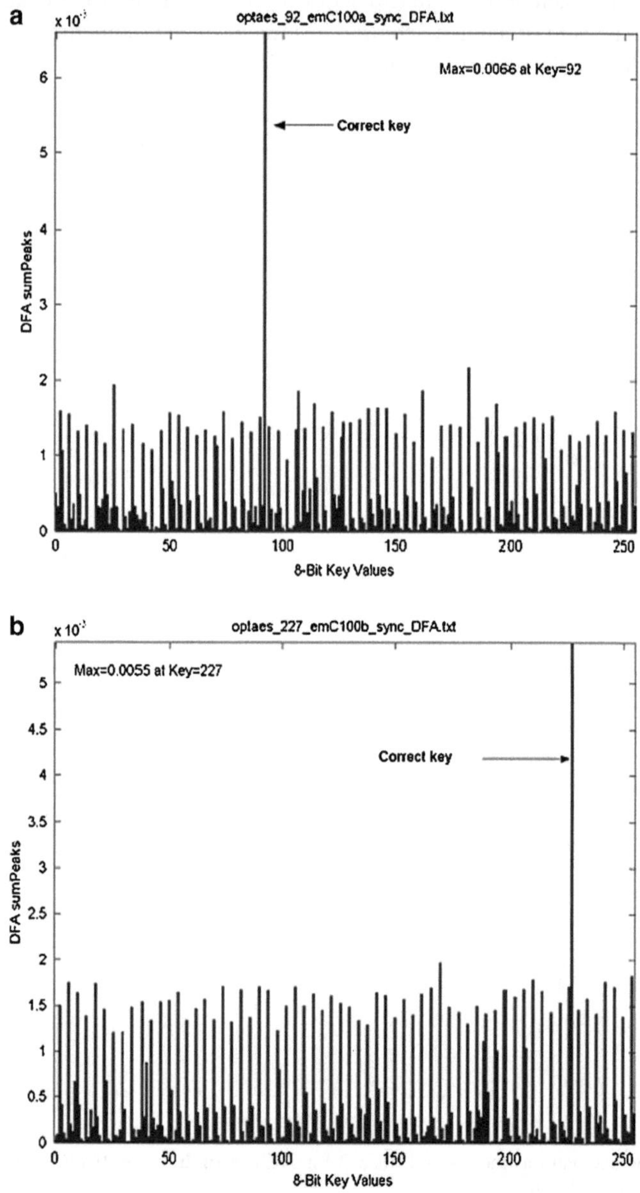

Fig. 8.36 All keys search using frequency-based DEMA results, for keys 92, 227, 61, 158 in (**a–d**)

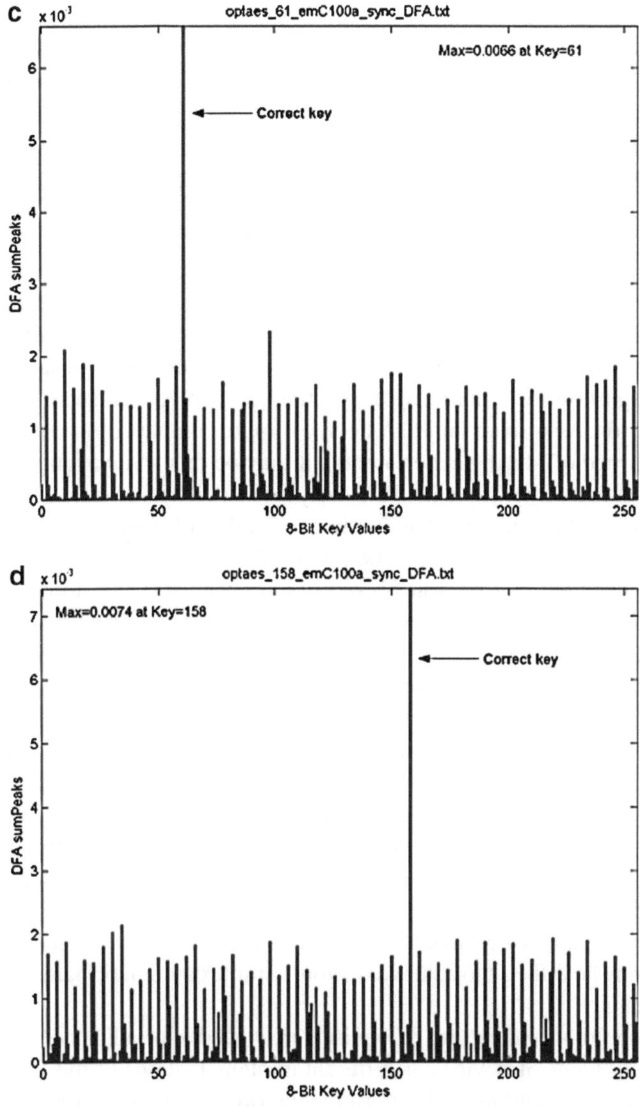

Fig. 8.36 (continued)

utilizing the frequency-based DEMA for four different keys. The frequency-based differential EM analysis was run on the extracted group 4 section N.

Further experiments were performed to determine how many EM traces were required before a successful differential analysis identified the correct key. Two separate EM acquisitions were used with the complete Java code. The first acquisition determined the correct key only after 128 EM traces (where the next highest key magnitude was 23% of the highest magnitude identifying the correct key). The second acquisition required only 64 EM traces in order to find the correct key (and the next highest magnitude 87% of the correct key magnitude). Utilizing only 32 traces, and in the second EM acquisition the correct key had the third highest magnitude.

The methodology, together with frequency-based DEMA, created a successful side channel attack on a real PDA running Java code. The attack is repeatable and works for all keys which were tried. The PDA attack is also expected to work with the same pattern recognition algorithm for different key sizes or number of rounds using the Java-based AES. Specifically the intermediate data, $t(0)$, was found to be attackable because it was written to and/or from memory.

In a more realistic embedded device the cipher source code or algorithm may be unknown, hence even more analysis may be necessary in order to find the approximate location of the attack point. Although other ciphers were not attacked on the PDA, it is likely that the methodology would remain the same, resulting in a new pattern recognition for identifying and extracting appropriate sections of the new cipher's rounds. Also since the PDA device is quite complex (and only the wireless radio was turned off during experiments), it may be that some acquisitions may be more difficult to analyze since the PDA may be in a different "state." In these cases, some traces may have to be deleted, a larger number of traces may be required, or the pattern recognition may have to be modified.

It is highly likely that section N used in the attack was the store of value $t(0)$. This is unlike previous research which had not considered attacks using $t(0)$. Only the S-box tables were considered viable attack points (Kocher et al. 1999; Agrawal et al. 2002; Itoh et al. 2002). Hence although the S-box tables were not directly attacked as in most previous research (Kocher et al. 1999; Agrawal et al. 2002; Itoh et al. 2002), an attack was still successful and viable. It is also interesting to note that further analysis did not successfully attack the S-box table accesses. It may be that the S-box tables were loaded all at once into cache earlier in the code, and hence on-chip cache memory accesses were more difficult to attack. The frequency-based DEMA was very successful once the area of attack was extracted. However time-based DEMA which was not successful in obtaining the correct key on the PDA, still requires excellent EM trace alignment in order to work. Excellent EM trace alignment may be difficult to achieve without complex automatic pattern recognition or other means.

The thresholding and pattern recognition may need to be adjusted for EM signals from different types of devices or different implementations of the Java code. The algorithms for thresholding and pattern recognition, providing extraction of regions in the EM, are relatively simple. This is partly due to the Java execution, where the high EM activities are often separated by long periods of low EM activity (such as the case for section N). Additionally, the simple extraction algorithm is largely a result of the frequency-based DEMA which does not require perfect alignment. It was assumed that the cipher implementation is the same in both the characterization device and the device under attack. However, it may be possible for the attacker to explore various implementations on the characterization device and determine which implementation more closely matches the EM signals from the device under attack. In cases where the source code is not available, this type of analysis on the characterization device would be crucial. However, if the device under attack utilizes assembly or a high-level language and the attacker does not have access to the necessary compilation tools, then an attack would be more difficult.

The proposed methodology may also be applied to power signals, however these are more difficult to obtain from the chip inside the PDA housing the respective processor. Power signals obtained from the battery of the mobile device may also be difficult since signals likely would be buried amidst the current drain of other components requiring power. For example on PDA devices, there are many different components including power management circuitry, radio circuitry, baseband processor, etc.

The methodology provides a repeatable approach for finding the cryptographic key. Unlike evaluation boards and despite interrupts and other interfering events, the PDA running the Java-based cipher provided opportunity for frequency-based differential attack largely due to the many data stores and loads to and from memory typical in a stack-based Java virtual machine implementation. The pattern recognition technique required a significant amount of man hours developing pattern rules and verifying them. There are two faster approaches for alignment. One uses a variation (Gebotys and White 2008a, b) of the phase-only correlation (Homma et al. 2006) to align traces. The other uses a phase substitution technique (Gebotys and White 2008b). Both techniques are much faster to implement than developing pattern matching rules. The phase-only correlation sometimes correlates incorrectly shifting the trace to align a larger EM characteristic while misaligning the attack region, thus adding some noise to the attack. The phase substitution provides perfect alignment of the traces (even thwarting countermeasures that reverse the order of operations). However, it adds amplitude noise to the traces. In both cases additional traces are required to deal with this added noise.

Given the attack potential of embedded systems, as demonstrated in this chapter with a real PDA device, clearly countermeasures are very important in order to protect the security of a cryptographic algorithm. The next chapter, Chap. 9, will examine some countermeasures focusing on the S-box type operation. To illustrate the wide range of issues involved in attacking embedded devices through the side channel, the next section will describe an experimental setup which supports an eavesdropping attack on contactless smartcards.

8.7 Experiments with Contactless Chip Cards

Here we consider the challenging problem of analyzing a class of contactless cards that do not have batteries and are totally reliant upon the card reader to provide them with power. Although contact smart cards have been attacked successfully, contactless cards pose several challenges to SCA. This section will illustrate that readily available equipment can be used to expose the EM signals of off the shelf commercial contactless smart cards. Although no practical attacks on the side channel of commercial untampered contactless cards have been announced, the EM side channel is accessible as will be illustrated.

Contactless cards were introduced briefly in Chap. 1. Compared to PDA devices, these cards have a smaller operating system and their hardware is typically contained within one chip. Contactless cards nevertheless remain a challenge largely since the side channel is heavily dominated by the carrier. Although there have been reports of contactless cards being read by malicious card readers, we assume that these types of attacks would be thwarted by card reader authentication. In these cases, only authenticated readers would be able to communicate with the contactless cards. Thus a possible attack on this secure scenario would involve eavesdropping on the communications. This section illustrates how to do this without contact with the card. This is an interesting attack on its own since any data being passed in the clear over a communications channel with an authenticated reader could easily be stolen. An even more interesting attack is to take this one step further. Assume that during a legitimate communication with an authenticated reader, a probe was able to acquire enough data from the contactless card to run a SCA and, thereby, extract the secret key from it. For example, when the card was communicating with the reader, an embedded probe was able to record the side channel of the card. Perhaps the card reader is malicious, and never authenticates with the contactless card, but instead feeds its responses (in reply to the contactless cards challenges) which are checked by the card with the secret key. The side channel of each processing of the response by the card is recorded and a SCA is mounted to extract the key from the card. This possibility will be discussed in this section.

Figure 8.37 shows a contactless card where the plastic has been filed away to reveal the single die connected to the antennae within the card. This happens to be what is called a dual interface card. This means that it can function as a contactless card or a contact card. We use the JCOP30 card, which is the Java card open platform

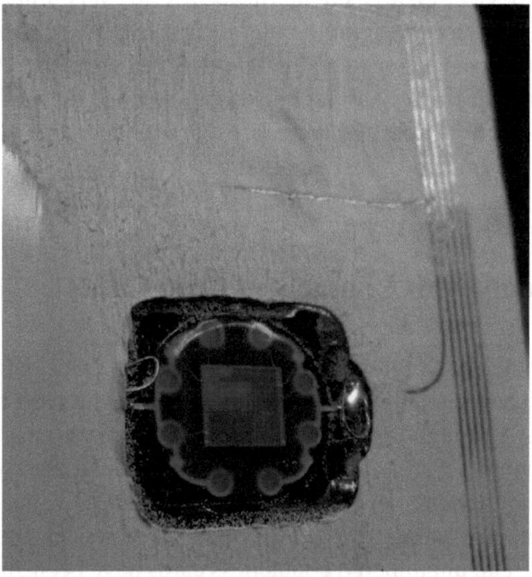

Fig. 8.37 Filing of contactless smart card revealing die within plastic

card supporting dual interface modes. It uses the Philips/NXP P8RF5016 circuit with IBMs proprietary Java card operating system in ROM (Philips 2003). It has 512 bytes of EEPROM for a transaction buffer, 750 bytes of RAM for a transient java heap, and 200 bytes for the Java stack. The processor is the 8-bit low power 80C51 with configurable clock speed. One can see the pins of the die connected to the pads for the contact mode of operation. As well one can see the connections from the die to the antennae. Even the patterning on the die is visible indicating the likely location of on-chip memory.

The first challenge in investigating this embedded system was triggering. Unlike previous research in this area, a wireless trigger was implemented. The ISO 14443 standard was used to find the trigger. In particular part 2 of this standard (14443–2 2001) indicated that communication with the contactless device starts with a "sequence of Z" which was a pause at the beginning of a bit duration. The pause after a long string of unmodulated carrier indicated a start of a new command transmission. The scope was programmed to trigger on this type of waveform. Specifically it triggered when the input signal remained below 50 mV for 2 μs. The trigger was also programmed to ignore trigger signals over the time period during which the relevant command was processed. This trigger technique should also work with all ISO 14443 type A devices that include RFID and contactless cards. The experimental setup for acquiring EM emanations from the contactless card is shown in Fig. 8.38. An example of the EM signals captured with the scope are shown in Fig. 8.39. Here one probe is used to acquire the EM trace as well as trigger the scope (Berkes 2008). In more recent experiments, positioning of a second probe provided more accurate triggering. The 13.56-MHz carrier is always present in the field to

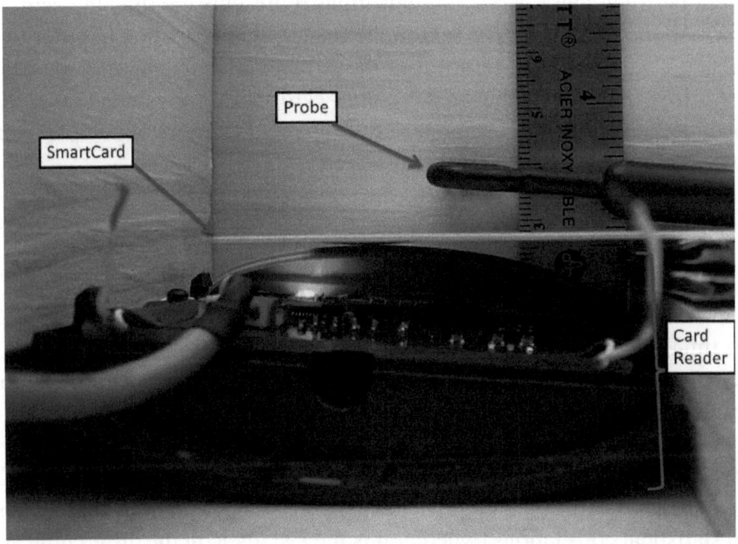

Fig. 8.38 Experimental setup for triggering and recording EM emanations of a contactless smartcard

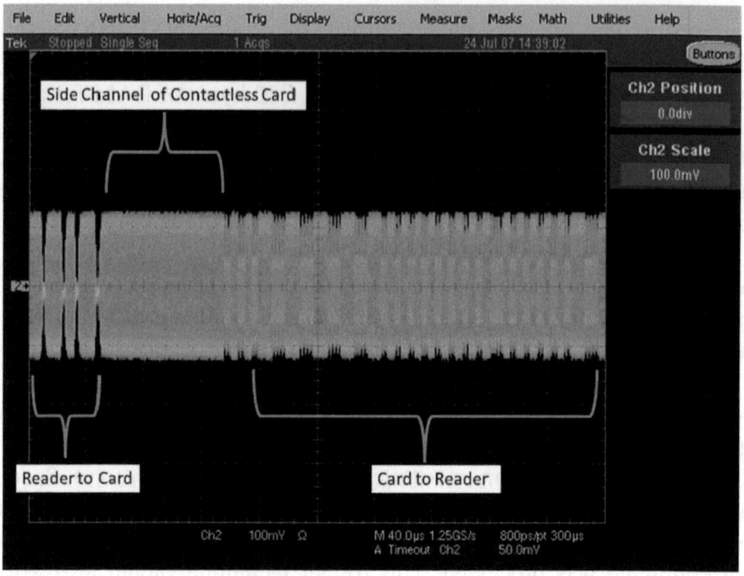

Fig. 8.39 Emanations from contactless smart card illustrating communications to and from card reader as well as the area of the side channel of the smart card. The signals from the reader to the card are modulated with 100% ASK. The load modulation was harder to see until the probe was positioned appropriately

supply power to the card. The possible area for SCA is delineated by the communications from the card reader and the communications to the card reader. The time interval in between these two communications represents the time the card is computing a response which will be sent to the reader. It appears that in order to obtain further side channel details, removal of the strong carrier signal may be necessary using realtime filtering (Riscure 2009), dual-card real time subtraction techniques (Brightsight 2009), or other techniques. Side channel analysis of complex commercial contactless cards requires further research.

8.8 The SCA Timeline and Future Technologies

A brief timeline of side channel analysis advances is presented in Table 8.3. Most research publications are not shown since the table would be too large. The focus of this timeline was to highlight some embedded devices that have been attacked using SCA, in addition to some new analysis techniques that have arisen. Some other research listings can be found at (side channel lounge) as well as many crypto conferences, such as CHES. Although, in general, the attack techniques are all based on general statistical correlations (whether it is the t-statistic in the differential analysis case or correlation coefficients) or probabilities (mutual information analysis).

Table 8.3 Timeline of side channel attacks on embedded systems

Year	Embedded system	Type of attack
1996	(Kocher 1996) Smart card	Timing attack
1999	(Kocher et al. 1999) announced in 1998, Smart card	Differential power analysis
	(Messerges et al. 1999) smart card	Multibit DPA
	(Fahn and Pearson 1999) smart card	Inferential power analysis
2000	(Quisquater et al. 2000) Smart card processor	Differential EM analysis
2001	(Agrawal et al. 2001) smartcard	SEMA, DEMA
	(Gandolfi et al. 2001) 8-bit microcontrollers, Comp128, RSA, DES	DEMA, DPA, SEMA, SPA
2002	(Gebotys et al. 2002) DSP processor platform, ECC	Simple and differential power analysis
	(Rao et al. 2002) SIM card in GSM phones	Partitioning attack
	(Chari et al. 2002) smartcard	Template attacks
2003	(Agrawal et al. 2003a) Palmpilot and RSA accelerator	Timing attack and simple EM analysis
	(Agrawal et al. 2003c) chip card	Multichannel attack
	(Mangard 2003) smartcard	Far-field EMA
2004	(Carlier et al. 2004) FPGA running AES	Square EM attack
	(Standaert et al. 2004) FPGA	SPA, DPA
	(Lemke et al. 2004) ATM163 microcontroller, IDEA	DPA
	(Brier et al. 2004) security chips	Correlation power analysis
	(Gebotys 2004) ECC on DSP Star*core Processor	SPA, DPA
2005	(Gebotys et al. 2005) PDA, ECC, and AES	Frequency based DEMA
	(Tromer 2004) RSA running on Intel Celeron 666 MHz	Acoustic analysis
	(De Mulder et al. 2005) FPGA running ECC	SEMA, DEMA
2006	(Oren et al. 2006) RFID	SPA
2007	(Fischer et al. 2007) FPGA, stream cipher	DPA
2008	(Gierlichs et al. 2008) AT90S8515 8-bit microcontroller	Mutual information analysis
	(Eisenbarth et al. 2008) Keeloq attack	DPA
2009	(Kizhvatov et al. 2009) AVR XMEGA	Correlation analysis

Although EM attacks on smart cards have been investigated (Agrawal et al. 2001; Gandolfi et al. 2001), EM or power attacks on other embedded systems have not been widely researched. Apart from far-field EM emanations from a Palm–Pilot and SSL accelerator (Agrawal et al. 2003a, b) and the attack on a RFID passive tag (Oren and Shamir 2006). In the later case of the RFID attack, the SPA of the receive antennae was used to extract the kill password, while the attacker sent data to the passive RFID tag. In Agrawal et al. (2003a) it was shown that AM demodulation could be utilized to extract EM signals at a distance from an RSA accelerator, as well as Palm–Pilot. They found that precise timing attacks could be performed at 50 in. away from the RSA device by listening to the AM demodulated EM signals on a receiver. EM leakage was shown to be better than the power leakage, thus leading to more information about the cryptographic operations. In Agrawal et al. (2003b) the researchers provided further details of the SEMA using AM demodulation of the RSA accelerator. They also provided a comparison of DPA to DEMA using a smartcard and further introduced the template attack again on a smartcard. Differential and simple EM attacks on an embedded low power highly parallel processor core was performed (Gebotys et al. 2002). This chapter covered differential EM analysis on a real PDA device, illustrating challenges in trace alignment and SEMA. Further in-depth reading on the subject of differential power analysis can be found in Mangard et al. (2007).

In summary launching SCA on real embedded devices calls for special alignment techniques, special triggering techniques, and more powerful attacks that demand greater knowledge of the location of attack points. Interrupted events in embedded systems produce noise and intermitten patterns that may never fully align. Since alignment is crucial for SCA, attack region identification along with windowing in on the attack region is crucial, to eliminate the many more false peaks created around the attack region by interrupts.

Our recent empirical results indicate that the amplitude of the EM side channel is reducing as newer technologies are adopted. This is likely due to the lower powered technologies where supply voltages are reduced despite higher clock frequencies. SCA with EM signals emanating from these newer technologies has become much more sensitive to noise than previous older technologies. We have found that newer technologies demand more powerful attack techniques such as correlation analysis, identification of EM attack characteristics along with hybrid techniques for aligning traces and focusing or windowing in on the region of attack. More importantly strong countermeasures will still be very important with future technologies. The next chapter will discuss the important topic of countermeasures.

References

14443–2 (2001) ISO/IEC Identification cards – Contactless integrated circuit(s) cards – Proximity cards – Part 2: Radio frequency power and signal interface

Agrawal D et al (2001) The EM side-channel... methodologies. Available At http://www.research.ibm.com/intsec/emf.html

Agrawal D et al (2002) The EM side-channel(s). In: Proceedings of CHES 2002, LNCS 2523, Springer, New York, pp 29–45

Agrawal D et al (2003a) Advances in side-channel cryptanalysis EM analysis and template attacks. In: RSA cryptobytes, Vol 6, No 1, Spring 2003, 20–32

Agrawal D et al (2003b) Power, EM and all that: is your crypto device really secure? Presentation at the 7th workshop on Elliptic Curve Cryptography (ECC 2003), Waterloo, Ontario, August 2003. Available At http://www.cacr.math.uwaterloo.ca/conferences/2003/ecc2003/rohatgi.ppt

Agrawal D et al (2003c) Multi-channel attacks. CHES 2003, LNCS 2779, Springer, Berlin Heidelberg, pp 2–16

Berkes J (2008) Side-channel monitoring of contactless Java cards, MASc Thesis, Department of Electrical and Computer Engineering, University of Waterloo

Brier et al (2004) Correlation power analysis with a leakage model. CHES 2004, LNCS 3156: 16–29

Brightsight (2009) Poster discussion at CHES 2009, http://www.brightsight.com
Carlier et al (2004) Electromagnetic side channels of an FPGA implementation of AES. IACR 145, http://eprint.iacr.org/2004/145.pdf
Carsten B (1998) Sniffer probe locates sources of EMI; EDN magazine, June 4, 1998
Chari S et al (2002) Template attacks. CHES2002, LNCS2523, 13–29
Chen X (2004) Simple power analysis threat in embedded devices, MASc Thesis, Department of Electrical and Computer Engineering, University of Waterloo
De Mulder E et al (2005) Electromagnetic analysis attack on an FPGA implementation of an elliptic curve cryptosystem. In: Proceedings of the IEEE international conference on computer as a tool (EUROCON), 1879–1882
Eisenbarth T et al (2008) Physical Cryptanalysis of KEELOQ code hopping applications, http://eprint.iacr.org/2008/058.pdf
Electro-Metrics Inc., *Instruction Manual: Near Field Probe Set Broadband Response Model EM-6992*, 2004
Fahn PN, Pearson PK (1999) IPA: a new class of power attacks, CHES 1999, LNCS 1717:173–186
Fischer W, Gammel BM, Kniffer O, Velten J (2007) Differential power analysis of stream ciphers, CT-RSA 2007, LNCS 4377, Springer, Berlin Heidelberg, pp 257–270
Gandolfi K et al (2001) Electromagnetic analysis: concrete results. In: Proceedings of CHES, LNCS 2162, Springer, New York, pp 251–261
Gebotys C, Gebotys R (2002) Secure Elliptic Curve Implementations: An Analysis of Resistance to Power Attacks in a DSP processor, Proceedings of CHES 2002, pp 98–113
Gebotys C, White BA (2008a) EM analysis of a wireless Java-based PDA, ACM transactions on embedded computing systems, Vol 7, No 4, Article 44, pp 1–28
Gebotys C (2004) Design of secure cryptography against the threat of power-attacks in DSP embedded processors, ACM Trans Embed Comput syst 3(1):92–113
Gebotys C (2006) A split-mask countermeasure for low-energy secure embedded systems, ACM Trans on Embedded Computing systems 5(3):577–612
Gebotys C, Ho S, Tiu CC (2005) EM analysis of Rijndael an ECC on a wireless java-based PDA, CHES 2005, LNCS 3659 GmbH, pp 250–265
Gebotys C, White BA (2008b) EM alignment using phase for secure embedded systems. Des Autom Embed Syst, Springer, Netherlands, 12(3):185–206
Gerke D, Kimmel B (2002) EDN designers guide to electromagnetic compatibility. Newnes, Oxford
Gierlichs et al (2008) Mutual information analysis – a generic side-channel distinguisher, CHES 2008, LNCS 5154, pp 426–442
Gladman B (2003) A specification for AES, the AES algorithm. Available At fp.gladman.plus.com/cryptography.technology/AES/aes.spec.311.pdf, 18–19
Ho S (2005) EM Analysis of ECC Computations on Mobile Devices, MASc Thesis, Dept of Elec and Comp Engineering, University of Waterloo, Waterloo ON, Canada
Homma N et al (2006) High-resolution side-channel attack using phase-based waveform matching. LNCS CHES 2006 187–200
Itoh K, Takenawa M, Torii N (2002) DPA countermeasure based on the masking method. In: Proceedings of ICISC 2001, LNCS 2288, Springer, New York, pp 440–456
Kizhvatov I (2009) Side channel analysis of AVR XMEGA Crypto Engine, ACM proceedings of Workshop on Embedded Systems Security (WESS), 2009
Kocher P (1996) Timing attacks on implementations of Diffie-Hellman, RSA, DSS, and other systems. LNCS 1109, Springer, New York, pp 104–113
Kocher P, Jaffe J, JunN B (1999) Differential power analysis. In: CRYPTO'99, Springer:388–397 (see also announcement in 1998 Introduction to DPA and related attcks http://www.cryptography.com/resources/whitepapers/DPA.pdf
Lemke et al (2004) DPA on n-bit sized boolean and arithmetic operations and its application to IDEA, RC6 and the HMAC construction. CHES2004, LNCS 3156, pp 1611–3349
Mangard S, Oswald E, Popp T (2007) Power analysis attacks – revealing the secrets of smartcards, Springer, New York

Mangard S (2003) Attacks on cryptographic ICs based on radiated emissions. In: Ostermann L (ed) Proceedings of austrochip 2003, 13–16
Messerges T et al (1999) Investigations of power analysis attacks on smartcards, proc of USENIX workshop on smartcard technology
Mills E (2009) Quote by B Jun in "Leaking crypto keys from mobile devices", CNET news, http://news.cnet.com/8301-27080_3-10379115-245.html
Oren Y, Shamir A (2006) Power analysis of RFID tags. Available At http://www.wisdom.weizmann.ac.il/~yossio/rfid
Philips/NXP Semiconductors. (2003) P8RF5016 – Secure Dual Interface Smart Card IC (Short Form Specification) Revision 1.4
Quisquater J-J et al (2000) A new tool for non-intrusive analysis of smartcards based on EM emissions. In Rump Session, Eurocrypt
Rao JR, Rohatgi P, Scherzer H, Tinguely S (2002) Partitioning attacks: or how to rapidly clone some GSM cards. IEEE Symp Secur Priv:31–41
Riscure (2009) Poster discussion at CHES 2009, http://www.riscure.com
Shamir A (2009) How Cryptosystems Are Really Broken. CS Distinguished Lecture presented at the University of Waterloo, October 2009
Side channel lounge http://www.crypto.ruhr-uni-bochum.de/en_sclounge.html
Standaert FX, Örs SB, Preneel B (2004) Power analysis of an FPGA: implementation of rijndael: is pipelining a DPA countermeasure? CHES 2004, Springer, 2004, LNCS, 3156, 30–44
Star*core (2000) SC140 DSP core reference manual (MNSC140CORE), Rev.1, 6/2000
Sylvester et al (1999) BACPAC – Berkeley Advanced Chip Performance Calculator, http://www.eecs.umich.edu/~dennis/bacpac/bacpac_models.html
Tromer E (2004) Hardware-based cryptanalysis. PhD Dissertation, A. Shamir Advisor, http://people.csail.mit.edu/tromer/phd-dissertation/
Tektronix Inc. (2003) *User Manual: Digital Phosphor Oscilloscopes TDS7254*

Chapter 9
Countermeasures

Although many embedded devices offer more resistance to bus probing attacks due to their compact size, susceptibility to power or electromagnetic analysis attacks must be analyzed. Most attacks on embedded systems involve attacks on the implementation of cryptographic methods as discussed in Chap. 8. The focus of this chapter is on techniques for resisting attacks which may occur even when secure implementations exist. Specifically resistance to side channel analysis will be discussed and analyzed with experimental results. The objective of a countermeasure is to make an attack much harder. Of course it would be ideal to have a countermeasure which prevents any attacks using the side channel. However, we do not have perfect models for the side channel to guarantee that no attack is possible. Nor can we guarantee the use of truly random values in our countermeasure. Thus the quality of a countermeasure typically can be measured by the average number of traces required in order to launch a successful attack on a specific platform.

From a chip point of view, recall that external memory accesses or external data transfers (those which cross the chip boundary) dissipate large amounts of power and hence exhibit well-defined side channels which in general are easy to detect. These are an important focus for side channel resistance. However, secondary to off-chip loads, stores, or data transfers are on-chip memory accesses. This power dissipation largely depends on the size of memory, however, given large on-chip caches standard in most processor-based platforms it is an important secondary focus for security. Less significant but perhaps important for attacks where the chip is decapsulated are switching of on-chip busses if the busses have high enough capacitances and possibly but to a lesser extend datapaths and registers. Thus side channel analysis is expected to be successful if critical data is transferred "unprotected" on or off the chip as well as possibly on-chip memory access. Hence in many secure systems, it is important to consider resistance to side channels for critical data transfer to and from the chip, as well as critical data involved in memory accesses on and off of the chip.

There are many proposed countermeasures for side channel resistance which can be used at the circuit/gate level, architecture level, software level, and system level. For example, there are proposed circuit implementations which involve technologies such as enhanced current balanced logic, ECBL (Khatibzadeh and Gebotys 2007), wave dynamic differential logic, WDDL (Tiri and Verbauwhede 2004),

C.H. Gebotys, *Security in Embedded Devices*, Embedded Systems,

DOI 10.1007/978-1-4419-1530-6_9,

sense-amplifier based logic, SABL (Tiri and Verbauwhede 2002), and adiabatic circuits, AC (Gebotys and Zhang 2003). In these approaches, the side channel leakage is reduced through symmetry at the expense of area (WDDL, SABL), or power (ECBL) or speed (AC). At the architecture level, techniques such as precharging the bus with random data and performing redundant parallel operations may be used. Other architectural techniques may involve the use of random clocks or random modulations in the case of contactless cards (Hancke 2007) in order to misalign acquired traces. At the software level, some obvious countermeasures involve inserting redundant instructions, removing data-dependent instructions such as loops or jumps, and randomly changing the order of operations. System level countermeasures include techniques such as detached power supplies (Shamir 2000). Suitable countermeasures generally depend upon the specific embedded platform of interest. For example preloading all tables into cache and making crypto functions un-interruptable may be good approaches in platforms where on-chip data processing is much more resistant to side channel analysis than off-chip data transfers. Three examples of countermeasures are provided in this chapter. The first example is independent of the algorithm and involves the insertion of random delays in order to misalign traces to thwart differential analysis. This is presented in Sect. 9.1. The second countermeasure is masking which attempts to decorrelate the data within the cryptographic algorithm and is discussed in Sect. 9.2. The final countermeasure to be discussed is at the algorithmic level, where components of the algorithm are modified with redundant operations in order to minimize leakage of the side channel to be discussed in Sect. 9.3.

9.1 Misalignment Countermeasures

Countermeasures developed to thwart differential analysis include inserting random delays. This may be achieved using a random number of NOPs at different points in the cryptographic algorithm. Other approaches include for example the use of random clocks, randomly inserting operations, or using more than one differently structured algorithm. Apart from frequency-based analysis and preprocessing with alignment techniques, time domain differential analysis, correlation analysis, and most other attacks require "perfect" alignment. This section will discuss the impact of misaligned traces on the attack.

In this experiment, the load of the S-box or attack at the output of the S-box within AES is used to illustrate the misalignment problem. Unlike the large misalignments naturally occurring in traces obtained from the PDA in Chap. 8, here we will provide an experiment illustrating the impact of even small delays. In this experiment, the access of data at the output of the S-box was delayed according to the last 2 bits of the data accessed. For example the number of inserted NOPs was 12, 9, 6, and 3 for the cases where the last 2 bits of data were 00, 01, 10, and 11. This resulted in 744 traces with each type of multi-NOP delay for a total of 2,976 traces. Using the experimental setup shown in Fig. 8.5, both power and EM traces were acquired in order to run this side channel analysis. The differential trace of a

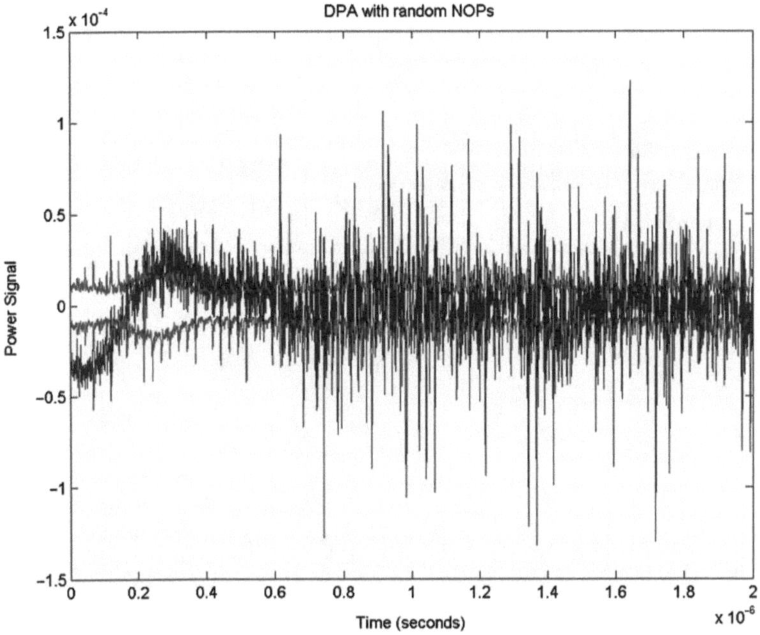

Fig. 9.1 Impact of random NOPs insertion on differential trace of a correct key guess

correct key guess is shown in Figs. 9.1 and 9.3 using power and EM measurements, respectively. These differential traces show lots of noise unlike aligned differential traces shown in Chap. 8. The all keys guess (e.g., plot of differential height vs. key guess, refer to Chap. 8) using DPA and DEMA techniques are shown in Figs. 9.2a and 9.4a. Neither attack was successful in determining the correct key.

Differential frequency-based analysis (DFA, see Chap. 8) is performed on these misaligned traces. This approach is able to find the correct key using the power traces and found the correct key in the top six keys using the EM acquisitions as shown in Figs. 9.2b and 9.4b respectively. Random delays have been reported as a countermeasure in SIM cards and other smart cards. However it is important to note that complex embedded systems will naturally have misaligned traces due to interrupts, cache misses, and other types of system activity. For example the correct and incorrect key differentials in the frequency domain are shown in Fig. 9.5a, b for the unaligned EM acquisitions acquired from a PDA device. The trace was divided into ten time segments (windows) and DFA was applied in each of these segments (Gebotys et al. 2005). This technique is referred to as differential power spectral analysis (DSA). Instead of performing differential frequency on the entire time interval acquired on the scope as described in Chap. 8, the time interval is divided into segments. The differential frequency analysis is generally applied separately to each segment. An additional difference from the DFA is that a Hamming window is applied. This is used to suppress the Gibbs' phenomena in spectral windowing. This variation of DFA can be used when there are small misalignments

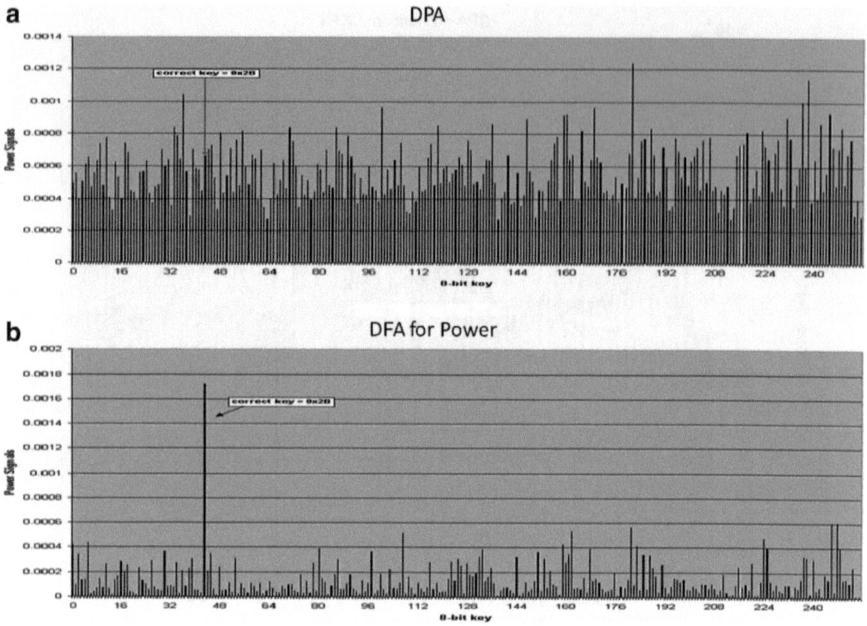

Fig. 9.2 Impact of insertion of random NOPs on all keys guess DPA in (**a**) and DFA in (**b**)

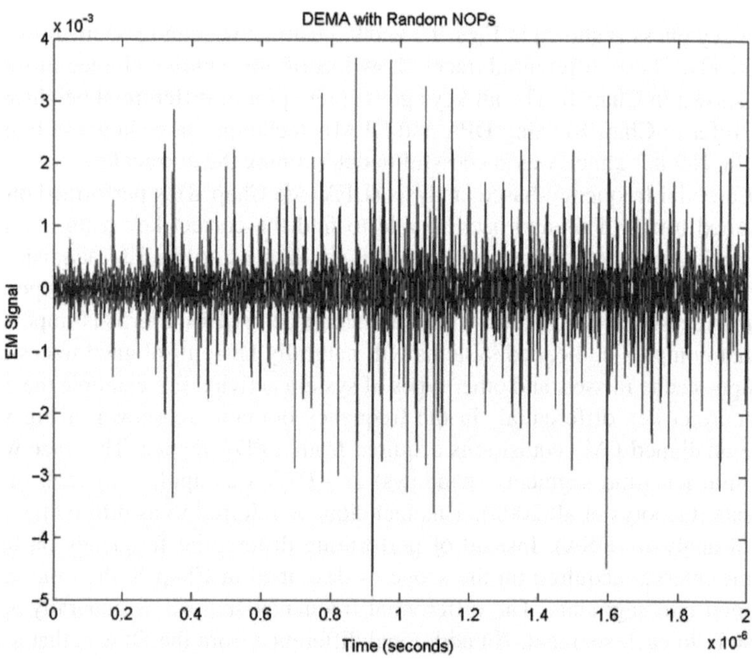

Fig. 9.3 Impact of insertion of random NOPs on correct key guess (**a**) DPA results and (**b**) DFA using power traces

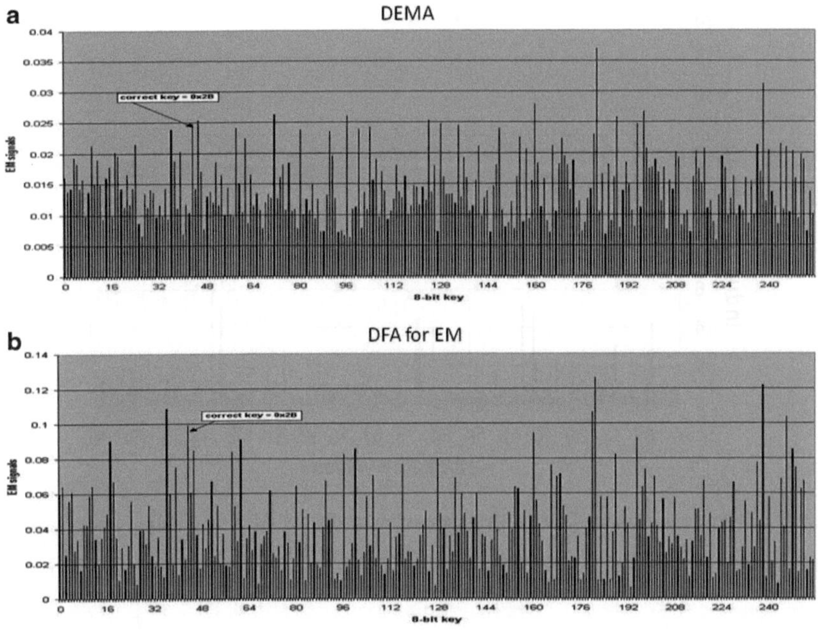

Fig. 9.4 Impact of insertion of random NOPs on all keys guess (**a**) DEMA results and (**b**) DFA using EM traces

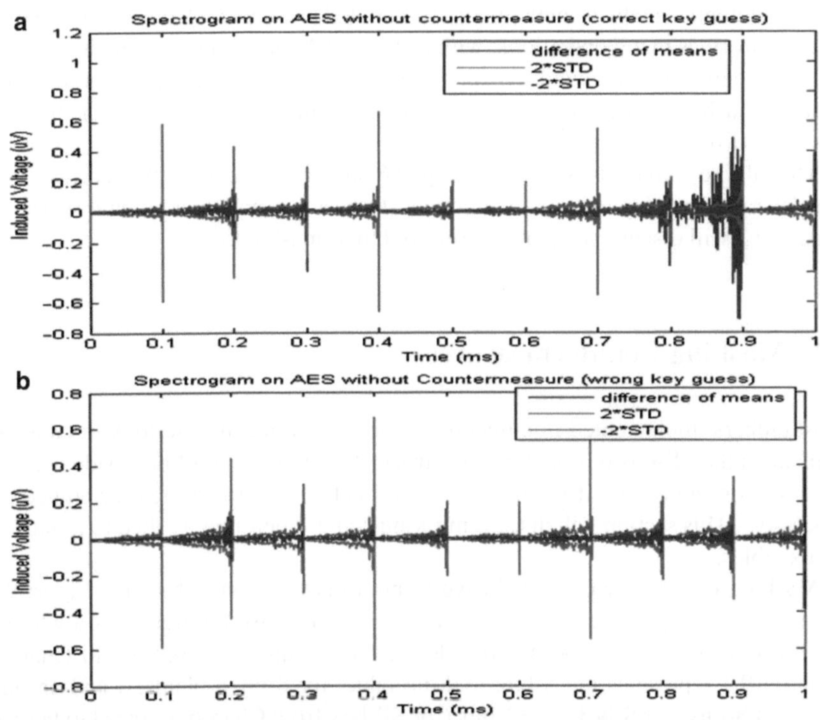

Fig. 9.5 DSA differential trace for correct key in (**a**) and wrong key in (**b**)

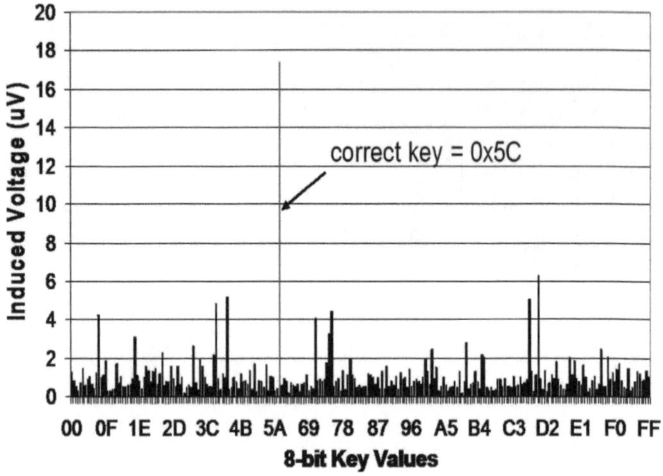

Fig. 9.6 All keys guess using DSA with standard deviations for AES on a PDA

and the time window contains more than a few different EM "activities." For example, if the scope is centered around one EM activity, such as a memory accesses, then analysis on the entire time frame will usually work well using the frequency-based differential analysis mathematics (illustrated in Chap. 8). However, if the time interval involves many operations, with differing EM activities, then it should generally be segmented into windows applying a differential spectral analysis (Gebotys et al. 2005a, b). It should be noted that the red (lighter) lines in Figs. 9.1, 9.3, 9.5 represent two standard deviations.

The all keys plot is shown in Fig. 9.6 clearly indicating the correct key. Besides random delay insertion techniques, there are many other countermeasures. Section 9.2 will discuss an approach referred to as masking.

9.2 Masking Countermeasures

Most ciphers implemented in embedded processors typically use lookup tables to implement their S-box or nonlinear constructs. Since the focus of most side channel attacks is on the S-box, it is important to provide countermeasures for embedded processors. This section will discuss masking countermeasures applied to lookup or S-box tables.

Masking of data involves exclusive-oring a mask, n, with the data. Figure 9.7 illustrates masking at the input of the S-box, which is also referred to as a whitening process (Messerges 2000). Consider the S-box attacks described in Chap. 8. Equation 9.1 represents the S-box look up table, for 8-bit p plaintext and an output, y, of 8-bits (for S-box) or 32-bits (for SP-box from Chap. 6, referred to here as

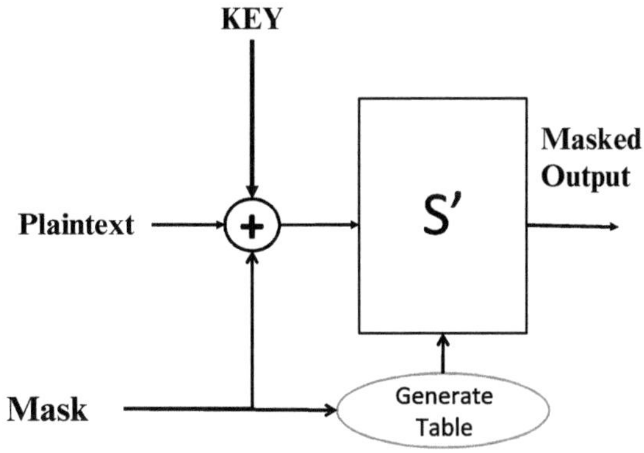

Fig. 9.7 Random masking countermeasure requiring table regeneration

S-box). Now if we mask the inputs and outputs of the table, with values n and m, respectively, as shown in (9.2), we have permuted the data in the tables storing them at new addresses and masked their outputs with one fixed output mask, m. The input data, key, and output of the S-box are p, k, y, respectively, in (9.1) and (9.2). We will refer to the S-box table as S when it is masked and S otherwise. Equation 9.1 is the normal S-box operation, whereas (9.2) uses a masked S-box, S, where a fixed input mask, n, and a fixed output mask, m, is utilized.

$$S(p \oplus k) = y \tag{9.1}$$
$$S'(p \oplus k \oplus n) = m \oplus y \tag{9.2}$$

Consider an attack at the input of the S-box. Recall from Chap. 8 that the differential spike at the input to the S-box (or output of exclusive-or operation in theory) is compared with the differential spike of loading the plaintext. If they have the same orientation then the key bit was a "0" otherwise the key bit was a "1." However, in the masking of Fig. 9.7, an attacker can no longer determine the key bits without the knowledge of the mask. Hence for a constant mask, the attack at the input to the S-box has been made harder. It is possible that hamming weights of the mask may reveal information about the mask. To make this countermeasure more secure, the masks should change randomly as shown in (9.3), where masks are a function of time m_t and n_t for time t.

$$S'(p \oplus k \oplus n_t) = m_t \oplus y \quad \text{for time } t. \tag{9.3}$$

Recall from Chap. 8 that in a differential attack a partitioning bit is chosen at the output of the S-box and computations on partitions of traces are performed for each key guess. With a constant random mask n in (9.2), the only difference is that the partitioned groups would be switched if the bit was flipped by the mask. However, since the differential attack computes the difference of means of groups, the masking has no impact. For example group "0" would contain all traces whose bit was a "0"

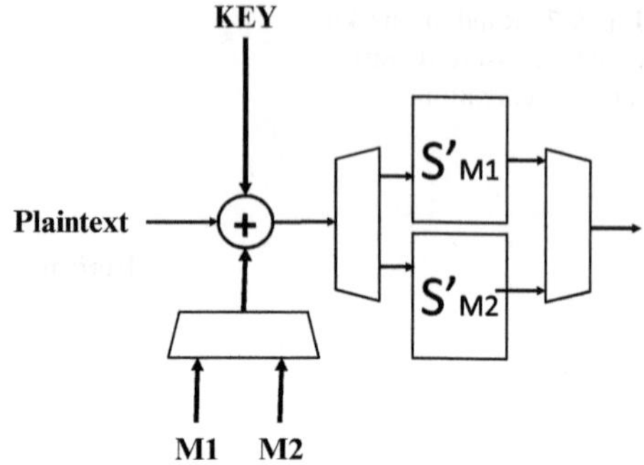

Fig. 9.8 Dual masking countermeasure using precomputed masked tables

if the bit position of mask n was a "0" or "1" if the mask bit position was a "1" and vica versa for the other group. Hence using a fixed constant mask as in (9.2), an attack at the output of the S-box is still viable.

If the input and output masks of the S-box changed randomly in value (Messerges 2000) as indicated in (9.3), then some traces placed in group "0" would have bit value "1" (when the current masking bit was a "1") and other traces in this same partition would have bit value "0" (when the current masking bit was a "0"). The same would happen for group "1." If the value of the mask changed randomly over time, in theory there would be an equal number of traces with a bit value of "1" and a bit value of "0" in each group. Thus the differential analysis, which relies upon all traces in each group to have the same bit value, would fail. Unfortunately changing the mask for each trace involves a very large overhead in time or storage. For example, every new random mask would require the S-box table to be recomputed, as shown in Fig. 9.7. Alternatively the masked S' tables could be precomputed (Itoh et al. 2002) as shown in Fig. 9.8 for two masks. This would require x masked S-boxes for x different masks, which may require significant amounts of memory. Thus this approach may not be suitable for memory-constrained embedded devices. Some researchers have suggested using only two masks with precomputed tables (Itoh 2002) as shown in Fig. 9.8. Alternative approaches to protecting the S-box are discussed in Sect. 9.2.1 along with a description of higher order attacks.

9.2.1 S-Box Output Masking

An alternative to the overheads of table recomputation and table regeneration is the table masking countermeasure (Gebotys 2006a). This table masking countermeasure stores randomly masked S-box data in a masked S-box table, S'. However, unlike the masking in (9.2), each addressed data in the table uses a different random mask. Specifically (9.4) illustrates the new masked S-box table, where the table output and input masks are represented by the function $r()$ and n.

$$S'(p \oplus k \oplus n) = r(p \oplus k) \oplus S(p \oplus k), \tag{9.4}$$

$$M(p \oplus k \oplus n) = r(p \oplus k) \oplus m. \tag{9.5}$$

The function $r(x)$ represents random data which is different for each table address (x). A second corresponding table (called the mask table, M) is used to store a corresponding mask for each address. The definition of the mask table is shown in (9.5) such that the exclusive-or of the masked S-box table and the mask table is data masked with a fixed value or $S'(x \oplus n) \oplus M(x \oplus n) = S(x) \oplus r(x) \oplus r(x) \oplus m = S(x) \oplus m$, for all x. The final mask, m, is split into two values, $r(x)$ and $M(x)$, and $S(x) \oplus m$ alone is never computed in the implementation in AES. Even if $S(x) \oplus m$ is computed, there may still be security with this counter measure since the output of an exclusive-or operation may not be attackable (such as the case on embedded 32-bit processors) and the memory accesses are masked, specifically the memory access of S and M. Also the subsequent exclusive-or operation $(S' \oplus M)$ is typically not attackable as discussed in Sect. 8.4 (unless strong invasive attacks using a decapsulated chip are launched). Hence this countermeasure protects the output of the S-box using 256 different masks (one per address of S-box).

The cryptographic algorithm, once incorporated with the countermeasure, does not need to be changed even when new masked tables are computed. The masking of tables is performed before the cryptographic tables are downloaded to the device, hence an attacker does not have access to m (nor can the attacker obtain the Hamming weight of m from power measurements). Since tables are generated only once, this value m along with the mask of the round keys is used to precompute tables before the cryptographic tables are downloaded to the device. Figure 9.9 illustrates the computations performed on the data accessed in AES from the mask tables [M in (a) or $M1, \ldots, M4$ in (b)] before they are merged with the data accessed from the masked S-box tables. The light box indicates the original S-box computations required with no countermeasure assuming $S' = S$.

For example, recall from Chap. 6, in the table method of AES (used for fast implementation on 32-bit processors), all data accessed from the S-box tables are exclusive-ored together and then exclusive-ored with the round key to produce a value $t0$ of (9.6). Using the table masking countermeasure this same computation is performed as shown in (9.7) to produce $t0a$. Next all corresponding data accessed from the mask table are exclusive-ored together to produce a value $t0b$. Finally these two values are exclusive-ored together. This is shown in Fig. 9.9a where one mask table, M is used.

$$t0 = S1\left(\{S0\}_{b3}\right) \oplus S2\left(\{S1\}_{b2}\right) \oplus S3\left(\{S2\}_{b1}\right) \oplus S4\left(\{S3\}_{b0}\right) \oplus rk_i, \tag{9.6}$$

($where\{w\}_b$ refers to byte b of the 32-bit word w).

$$\begin{aligned} t0a &= S1'\left(\{s0 \oplus n\}_{b3}\right) \oplus S2'\left(\{s1 \oplus n\}_{b2}\right) \oplus S3'\left(\{s2 \oplus n\}_{b1}\right) \\ &\quad \oplus S4'\left(\{s3 \oplus n\}_{b0}\right) \oplus (rk_i \oplus n) \\ t0b &= M\left(\{s0 \oplus n\}_{b3}\right) \oplus M\left(\{s1 \oplus n\}_{b2}\right) \oplus M\left(\{s2 \oplus n\}_{b1}\right) \oplus M\left(\{s3 \oplus n\}_{b0}\right) \\ t0 &= t0a \oplus t0b \end{aligned} \tag{9.7}$$

The value $t0$ is input to the next set of tables, which may also have masked inputs. Alternatively the computation would follow (9.8) using four mask tables, $M1$, $M2$, $M3$, $M4$. Figure 9.9b illustrates the computations required in the scheme for AES, where $S1$, $S2$, $S3$, $S4$ are the four masked S-box tables and $M1$, $M2$, $M3$, $M4$ are the mask tables.

With one mask table only ($M1 = M2 = M3 = M4$) as shown in Fig. 9.9a, if tables are generated with one value of m, then $t0$ will be unmasked (since the four m's cancel) or masked with the round key mask which would be n, ready for input to the masked tables.

In order to produce a masked value at the output of the masked tables independent of the round key mask, at least one or all of the masked S-box tables can be defined as $S_j(x) = S_j(x) \oplus r(x) \oplus m_j$, where m_j is a random mask for table j. In this approach the final mask on $t0$ would be $m_1 \oplus m_2 \oplus m_3 \oplus m_4$ exclusive-ored with the round key mask creating a final mask of n, in AES since there are four S-box tables ($j = 1, \ldots, 4$), the computations would follow (9.8).

$$\begin{aligned} t0a &= S1'(\{s0 \oplus n\}_{b3}) \oplus S2'(\{s1 \oplus n\}_{b2}) \oplus S3'(\{s2 \oplus n\}_{b1}) \\ &\quad \oplus S4'(\{s3 \oplus n\}_{bo}) \oplus (rk_i \oplus \text{mask}) \\ t0b &= M1(\{s0 \oplus n\}_{b3}) \oplus M2(\{s1 \oplus n\}_{b2}) \oplus M3(\{s2 \oplus n\}_{b1}) \\ &\quad \oplus M4(\{s3 \oplus n\}_{bo}) \\ t0 &= t0a \oplus t0b \\ &\text{where } n = m_1 \oplus m_2 \oplus m_3 \oplus m_4 \oplus \text{mask} \end{aligned} \tag{9.8}$$

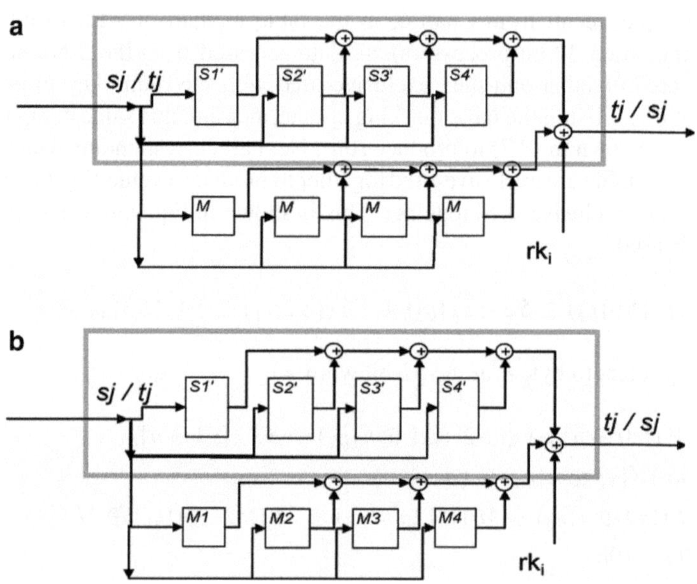

Fig. 9.9 Multiple masked table countermeasure in AES with one table in (**a**) or four tables in (**b**)

If the set of $r(x)$ is the same set used for all S-boxes then only one $M(x)$ table needs to be stored. However, if storage is not an issue in the embedded system, a different set of $r(x)$ for each S-box could be used, to increase the number of different random masks. For example a separate mask table, $M_i(x)$, for each S-box, $S_i(x)$, could be implemented. Furthermore dynamic updating of random masks $r_i(x)$ is easily supported in this scheme. For example after access of data from both M_i (x) and $S_i(x)$, a remasking of these two data values can be computed by exclusive-oring them with another new random value. Then these newly masked values can be stored back into the original $M_i(x)$ and $S_i(x)$ tables. This procedure for dynamically updating or refreshing random masks (during the en-/decryption) is shown in Fig. 9.10a. This

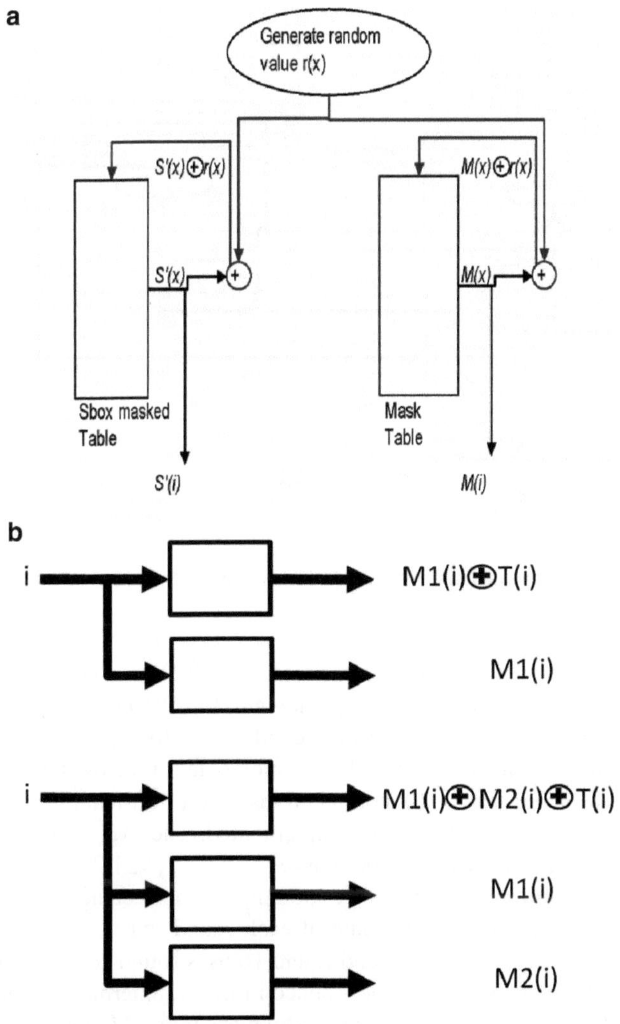

Fig. 9.10 Remasking at output of S-box and M table (**a**) and extension to split masks in (**b**)

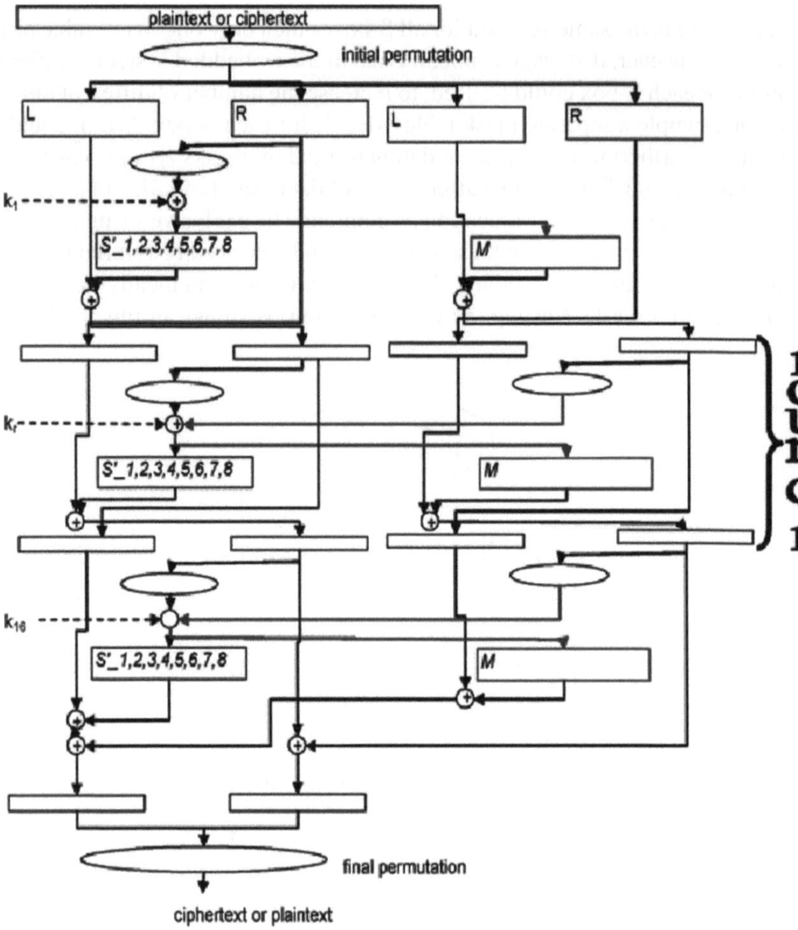

Fig. 9.11 Table masking countermeasure in *dashed box* for DES

countermeasure using one or more mask tables, $M(x)$, can be extended to more split masks at the output of the S-box as shown in Fig. 9.10b.

A similar implementation for DES and other cryptographic algorithms which use look up tables is also possible. The implementation of the table masking for DES is shown in Fig. 9.11. The computations in the left-hand column represent normal DES with no countermeasure implemented. The oval shapes represent the expansion-permutations in DES. The masked S-boxes ($S\_1,2,3,4,5,6,7,8$) provide both S-box substitutions and P-box permutations. The rectangular boxes are just used to illustrate input and output data of each round (where round 1 uses round key k_1, round r uses round key k_r and round 16 uses round key k_{16}, shown on the left). In DES there are 8 S-box tables, hence in this countermeasure eight masked S-box tables ($S\_1,2,3,4,5,6,7,8$) and one extra mask table, M, are used. Initially for round 1, it is assumed that the round key k_1 is masked (where the mask is equivalent to the input table mask of the masked S-box tables, which is our fixed mask m). The initial L and R boxes of the right-hand column are filled with zeros initially. Subsequent rounds repeat the operations shown in round r.

9.2.2 Evaluating Countermeasures and Higher Order Analysis

Security analysis should be performed on any countermeasure. The analysis is typically based upon assumptions of the side channel analysis model. Real empirical support for the countermeasure is also very important. Experimental results should not only verify that a differential attack is thwarted but also verify that the countermeasure is strong enough that a higher order differential analysis is very difficult to perform due to the large number of side channel traces required. How many traces should be required? In high security devices, the answer to this question would be a number greater than that defining the cryptoperiod of the key. Specifically if a successful higher order analysis requires n traces, then the cryptoperiod of the key is a period of time defined by at most $(n - m)$ uses of the encryption algorithm, where $m \gg 0$. This section will discuss and demonstrate the verification of the countermeasure described in Sect. 9.2.1.

Differential analysis defined in Chap. 8 is also called a first-order differential analysis. This is largely because only one "sample" in each trace is needed at the attack point. Using the table masking countermeasure (Sect. 9.2.1), where masks randomly change, a first-order fixed plaintext (e.g. attacker cannot control the plaintexts) differential attack is generally thwarted. A more difficult attack referred to as a second-order differential attack (Messerges 2000) may be launched. It is more difficult for two reasons. The first is that two samples are required, one is the sample of the mask and the other is the sample of the masked data. This is difficult since typically attackers do not know the location of this data, hence all samples must be considered. The second reason is that many more traces are required in order to launch this second-order attack.

Countermeasures should also be evaluated for chosen plaintext attacks, where unlike the fixed plaintext attack, the attacker can control the plaintexts. In this case the attacker may use plaintexts which modify only one byte of the 128-bit input. In the table method of AES, only one SP-box output of round 1 (which corresponds to the input plaintext byte which is changing) will have different outputs for each different plaintext. Thus the table masking countermeasure will end up with an effective constant mask at the input of SP-boxes of round 2, or effectively the output of the exclusive-or (*t0* in (9.6) or (9.7)). This enables a first order attack (Gebotys 2006) at the output of the exclusive-or operation. Countermeasures should also be analyzed for ciphertext attacks. For example an attacker may be able to record the ciphertexts and launch an attack at the input of the final S-box (specifically the state inputs to *Te4* in Fig. 8.23) used in the last round of the cipher. This attack may be viable on some platforms using this table masking countermeasure if the output of exclusive-or operations are attackable such as in (Coron et al. 2009) where an 8-bit processor Atmel ATmega16 was utilized. As previously discussed in Sect. 8.4, these attacks of the exclusive-or operation are typically very difficult on 32-bit embedded processors and so too are attacks of the correlated 8-bit address input to the SP-boxes which have 32-bit outputs. For example attacks on the output of exclusive-or operations were not possible in previously examined 32-bit embedded processors and PDAs (Gebotys 2006). In the table masking countermeasure, a second-order differential analysis can be thwarted by using more than one mask table [for example using $M_1(x)$ and $M_2(x)$, in place of $M(x)$, and $m = r(x) \oplus M_1(x) \oplus M_2(x)$]. Thus, n mask tables can be created by further splitting the masks. The security of this countermeasure scales with the number of tables. For n mask tables (M_1, $M_2, \ldots, M_n$) and one set of masked S-box tables, (S), a (n)th order DEMA attack is thwarted [where $m = r(x) \oplus M_1(x) \oplus M_2(x) \oplus \cdots \oplus M_n(x)$, for all x, thus splitting mask m into $n + 1$ masks]. For example in AES, a second-order DEMA can be thwarted with a storage overhead of two extra mask tables (M_1, M_2 where $m = r(x) \oplus M_1(x) \oplus M_2(x)$, for all x). Given one set of S-box tables accounts for 4 KB, the masked S-box and mask table memory requirements for thwarting a second-order DEMA would increase to 12 KB, which should still be practical for a small wireless device such as a PDA, where typically over 128 MB is available. The higher order attack typically will require many more EM traces and thus provides an increase in difficulty of launching the attack. Note that once again, even with n tables, to avoid a nth order DEMA, the exclusive-or of the masked S-box table and mask table, specifically $S(x) \oplus M_1(x) \oplus M_2(x) \oplus \cdots \oplus M_n(x) = m \oplus S(x)$ is never computed.

Appendix A presents two attack statistics, one where m is known and the other for m unknown. Furthermore if m changes for different iterations of the encryption algorithm, then the statistic presented in Appendix A does not work. In fact it is not clear how to formulate a statistic for this case, given that each trace may have been generated with a different value of m. The value of m can be changed after a large number of algorithm executions (e.g., a number less than the number of traces required for a successful second-order DEMA), by creating a random number, r, and exclusive-oring the mask table with this value ($M(x) \oplus = r$, for all x).

If an attacker has access to the masked tables (for example by dumping the memory contents of the device) and m does not change, it may be possible to derive the fixed values in the masked S-boxes and then perform a first-order attack to obtain the key (assuming the key is not stored in the accessible memory). For example, if the attacker exclusive-or's the masked S-box table with the mask table(s), the resultant value, $M(x) \oplus S(x) = m \oplus S(x)$, is still masked with m, unlike (18) where access to the tables immediately reveals the location of the S-box values and random values through exclusive-oring [where $M(x) \oplus S(x) = S(x)$]. Next the attacker must exclusive-or two values together, $(m \oplus S(x)) \oplus (m \oplus S(y)) = S(x) \oplus S(y)$, for two values of y, and compare these with precomputed pairwise exclusive-ors generated by the attacker with the known S-box values. When both matches occur [i.e., $S(x) \oplus S(y)$ and $S(x) \oplus S(z)$] the value of $S(x)$ is determined in the $S(x)$ masked S-box table. With this knowledge, the value of m can be determined and finally used to locate the remaining S-box values in the masked S-box. Given the location and value of masked S-box values, a first-order DEMA can be launched to obtain the secret key. However, if m changes, as described in the previous paragraph, then even this attack is thwarted.

The countermeasure is demonstrated using the ARM7TDMI processor evaluation board setup in Chap. 8. The AES cipher was attacked using a total of 2,976 EM traces. The first-order DEMA without any countermeasure is shown in Fig. 9.12a. The correct key is clearly evident. In Fig. 9.12b the table masking countermeasure has been implemented. The key has not been found hence the countermeasure has successfully resisted the DEMA attack.

A second-order differential EM analysis was also performed on AES, implemented with the proposed countermeasure, in order to see if the correct guessed key would be apparent. In the proposed countermeasure, the S-box was transformed into a masked S-box and one mask table. In the analysis, three acquisitions of the same set of plaintext and same set of masked S-box tables and mask tables were performed. Each acquisition produced 6,000 EM traces, 3,000 were used to obtain samples of the masked S-box and the other 3,000 were used to obtain the corresponding samples of the mask tables. In this experiment we assume that the attacker knows the location of samples of power or EM to be used to launch a higher order analysis. Different bits at the output of the masked S-box load were used to perform the partitioning. Unless otherwise stated, the probability difference, $\Lambda_{m=0}$ or $\Lambda_{m=1}$ (see Appendix A), was used to illustrate the results when the value of m was known to the attacker (known m). The all keys guess analysis for known m is shown in

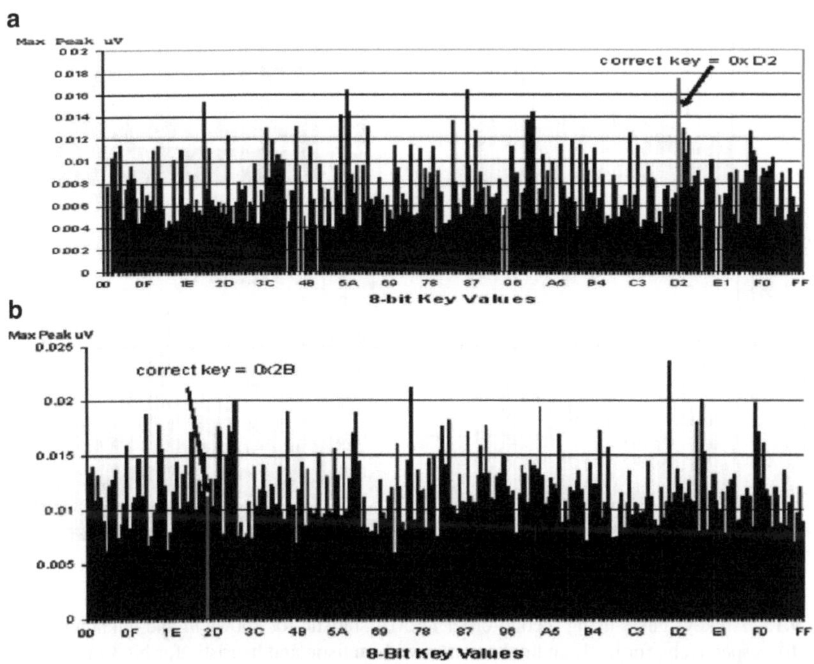

Fig. 9.12 DEMA without (**a**) and with the masking table countermeasure (**b**) at output of S-box

Fig. 9.13a using the least significant bit at the output of the S-box table for partitioning. In all cases, the correct key is at 00 (the correct key has all bits equal to zero, so if the analysis is successful, then the highest peak in the bar chart would be at the far left at 00, identifying the correct key). The absolute probability difference or statistic Λ was used to illustrate an attack when the value of m is unknown (*unknown m*). Assuming the value of m is unknown, the all keys guess analysis is shown in Fig. 9.13b for the same set of EM traces and the least significant bit as a partitioning bit. Figure 9.13c also illustrates the successful results, using the same set of EM traces with unknown m, but with bit 3 as the partitioning bit. The ad hoc second-order DPA bias statistic (Messerges 2000), also called the heuristic in Appendix A, was also used to plot the all keys guess and is shown in Fig. 9.13d for the same set of EM traces, also using bit 3 to partition. The absolute probability difference or statistic Λ (Fig. 9.13c) identified the correct key, whereas the heuristic (Fig. 9.13d) did not identify the correct key with the highest peak (since the correct key was the second highest peak). Using a total of 1,400 EM traces, the all key guess analysis with unknown m was performed many times, each using a different partitioning bit. Only 8 out of 32 partitioning bits revealed the correct key. When the total number of EM traces was increased to 2,800 and 8,000, the number of partitioning

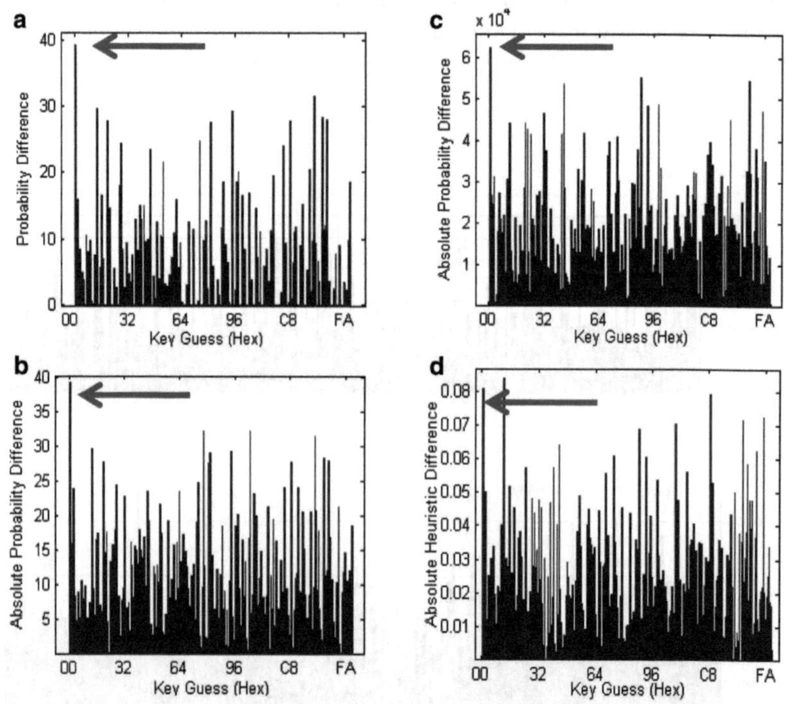

Fig. 9.13 All keys guess using second-order DEMA statistic for known m and unknown m in (**a**) and (**b**), respectively, for bit 1, and unknown m with statistic and heuristic for bit 3 in (**c**) and (**d**), respectively

bits which revealed the correct key increased to 10 and 13 out of 32, respectively. However, the analysis showed that the key most frequently identified as having the highest peak (over all bit positions) was the correct key. Analysis using dynamically changing values of m, with over 5,600 EM traces, using the absolute probability difference or statistic Λ, revealed the correct key in only 2 out of 32 analyses.

The same two sets of EM traces were also used to analyze the convergence time for a second-order DEMA. In this experiment, the analysis is performed to determine the value of m [analogous to the attack on the whitening process or key-xoring countermeasure (Messerges 2000), except the analysis is done at the output of the S-box and unlike (Messerges 2000) where the key is determined, here we determine the value of m]. The y-axis of the graphs in Fig. 9.14 represents the number of bits of m which were incorrectly predicted using the statistic in Appendix A. Again a different set of acquisitions was also run with different random masks, and the results are shown in Fig. 9.14b. In Fig. 9.14a all 31 bits out of 32 bits of m (at the output of the S-box) were determined by the second-order DEMA. In Fig. 9.14b only 24 bits out of 32 bits of m are correctly determined using this analysis. In Fig. 9.15 over 8,000 EM traces were involved in the computations and Fig. 9.14a uses a different set of masked S-box and mask tables than Fig. 9.14b.

A third-order DEMA (see Appendix B) was also performed using the table masking countermeasure. In this experiment, approximately 6,000 EM traces using different random masks were acquired. Some results using unknown m are shown in Fig. 9.15a, b, and c with same EM data but using bit 1, 2, and 3, respectively, as a partitioning bit. In all cases, the correct key is at 00 (the correct key has all bits equal to zero, so if the analysis is successful, then the highest peak in the bar chart would be at the far left at 00, identifying the correct key). Figure 9.15a correctly identifies the correct key, however, Fig. 9.15b, c do not. Figure 9.15c is close, since the correct key is the fourth highest peak. Using over 5,000 EM traces, only 1 out of 32 analyses was able to find the correct key as the highest DEMA peaks. The histogram of these keys with the highest peaks resulting from the 32 analyses did not indicate the correct. In the second set of EM acquisitions, obtained from the same set of data, none of the analyses revealed the correct key.

Figure 9.16 illustrates the convergence time for a third-order differential EM attack on the value of m. In both cases, different random masks are used in addition to different acquisitions. However, in both cases 15 out of 32 bits of the value of m remain incorrectly determined by the analysis. Both figures, use the same value of m but each involve three different sets of acquisitions and different masked S-box tables and mask table, each having a total of 6,000 EM traces (since the x-axis reports the number of EM traces in each partition, for partition bit being a 0 and being a 1). It is interesting to note that the number of incorrect bits of m (y-axis) in Fig. 9.16 do not seem to decrease, but instead appear to converge at 15 incorrect bits.

Further security may be attained by making higher order differential analysis more difficult. For example consider a second-order differential analysis launched on the mask table and masked S-box table. It may be likely that the sample of the mask table and a second sample of the masked S-box table are readily available in the implementation, hence providing aid to the attack. A more secure

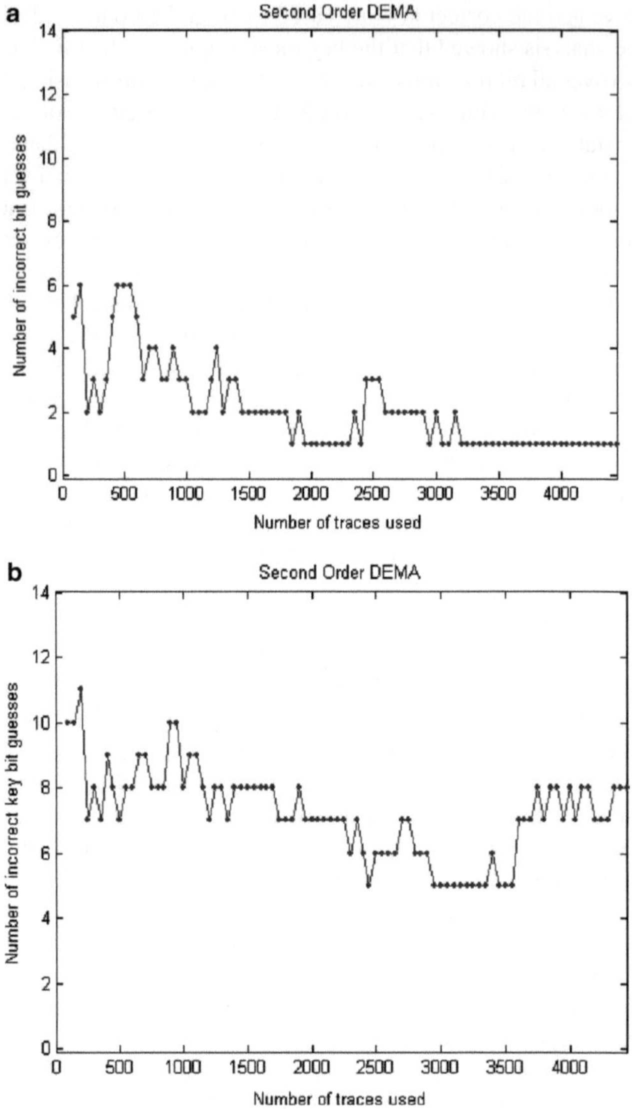

Fig. 9.14 Number of incorrect bits of m vs. number of EM traces using second-order DEMA using different random masks

implementation would be to apply table splitting. Previously we used two tables, S' and M, defined by (9.4). Table splitting also uses two tables which replace S' and M with $M1$ and $M2$. They are defined using (9.9) or alternatively using (9.10). In these equations the term $\{w\}_{x...y}$ represents *bit x* to *bit y* of w and $||$ represents bit concatenation. For example, the masked S-box table data be represented by a 32-bit quantity $S'(p \oplus k \oplus n) = \{S'\}_{31...0}$ and the associated mask table data

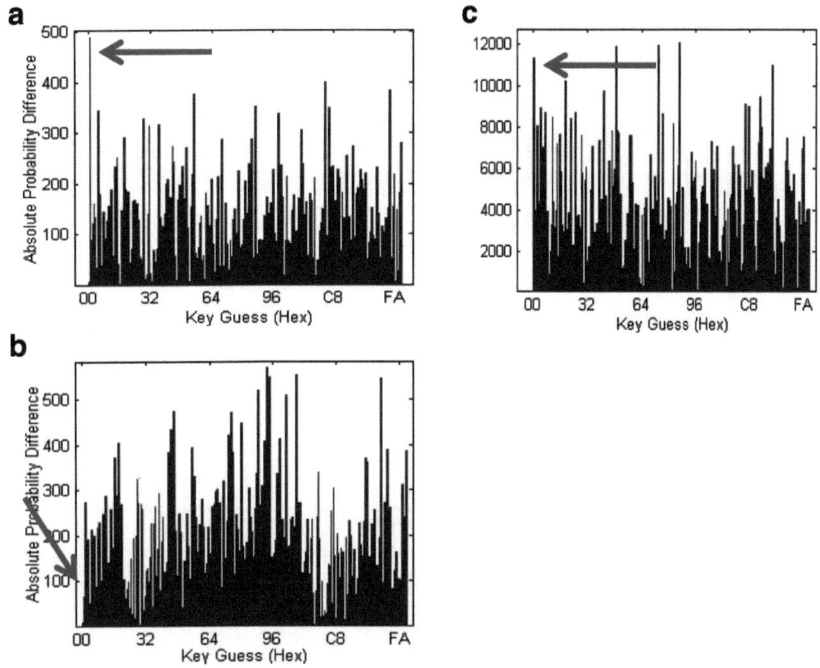

Fig. 9.15 The third-order DEMA, showing all keys guess for unknown m using bit 1, 2, 3 in (**a**), (**b**), (**c**) illustrating correct, incorrect, and incorrect key guesses, respectively

by a 32-bit value $M(p \oplus k \oplus n) = \{M\}_{31\ldots0}$. By splitting the tables, one table would access the half word of S' concatenated with the half word of M, for example $\{S'\}_{15\ldots0}\,\{M\}_{15\ldots0}$ or $\{S'\}_{31\ldots16}\,\{M\}_{150}$. In the first case, $\{S'\}_{15\ldots0}\,\{M\}_{15\ldots0}$, a second-order attack would have to be launched with only one sample per trace. In the second case, $\{S'\}_{31\ldots16}\,\{M\}_{15\ldots0}$, the sample of a share of data would be made noisy by the other bits containing masking information of another data share.

$$M1(p \oplus k \oplus n) = (\{S'(p \oplus k \oplus n)\}_{15\ldots0} \parallel \{M(p \oplus k \oplus n)\}_{15\ldots0}),$$
$$M2(p \oplus k \oplus n) = (\{S'(p \oplus k \oplus n)\}_{31\ldots16} \parallel \{M(p \oplus k \oplus n)\}_{31\ldots16}), \quad (9.9)$$
$$M1(p \oplus k \oplus n) = (\{S'(p \oplus k \oplus n)\}_{15\ldots0} \parallel \{M(p \oplus k \oplus n)\}_{31\ldots16}),$$
$$M2(p \oplus k \oplus n) = (\{S'(p \oplus k \oplus n)\}_{31\ldots16} \parallel \{M(p \oplus k \oplus n)\}_{15\ldots0}). \quad (9.10)$$

The energies of previously researched countermeasures (Messerges 2001; Itoh et al. 2002) were computed to compare with the proposed countermeasure energy. The evaluation board containing the 32-bit ARM7TDMI RISC processor core on one chip separate from the memory along with a current meter was used to obtain current measurements. All the energy measurements were obtained from a current meter measuring the current drawn from the supply pin of the chip core, while the processor was executing the algorithm in an infinite loop. These energies

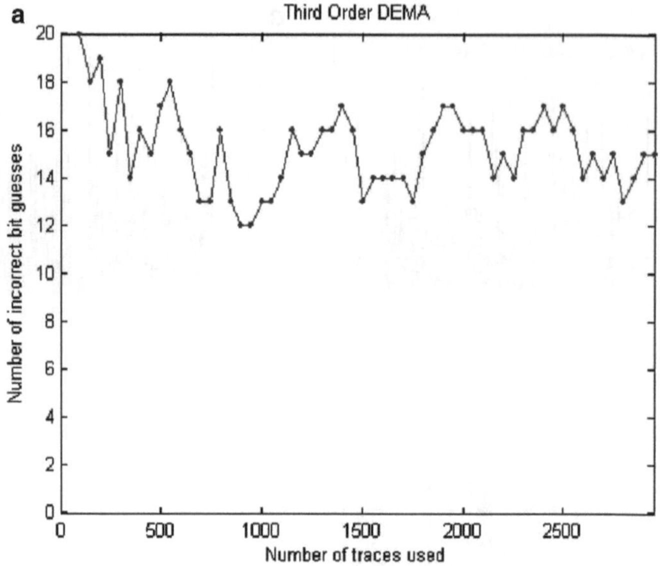

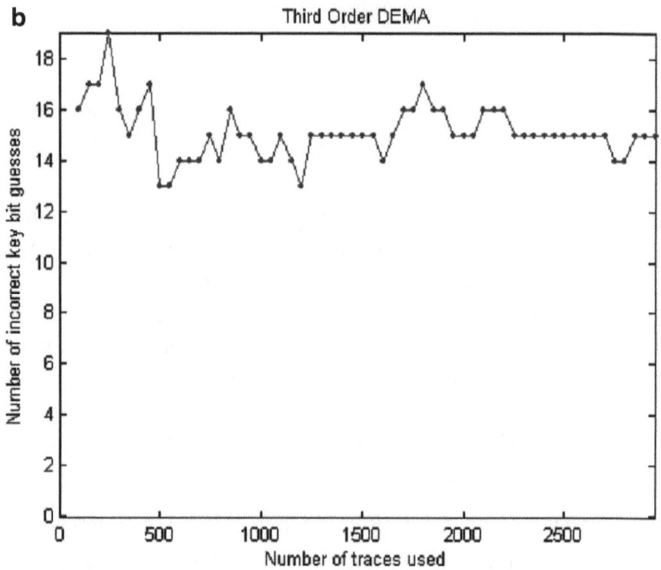

Fig. 9.16 The third-order (split mask) DEMA for determining the value of m using different sets of randomly generated masks and data acquisitions in (**a**) and (**b**)

reflected the processor core energy consumption only and not the input/output buffer power or memory power. Several different countermeasures were implemented in the ARM7TDMI in order to compare energies as shown in Table 9.1. The proposed countermeasure using only one additional table (thwarting a first-order DEMA) is

Table 9.1 Comparison of Processor Energy

	AES	M	M_1, M_2	Reference: Messerges (2001)
$\|$S-box$\|$	5	6	7	5
$\|ld/st\|$	160	320	480	2,048
I (mA)	4.46	4.41	4.40	4.17
E_p (mJ)	0.33	0.57	0.81	2.92
E times	1	1.7	2.4	8.9

shown as M, whereas the proposed countermeasure using two extra tables (thwarting a second-order DEMA, thus providing increased security) is shown as M_1, M_2. The total number of 256 by 32-bit tables (denoted as $|S\text{-box}|$) used by the AES algorithm, the number of memory accesses ($|ld/st|$ for *load/store*) and the average current measured from the processor while it was executing the AES ($I(mA)$) algorithm is illustrated in Table 9.1. The original AES algorithm, optimized for ARM7TDMI is shown in column 2 as *AES*. Finally the energies (current multiplied by the supply voltage and latency of each implementation) are given in the table in mJ (E_p). Even the two table implementation of the proposed countermeasure [for example using $M_1(x)$ and $M_2(x)$, in place of $M(x)$], $P2$, is more energy efficient than the countermeasure in Messerges (2001), where table regeneration is required in order to change the random mask. The proposed countermeasure (M) provided 1.7 times increase in energy (*E times*) over AES with no countermeasure (*AES*) and 5.2 times less energy than the countermeasure reported in Messerges (2001), where table regeneration is required when a new mask is applied.

Since Table 9.1 did not represent the energy dissipation of the memory and since it is well known that memory energy dissipation is significant and often dominates, an analysis was performed with SRAM models from Liao et al. (2002). A comparison with the previously researched countermeasure which stores one set of S-box tables per mask (Itoh et al. 2002) is given in Table 9.2 which illustrates the impact of memory size on power dissipation. The number of different masks supported by each countermeasure is shown as $|masks|$; the order of the DEMA which may be required to attack the countermeasure (*HODEMA*) and the dynamic power dissipation of each SRAM (P_m (mW)) which would hold all the tables used in the AES algorithm for each countermeasure are given. The proposed countermeasures (M and M_1, M_2) require significantly less memory when supporting the same number of masks. For example with (1/5)th the number of masks (last column), countermeasure in Itoh et al. (2002) dissipates up to 10.4 times more energy than the proposed countermeasure.

9.2.3 Address Masking of S-Box

As discussed earlier, by masking the input address to the S-box, one can thwart the previous fixed plaintext differential attack described in Chap. 8 at the input to

Table 9.2 Comparison of SRAM dynamic power

	AES	M	M_1, M_2	Reference: Itoh et al. (2002)	Reference: Itoh et al. (2002)
$\|masks\|$	0	256	256	2	50
HODEMA	1	2	3	1	1
P_m (mW)	2.2	4.9	7.9	3.2	51.4
P times	1	2.2	3.5	1.4	22.7

lookup tables. However, if an attacker gains information regarding the input mask, they could use this information to determine the key. Any bits learned about the mask will lead to bits of the key using differential analysis. Information concerning the mask may be obtained from the side channel traces, such as hamming weight, etc. However, there are more powerful techniques which the attacker can use to attempt to obtain the key at the masked input to the S-box. These techniques are referred to as higher order differential analysis. A countermeasure to thwart this type of attack involves increasing the number of mask values and changing their values. This section will describe this general approach and discuss an implementation in AES.

In Fig. 9.17a, the "table look-up" has a masked input. Whenever a new random mask is used, a new table must be regenerated. However, in Fig. 9.17b, a number of splitmasks are used (Gebotys 2006b). When new masking is required, a new set of split masks are generated, however, table regeneration is not required, since the exclusive-or of the set of random masks does not change. For example, Fig. 9.17b has three masks as opposed to Fig. 9.17a where only one mask is used. However, the same S-box can be used in both cases if the input masks to the S-box are the same, or specifically Mask $= M1 \oplus M2 \oplus M3$, where Mask is from (a) and $M1$, $M2$, $M3$ is from (b). This concept of splitting masks, where Mask in (a) is split into three masks in (b) can be generalized to splitting a mask into m shares. Let ni_t represent the random mask i used at time t, where the cryptographic application can be described as:

$$Output = masked\ table(plaintext \oplus n1_t \oplus \cdots \oplus nm_t \oplus key),$$

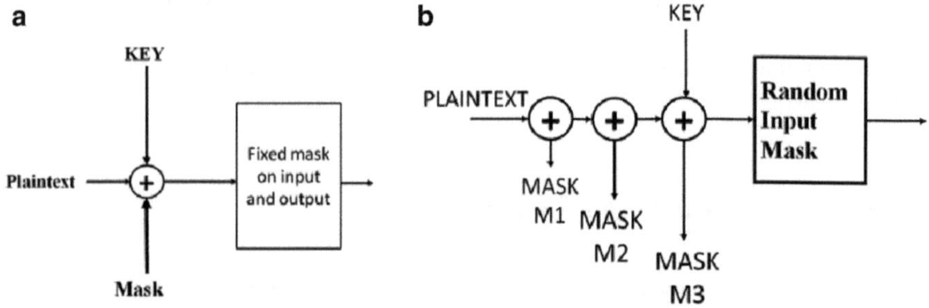

Fig. 9.17 General single mask in (**a**) and multiple masks in (**b**) at the input of the table

where

$$masked\ table(plaintext \oplus n1_t \oplus \cdots \oplus nm_t \oplus key) = table(plain\ text \oplus key).$$

After a new set of masks are generated, the cryptographic application can be described as:

$$Output = masked\ table(plaintext \oplus n1_t + 1 \oplus \cdots \oplus nm_t + 1 \oplus key),$$

where

$$n1_t \oplus \cdots \oplus nm_t = n1_t + 1 \oplus \cdots \oplus nm_{t+1}.$$

A second DPA (Messerges 2000) involves two power samples, specifically a power sample of the random mask and a power sample of the plaintext exclusive-ored with the key and random mask. An nth order DPA would involve n power samples, specifically a power sample of each $(n-1)$ random masks and a power sample of the plaintext exclusive-ored with the key and random masks. The implementation of the split-mask countermeasure (using two split masks) is shown in more detail in Fig. 9.18. The original secret or master key, *key*, is immediately masked after it is received or derived. Key mask m is exclusive-ored with the master key and the masked key, *mkey*, is stored (the unmasked master key is not stored or ever used again). Stored values are indicated in Fig. 9.18 using a black rectangle. The key mask, m, is a randomly selected fixed value and it is unchanged for different plaintexts. Key mask m is exclusive-ored with the two table input masks, $min1$, $min2$, and the result is also stored. The result is exclusive-ored with $r1$, where $r1$ is

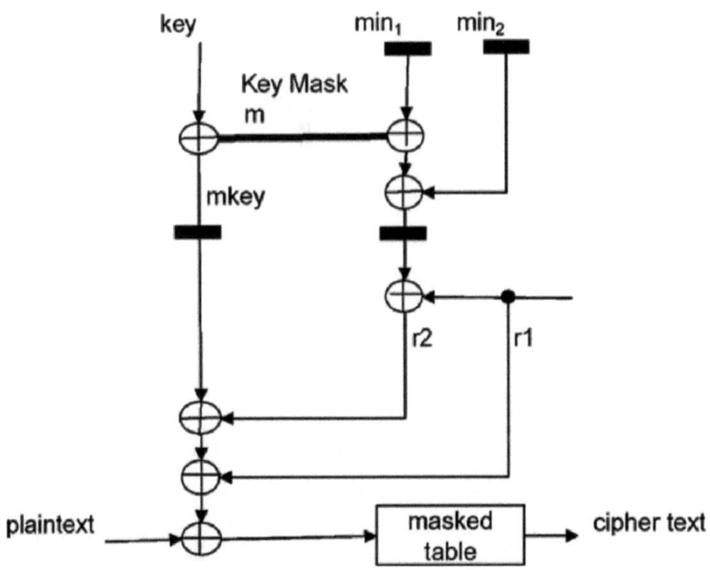

Fig. 9.18 Key, table, and split masks shown as m, $min1$, and $min2$ and $r1$ and $r2$

a random value generated for each new plaintext. The exclusive-or result is defined to be the value $r2$ (where mathematically $r2 = m \oplus min1 \oplus min2 \oplus r1$). The result of the masked key exclusive-ored with $r1$ and $r2$ is then exclusive-ored with the plaintext and input to the masked table.

Mathematically, the cryptographic application in Fig. 9.18 can be described as:

$$\begin{aligned} output &= masked\ table(plaintext \oplus r1 \oplus r2 \oplus mkey) \\ &= masked\ table(plaintext \oplus r1 \oplus (m \oplus min1 \oplus min2 \oplus r1) \oplus (m \oplus key)) \\ &= masked\ table(plaintext \oplus min1 \oplus min2 \oplus key). \end{aligned}$$

The masks whose names start with "*m*" (i.e., m, $min1$, and $min2$) are generated only once per key. The other masks whose names start with "*r*" (i.e., $r1$, $r2$) are randomly generated with every plaintext. The split masks $r1$ and $r2$ are exclusive-ored together to form a final fixed mask for the table input. These two masks can be randomized for each invocation of the cryptographic process. For example, one can generate a random $r1$ value and use the computation shown in Fig. 9.18 in order to generate a new $r1$, $r2$ pair. However, to avoid table regeneration, the final fixed

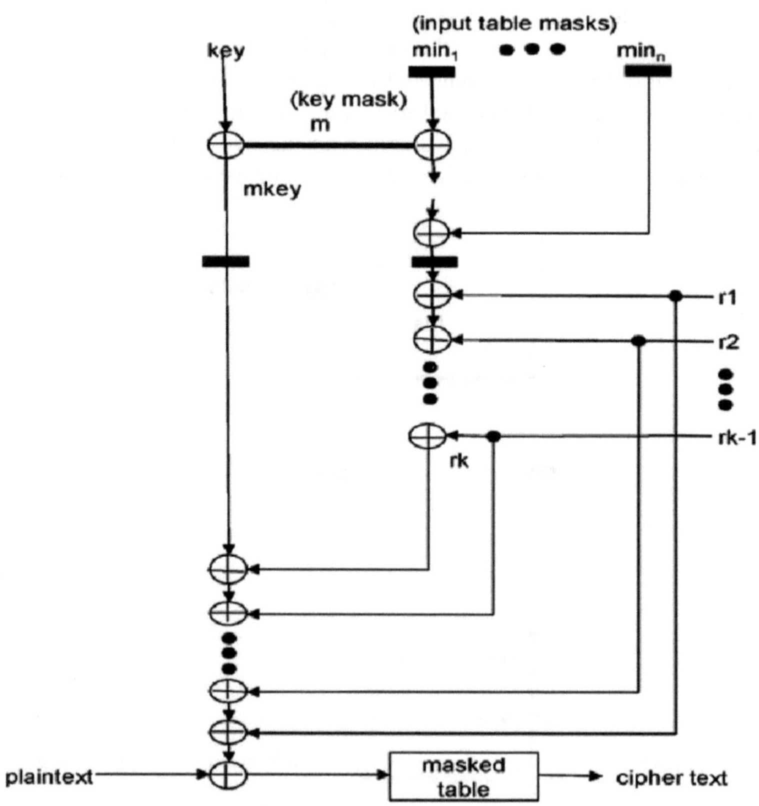

Fig. 9.19 Key, table, and split masks shown as m, $min1, \ldots, min\ n$, and $r1, \ldots, rk$

mask for the table input is not changed. The masked table (shown as table look-up in Fig. 9.18) is generated from an original unmasked table, *table*(i), using a two-part mask comprising $min1$ and $min2$. The table is defined as: *table*(i) $=$ masked table($i \oplus min1 \oplus min2$). Figure 9.19 is a further generalization of the split mask countermeasure (using n split masks). The input table masks are split into n masks and there are also k random masks used per plaintext. In this application, the masked key, n input table masks, and the final masked combined value need to be stored. The table does not need to be regenerated. In all schemes described, it is also possible to employ the proposed split-mask countermeasure in conjunction with some previously researched countermeasures, for example, where table regeneration is required (Messerges 2001) or where several masked tables (Itoh et al. 2002) are used. The combination of both countermeasures can be used to provide increased security. The problem of designing a countermeasure which supports changing input mask for the S-box (as well as output mask), yet avoids table precomputation or large memory requirements of multiple masked tables is a difficult one.

The security of the countermeasure implementation is also an important consideration. Consider using a number of masked precomputed tables in an embedded system where memory is not constrained. In general the side channel can be monitored to identify when a cache miss occurs. This may sometimes be used to provide useful information to the attacker, especially if the attacker has the specifics on cache line size and cache size. If the embedded system has a very small cache size such that only one S-box table can be loaded at a time, attackers may be able to group traces into possible sets of traces with a constant mask by identifying when a cache miss occurs. Alternatively the program structure may leak information to the side channel concerning jumps to load a new masked table. Thus the implementation of the countermeasure must also be secure. Section 9.3 will discuss the use of parallelism for side channel countermeasures, in particular a public key cryptographic application is considered.

9.3 Countermeasures for Public Key Cryptography

Both simple and differential power analysis results on a highly parallel DSP processor core (the Star*core SC140) will be presented in this section. The elliptic curve point multiplication algorithm will be analyzed for simple power analysis (SPA). Recall from Chap. 5 that the elliptic curve point multiplication uses the secret key as the scalar involved in the multiplication. The value of the scalar determines the number of point doublings and additions. Thus if an attacker can record the series of point additions and doublings, they can potentially determine the value of the secret key. This section will describe a countermeasure which resists this attack by designing the point summation to look the same as a point doubling operation. The SPA-resistant ECC was implemented in both binary and prime fields. The algorithm was coded using both "C" and assembly [to take full advantage of the high memory bandwidth (Gebotys 2004) and complex functional units of Star*Core

DSP processor core]. Small keys are used to illustrate the design for security methodology. In prime fields using a maximum sized key for 192-bit fields, point multiplication can be achieved in less than 3 ms running SC140 at 300 MHz. All power traces were obtained by executing the cryptographic algorithms on the SC140 at 100 MHz (for illustration purposes, although 300-MHz power traces showed almost identical power variation). The equipment setup shown in Chap. 8 ideally should be used to capture traces. However, the results which will be shown in this chapter were produced using a pattern generator which synchronized the processor to the oscilloscope. Also a high-speed oscilloscope without multiframe capability was used to capture the power traces. Hence only one trace could be acquired at a time. Better equipment was not available at the time during which we performed these experiments. The SC140 has four multiplier–accumulator units, high memory bandwidth (Gebotys 2004), and a VLIW architecture (Motorola/Lucent 1999) with a supply voltage of 2 V. It has a 16-bit instruction, 40-bit data registers, and a 128-bit data bus, where four 32-bit words can be loaded or stored from/to memory in one cycle. The SC140 architecture is illustrated in Fig. 9.20.

The countermeasure for public key cryptography makes the power trace of the double function and the power trace of the sum function look similar in order to thwart an SPA attack. It first modifies each function so that they use the same order of operations (inserting redundant operations where necessary) and the same clock cycle timing where possible. Redundant operations utilize real data variables in the code and create new temporary variables so that their power traces will vary in each execution. Next the methodology ensures that the code in between the call to the sum and double functions is also identical. This is important so that the attacker cannot identify the start of a double function vs. the start of a sum function.

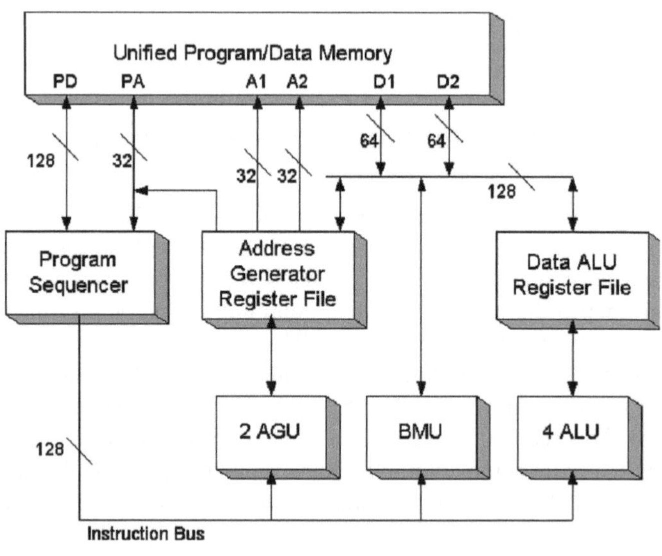

Fig. 9.20 Star*core SC140 architecture

The application of the methodology will be illustrated for binary fields and prime fields. For example in binary fields, both point doubling and summation functions perform 1 field inversion, 2 field multiplications, and 1 or 2 field squarings (see Table 5.5) along with other shorter loops and shorter operations such as modular reduction (performed after field multiplication and field squaring). The order of the major field operations (or operations requiring significant number of cycles such as inversion, multiplication, and squaring) is chosen to be the same and then through adding redundant operations the point summation is designed to be "equivalent" to point doubling. Here "equivalent" means that in each processor clock cycle, the processor performs the same instructions whether the application is point doubling or point summation (however, the data will be different). Table 9.3 identifies the minimum number of redundant operations added in the binary field point multiplication by "//".

Next the detailed "C" compilation of routines is performed. In-lined assembly is added to improve the performance of these key routines. A final check of the code at the assembly level is performed to ensure that cycle counts are the same, where possible. Finally, dynamic power is measured to verify that the point doubling, point summation, and point multiplication power traces are resistant to SPA. Any final timing differences are resolved through redundant operation insertion.

The prime field ECC point multiplication was implemented with Jacobi projective coordinates (Chudnovsky and Chudnovsky 1986), as previously discussed in Chap. 5. However, the modification of sum and doubling routines for security against power attacks was more complex (Gebotys 2004) than that for the binary fields. The modified list of operations is shown in Fig. 9.21. The operations in column 1 are for point doubling [output point is $(x12, y12, z12)$] and point addition operations are in columns 2 and 3 [output point is $(x3, y3, z3)$]. The underscored operations are the redundant operations added for resistance to a side channel attack. The number of redundant operations and nonoptimal operations (coefficient multplies implemented as full multiplications) were minimized. Once the point doubling and summing routines are modified to be "equivalent" from a power point of view, it is important to make sure that the attacker is unable to identify any differences in the higher level algorithm as well.

Table 9.3 Redundant operations inserted into sum and double binary field routines

Double routine	Sum routine
Inversion	Inversion
Multiplication	Multiplication
Loop1	//LOOP1//
Squaring	Squaring
Loop2	Loop2
XOR and//Loop4//	Loop4 and //XOR//
Multiplication	Multiplication
Squaring	//Squaring//
//Loop3//	Loop3

DOUBLE	SUM 1	SUM 2
b1 = y ^2	z2s = z2 ^2	hs = h ^2
e1 = z ^2	z1s = z1 ^2	om = j ^2
b2 = b1 * b1	z2c = z2s * z2	al = th * hs
b = b2<<3	__al=y2<<3	_th=y2<<3
__z2=y2 * x2	f = z1s * x2	la = h * hs
z31 = y * z	g = z2s * x1	th = la * i
e2 = x - e1	___th=x1-z1s	x3 = om - al
e3 = x + e1	___ga=x1+z1s	___be=al + om
e = e2 * e3	z1c = z1s * z1	z3 = ga * h
z12 = z31<<1	___om=g<<1	___al=hs<<1
c1 = e<<1	___ga=z1c<<1	___be=z3<<1
c = c1 + e	th = f + g	___om=be+z3
f1 = b1<<2	___al=z2s<<2	___ga=hs<<2
a = f1 * x	ga = z1 * z2	la = hs * g
f3 = a<<1	___la=ga<<1	___al=la<<1
d1 = c * c	i = y1 * z2c	___be=om *om
x12 = d1 - f3	___al=i-la	om = la –x3
y31 = a - x3	h = f - g	___ga=om-al
y32 = y31 * c	om = y2 * z1c	al = om * j
y12 = y32 - b	j = om - i	y3 = al - th

Fig. 9.21 SPA-resistant operations for point doubling and summation

Original Point Multiplication Code:

```
while (bit_count > 0)
        {
         point_dbl(&r, &temp, &curv);
         bit_count--;
         switch (NAF[bit_count])
         {
           case 1: point _sum (&p, &temp, &r, &curv);
                   break;
           case -1: point _sub (&temp, &p, &r, &curv);
                   break;
           case 0: copy_point (&temp, &r);
                   break;
         }
        }
```

Fig. 9.22 Original code for elliptic curve point multiplication from Rosing (1999)

The original point multiplication code, taken from Rosing (1999), is shown in Fig. 9.22. Each loop iteration performs a point double and then depending upon the bit of the NAF a point sum, point subtract, or a copypoint (the point is copied to another variable) is performed. The current routine indicates that the point doubling

```
while (bit_count > 0)
        {
        bit_count--;
         switch (NAF[bit_count])      {
           case 1: if (I−0) { sub(p,negpy);
                      point _dbl (&r, &temp,  &curv);
                      copy point (&temp, &negp);
                      I−1; bit_count++; }
                      elseif (I=1){ sub(p,negpy);
                       point _sum (&p, &temp, &r, &curv);
                      copy point (&temp, &negp);
                      I−0;}
                      break;
           case -1: : if (I=0) { sub(p,negpy);
                       point _dbl (&r, &temp,  &curv);
                      copy point (&temp, &negp);
                      I−1; bit_count++; }
                      Elsif(I−1){ sub(p,negpy);
                       point _sum(&temp, &negp, &r, &curv);
                      copy point (&temp, &negp);I=0;}
                      break;
           case 0: : if (I  0) { sub(p,negpy);
                       point _dbl (&r, &temp,  &curv);
                      copy_point (&temp, &negp);
                      }
        break;
    }
    }
```

Fig. 9.23 Modified code for point multiplication

could be identified distinct from the point summing (or subtracting) due to its location in the loop vs. the point summing further delayed by the switch statement. Furthermore, in some loop iterations a point summation would not be calculated, only a copypoint. Thus a timing analysis of the power trace could reveal when a double was being performed in the unmodified point multiplication code. To modify the code above to make it secure against power-attacks, the point doubling routine was first moved into the switch statement. Next a temporary variable, I, was used to ensure that both double and sum were handled by a switch statement within the loop. Finally, redundant code was added to make sure that each loop iteration performed the "equivalent" operations (where point double and point sum are now considered "equivalent"). The redundant operations involved a number of small loops and changing the point subtraction into a subtract and a point double (since if $P = (x, y)$, then $-P = (x, y)$ from, performed by *sub*() routine below). These modifications were done to ensure that the performance degradation was minimal. The redundant additional operations are underlined in Fig. 9.23. A minor modification was made to extend this approach for prime fields, where two sum functions are used.

Figure 9.24 shows the original unmodified binary field point doubling and point summation routines running on SC140 after field inversion (inv). The field

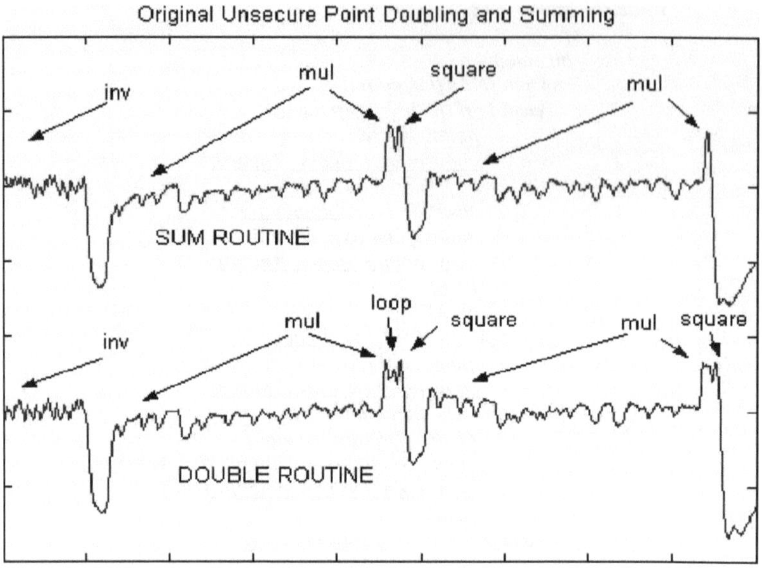

Fig. 9.24 Power traces of summing and point doubling algorithms at *top* and *bottom*, respectively

multiplications (mul), field squaring (square), and important loops (loop) are identified. This identification was obtained from cycle count information and separate RMS current magnitude values for field operations. In Fig. 9.24, the unmodified point summing had two peaks unlike the unmodified point doubling that had three peaks (middle peak due to a loop that was not present in the sum routine). When the point summing was modified by including a redundant loop operation in between the field multiplication and the field squaring, the power "peaks" could not be used to differentiate the two point routines, as shown in Fig. 9.25.

Figure 9.26 illustrates the SPA-resistant binary field power traces run for different keys, key = \$0b and key = \$15. The traces look identical; however, the differences of these two traces do reveal some very small timing differences. For the key \$0b, the binary fields SPA resistance was achieved at a negligible energy overhead and negligible performance overhead (in both cases less than 1%, i.e., 335 clock cycles difference); however, 11% overhead in code size was incurred (the ratio of SPA-resistant code with 6,546 bytes over 5,874 bytes).

Figure 9.27 illustrates the SPA-resistant trace and the original trace (where it can be clearly seen that the sum routine is approximately twice the latency of the double routine), both shown for key \$15 using prime fields. The key \$15 performs two doubles, a sum, two doubles and a sum, which can easily be determined from an SPA attack of the original power trace at the bottom of Fig. 9.27, but cannot be determined from the top SPA-resistant trace. The SPA-resistant power trace incurred a 49% energy overhead, along with 38% performance overhead and 26% code size overhead. For other keys such as \$0b the overhead in performance and code size was 26%.

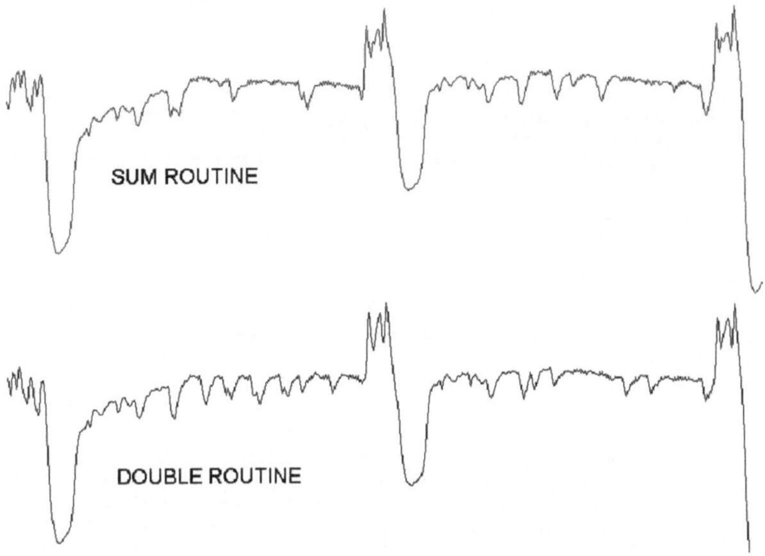

Fig. 9.25 Power traces of modified secure point summing (*top*) and doubling (*bottom*)

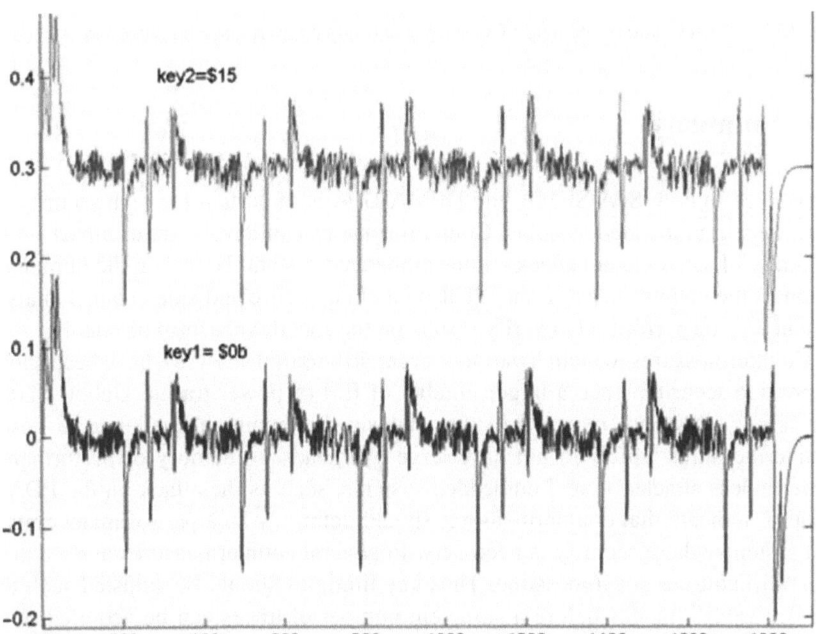

Fig. 9.26 Binary fields SPA-resistant traces for key $15 and $0b at *top* and *bottom*

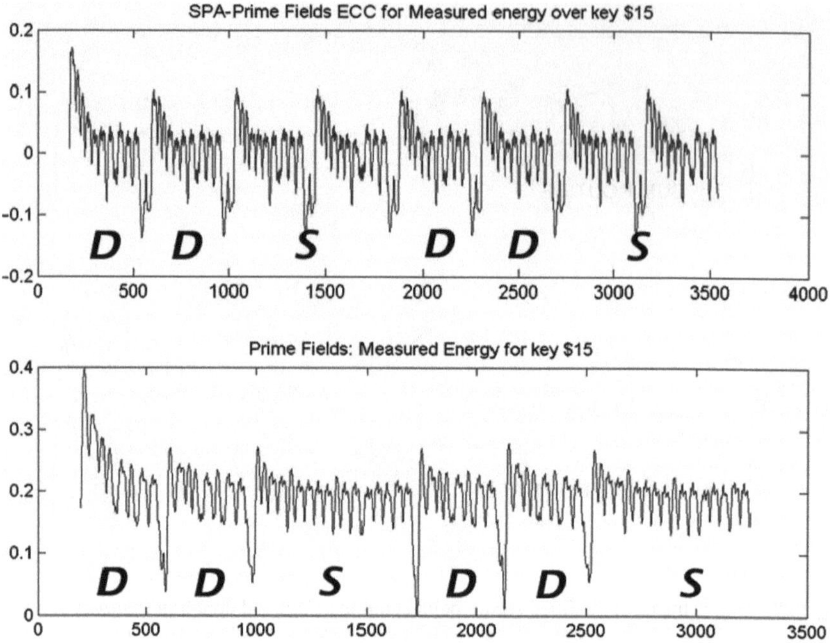

Fig. 9.27 SPA-resistant prime field ECC at *top* and original power trace for same key

9.4 Summary

Resistance to both SPA/SEMA and DEMA/DPA is important for high security applications in embedded systems. Countermeasures can have a great impact on the difficulty of side channel attacks of the embedded system. However, the implementation of the countermeasure should also be made secure and side channel analysis should also be performed to verify results on the specific platform of interest.

Countermeasures, which thwart low-order differential analyses, have been shown to provide security, since a larger number of EM or power traces (Gebotys 2004; Messerges 2000) are required in the analysis. It remains a challenge to design countermeasures which do not have large overheads in memory or performance. Nevertheless attacks on real embedded systems, such as the attack on the PDA in Chap. 8, indicate that countermeasures should definitely be a part of many embedded systems where security is necessary. In general countermeasures make attacks more difficult but not impossible. Thus key lifetimes should be adjusted such that the key value changes before a sufficient number of traces can be acquired in order to launch a successful attack. Chapter 10 discusses the impact of reliability and testability on security functions. Higher order (nth order) differential attacks (Messerges 2000) also have been illustrated in this chapter as an attack against some countermeasures. They are an extension of the first-order differential analysis which involves using joint statistics on multiple points within side channel traces.

Appendix A: Second-Order S-box Differential Analysis Derivation

The following partial code will be used to derive the second-order S-box DPA. We assume that the power (or EM) sample of the load of the corresponding mask table $[M(x)]$ is available (or random mask used in full random masking case) as well as the power (or EM) consumption of the masked S-box table load $[S(x)$, where x is an 8-bit quantity, in AES $x = \text{mask} \oplus \text{plaintext} \oplus \text{secret\_key} = m \oplus p \oplus k)$.

$$\begin{aligned} &\ldots \\ &8.\quad \text{load}\, S'[(m \oplus p \oplus k)_{0\ldots7}] \\ &;\text{where}\, S'[m \oplus p \oplus k] = S[k \oplus p] \oplus r_i \\ &a.\quad \text{load}\, M[(m \oplus p \oplus k)_{0\ldots7}] \end{aligned}$$

Let the kth normalized power consumption of the random data (line a.) $(M((m \oplus p \oplus k)_{0\ 7}))$ be represented by $b1_k$, and the normalized power consumption of the masked output of the S-box table $(S((m \oplus p \oplus k)_{0\ 7})$, on line 8) be c_k. Let $x = (m \oplus p \oplus k)_{0\ 7}$. Let $M((m \oplus p \oplus k)_{0\ 7})_i$ and $S((m \oplus p \oplus k)_{0\ 7})_i$ be the ith bit of $M((m \oplus p \oplus k)_{0\ 7})$ and $S((m \oplus p \oplus k)_{0\ 7})$, respectively. The distribution for these values is assumed to be Gaussian with mean of zero and standard deviation of one (due to normalization, $N(\mu, \sigma) = N(0, 1)$), $f_{b1}(b1_k) \approx N(0, 1)$, and $f_c(c_k) \approx N(0, 1)$. Let $f_{b1}^- = f_{b1}(b1_k | M[x]_i = 0) \approx N(-\varepsilon/2, \sigma)$ represent the distribution of the power consumption values of $b1_k$ such that the ith bit of $M(x)$ is zero. Assume that it is also a Gaussian distribution. Similarly let the following distribution of power consumption exist $f_c^- = f_c(c_k | S'[x]_i = 0) \approx N(-\varepsilon/2, \sigma)$. Let $f_{b1}^+ = f_{b1}(b1_k | M[x]_i = 1) \approx N(+\varepsilon/2, \sigma)$ represent the distribution of the power consumption values of $b1_k$ such that the ith bit of $M(x)$ is one. Similarly let the following distributions of power consumption exist $f_c^+ = f_c(c_k | S'[x]_i = 1) \approx N(+\varepsilon/2, \sigma)$. Next one can calculate the following joint conditional probability distributions of $b1_k$ and c_k (and $b1_k$ is equally likely to be a 0 or 1, and where the final mask, m, is equally likely to be a zero or a one):

$$f_{b1,b2,c}(b1_k, c_k | m \oplus S[x]_i = 1) = \frac{1}{2}\left(f_{b1}^- f_c^+ + f_{b1}^+ f_c^-\right),$$

$$f_{b1,b2,c}(b1_k, c_k | m \oplus S[x]_i = 0) = \frac{1}{2}\left(f_{b1}^- f_c^- + f_{b1}^+ f_c^+\right).$$

Next one can substitute using the normal Gaussian distribution (Messerges 2000) to obtain:

$$\Pr(\Psi | m \oplus S[x]_i = 1) \approx \prod_{k=0}^{N-1} \left[2\cosh\left(\frac{\varepsilon}{2\sigma^2}(b1_k - c_k)\right)\right],$$

$$\Pr(\Psi | m \oplus S[x]_i = 0) \approx \prod_{k=0}^{N-1} \left[2\cosh\left(\frac{\varepsilon}{2\sigma^2}(b1_k + c_k)\right)\right].$$

So far our derivation is identical to that in (Messerges 2000), however, here it is applied to our masked S-box scheme DPA. Now we will extend these results further to derive a statistic, which we can use in our analysis. The statistic will be used to determine the estimated probability of a correct key guess. We assume that the user has control over the input plaintexts; hence based upon the attackers' guess of 8 key bits, the attacker can determine the output of the S-box table. Using the following notation, let

$$\delta 1(key) = \{k \,|\, S[key \oplus p]_i = 1, \text{ or } S'[key \oplus p]_i \oplus M[key \oplus p]_i = 1, \text{where } m_i = 0\}$$

and

$$\delta 0(\text{key}) = \{k \,|\, S[key \oplus p]_i = 0, \text{ or} S'[key \oplus p]_i \oplus M[key \oplus p]_i = 0, \text{where } m_i = 0\}$$

to support partitioning of power traces (based upon the users input plaintext and 8 key bits guess), we obtain the following statistics:

$$\begin{aligned}
\Delta_{m_i=0} &\approx \prod_{k=0|k\in\delta 1}^{N-1} \left[2\cosh\left(\frac{\varepsilon}{2\sigma^2}(b1_k - c_k)\right)\right] \\
&\quad \times \prod_{k=0|k\in\delta 0}^{N-1} \left[2\cosh\left(\frac{\varepsilon}{2\sigma^2}(b1_k + c_k)\right)\right], \\
\Delta_{m_i=1} &\approx \prod_{k=0|k\in\delta 0}^{N-1} \left[2\cosh\left(\frac{\varepsilon}{2\sigma^2}(b1_k - c_k)\right)\right] \\
&\quad \times \prod_{k=0|k\in\delta 1}^{N-1} \left[2\cosh\left(\frac{\varepsilon}{2\sigma^2}(b1_k + c_k)\right)\right],
\end{aligned}$$

which represents the estimated probability that the partitioning of power traces according to the key guess is correct, given the value of the jth bit of m, m_j. Alternatively one could also use the following statistic representing the difference of estimated probabilities (probability of the partition being correct minus the probability of the partition being incorrect) given as the two formulas below depending upon the bit value of m:

$$\begin{aligned}
\Lambda_{m_i=0} \approx &\prod_{k=0|k\in\delta 1}^{N-1} \left[2\cosh\left(\frac{\varepsilon}{2\sigma^2}(b1_k - c_k)\right)\right] \\
&\times \prod_{k=0|k\in\delta 0}^{N-1} \left[2\cosh\left(\frac{\varepsilon}{2\sigma^2}(b1_k + c_k)\right)\right] \\
&- \prod_{k=0|k\in\delta 0}^{N-1} \left[2\cosh\left(\frac{\varepsilon}{2\sigma^2}(b1_k - c_k)\right)\right] \\
&\times \prod_{k=0|k\in\delta 1}^{N-1} \left[2\cosh\left(\frac{\varepsilon}{2\sigma^2}(b1_k + c_k)\right)\right]
\end{aligned} \qquad \text{for known } m,$$

$$\Lambda_{m_i=1} \approx \prod_{k=0|k\in\delta 0}^{N-1}\left[2\cosh\left(\frac{\varepsilon}{2\sigma^2}(b1_k-c_k)\right)\right] \times \prod_{k=0|k\in\delta 1}^{N-1}\left[2\cosh\left(\frac{\varepsilon}{2\sigma^2}(b1_k+c_k)\right)\right] - \prod_{k=0|k\in\delta 1}^{N-1}\left[2\cosh\left(\frac{\varepsilon}{2\sigma^2}(b1_k-c_k)\right)\right] \times \prod_{k=0|k\in\delta 0}^{N-1}\left[2\cosh\left(\frac{\varepsilon}{2\sigma^2}(b1_k+c_k)\right)\right] \quad \text{for known } m.$$

However, the difficulty lies in the fact that the attacker does not know the value of m. Hence a new statistic is required, as follows:

$$\Lambda \approx \left| \prod_{k=0|k\in\delta 0}^{N-1}\left[2\cosh\left(\frac{\varepsilon}{2\sigma^2}(b1_k-c_k)\right)\right] \times \prod_{k=0|k\in\delta 1}^{N-1}\left[2\cosh\left(\frac{\varepsilon}{2\sigma^2}(b1_k+c_k)\right)\right] - \prod_{k=0|k\in\delta 1}^{N-1}\left[2\cosh\left(\frac{\varepsilon}{2\sigma^2}(b1_k-c_k)\right)\right] \times \prod_{k=0|k\in\delta 0}^{N-1}\left[2\cosh\left(\frac{\varepsilon}{2\sigma^2}(b1_k+c_k)\right)\right] \right| \quad \text{for unknown } m.$$

This last statistic, Λ, is independent of the value of m. However, it assumes that m is some fixed value for all the power or EM traces acquired and that the probability of the partitioning corresponds to m_j is higher than the probability of the partitioning corresponds to $\overline{m_j}$. If the attacker does not know the value of m the attack is more difficult since all probabilities are positive according to Λ, unlike the case where an attacker could rule out cases where $\Lambda_{m_j=0}$ is negative if the attacker knows that $m_j = 0$. One could also use the heuristic from Messerges (2000), as follows:

$$\text{heuristic} = \left| \frac{\sum\limits_{k=0|k\in\delta 0} |b1_k - c_k|}{N} - \frac{\sum\limits_{k=0|k\in\delta 1} |b1_k - c_k|}{N} \right|.$$

Appendix B: Derivation of Third-Order S-box Differential Analysis Statistic

The third-order statistic can be derived using the same approach as illustrated in Appendix A. It will be used for the case where three tables are in use, specifically

$S(x)$, $M_1(x)$, and $M_2(x)$ [where $m = r(x) \oplus M_1(x) \oplus M_2(x)$, or $m \oplus S(x) = S(x) \oplus M_1(x) \oplus M_2(x)$].

Line 1. Load $S(x)$
Line 2. Load $M_1(x)$
Line 3. Load $M_2(x)$

Let the kth normalized power consumption of the masked S-box (*line* 1), the first mask table (*line* 2), and the second masked table (*line* 3) be represented by c_k, $b1_k$, and $b2_k$, respectively. The distribution for these values is assumed to be Gaussian with mean of zero and standard deviation of one [due to normalization, $N(\mu, \sigma) = N(0, 1))$, $f_{b1}(b1_k) \approx N(0, 1)$, $f_{b2}\ (b2_k) \approx N(0, 1)$, and $f_c(c_k) \approx N(0, 1)$]. Continuing with the same assumptions in Appendix A we obtain the following statistic:

$$\begin{aligned}
&f_{b1,b2,c}(b1_k, b2_k, c_k | S[x]_i \oplus m_i = 1) \\
&= \frac{1}{4}\left(f_{b1}^- f_{b2}^- f_c^+ + f_{b1}^+ f_{b2}^+ f_c^+\right) + \frac{1}{4}\left(f_{b1}^- f_{b2}^+ f_c^- + f_{b1}^+ f_{b2}^- f_c^-\right) \\
&f_{b1,b2,c}(b1_k, b2_k, c_k | S[x]_i \oplus m_i = 0) \\
&= \frac{1}{4}\left(f_{b1}^- f_{b2}^- f_c^- + f_{b1}^+ f_{b2}^+ f_c^-\right) + \frac{1}{4}\left(f_{b1}^- f_{b2}^+ f_c^+ + f_{b1}^+ f_{b2}^- f_c^+\right).
\end{aligned}$$

Next one can substitute using the normal distribution:

$$\eta(x) = \frac{1}{\sigma\sqrt{2\pi}} \mathrm{e}^{-\frac{1}{2}\left(\frac{x-\mu}{\sigma}\right)^2}$$

and factor out the constant $1/\sigma\sqrt{2\pi}$ to obtain the following expression for the probability:

$$\begin{aligned}
&\Pr(\Psi | S[x]_i \oplus m_i = 1) \\
&\approx \prod_{k=0}^{N-1} \left[\begin{array}{l} \left(\mathrm{e}^{-\frac{1}{2\sigma^2}\left((b1_k+\frac{\varepsilon}{2})^2+(b2_k+\frac{\varepsilon}{2})^2+(c_k-\frac{\varepsilon}{2})^2\right)}\right) + \left(\mathrm{e}^{-\frac{1}{2\sigma^2}\left((b1_k-\frac{\varepsilon}{2})^2+(b2_k-\frac{\varepsilon}{2})^2+(c_k-\frac{\varepsilon}{2})^2\right)}\right) \\ + \left(\mathrm{e}^{-\frac{1}{2\sigma^2}\left((b1_k+\frac{\varepsilon}{2})^2+(b2_k-\frac{\varepsilon}{2})^2+(c_k+\frac{\varepsilon}{2})^2\right)}\right) + \left(\mathrm{e}^{-\frac{1}{2\sigma^2}\left((b1_k-\frac{\varepsilon}{2})^2+(b2_k+\frac{\varepsilon}{2})^2+(c_k+\frac{\varepsilon}{2})^2\right)}\right) \end{array}\right].
\end{aligned}$$

For the rest of the analysis let ε represent $\varepsilon/2\sigma^2$

$$\begin{aligned}
&\Pr(\Psi | S[x]_i \oplus m_i = 1) \\
&\approx \prod_{k=0}^{N-1} \left(\mathrm{e}^{-(\varepsilon b1_k + \varepsilon b2_k - \varepsilon c_k)}\right) + \left(\mathrm{e}^{-(-\varepsilon b1_k - \varepsilon b2_k - \varepsilon c_k)}\right) + \left(\mathrm{e}^{-(\varepsilon b1_k - \varepsilon b2_k + \varepsilon c_k)}\right) \\
&\quad + \left(\mathrm{e}^{-(-\varepsilon b1_k + \varepsilon b2_k + \varepsilon c_k)}\right) \\
&\approx \prod_{k=0}^{N-1} \left[\left(\mathrm{e}^{-\varepsilon c_k}\left(\mathrm{e}^{-(\varepsilon b1_k - \varepsilon b2_k)} + \mathrm{e}^{-(-\varepsilon b1_k + \varepsilon b2_k)}\right)\right) + (\mathrm{e}^{+\varepsilon c_k}(\mathrm{e}^{-(\varepsilon b1_k + \varepsilon b2_k)}\right. \\
&\qquad \left. + \mathrm{e}^{-(-\varepsilon b1_k - \varepsilon b2_k)}))\right].
\end{aligned}$$

Next we substitute using the following trigonometric identity $\cosh(x) = (e^x + e^{-x})/2$ to obtain the following:

$$\begin{aligned}
&\Pr(\Psi|S[x]_i \oplus m_i = 1) \\
&\approx \prod_{k=0}^{N-1} \left(\mathrm{e}^{-\varepsilon c_k}\left(\mathrm{e}^{(-\varepsilon b1_k + \varepsilon b2_k)} + \mathrm{e}^{-(-\varepsilon b1_k + \varepsilon b2_k)}\right)\right) \\
&\quad + \left(\mathrm{e}^{+\varepsilon c_k}\left(\mathrm{e}^{-(\varepsilon b1_k + \varepsilon b2_k)} + \mathrm{e}^{(\varepsilon b1_k + \varepsilon b2_k)}\right)\right) \\
&\approx \prod_{k=0}^{N-1} (\mathrm{e}^{-\varepsilon c_k} \cosh(\varepsilon(-b1_k + b2_k)) + \mathrm{e}^{\varepsilon c_k} \cosh(\varepsilon(b1_k + b2_k))) \\
&\Pr(\Psi|S[x]_i \oplus m_i = 0) \\
&\approx \prod_{k=0}^{N-1} (\mathrm{e}^{-\varepsilon c_k} \cosh(\varepsilon(b_k 1 + b2_k)) + \mathrm{e}^{+\varepsilon c_k} \cosh(\varepsilon(-b1_k + b2_k))).
\end{aligned}$$

Now one can further use the following trigonometric identities $\cosh(x + y) = \cosh(x)\cosh(y) + \sinh(x)\sinh(y)$, $\sinh(-x) = -\sinh(x), \cosh(-x) = \cosh(x)$, and $\sinh(x) = (e^x - e^{-x})/2$, so the equation can be rewritten as:

$$\begin{aligned}
&\Pr(\Psi|S[x]_i \oplus m_i = 1) \\
&\approx \prod_{k=0}^{N-1} \begin{bmatrix} \mathrm{e}^{-\varepsilon c_k}(\cosh(\varepsilon b1_k)\cosh(\varepsilon b2_k) - \sinh(\varepsilon b1_k)\sinh(\varepsilon b2_k)) \\ +\mathrm{e}^{\varepsilon c_k}(\cosh(\varepsilon b1_k)\cosh(\varepsilon b2_k) + \sinh(\varepsilon b1_k)\sinh(\varepsilon b2_k)) \end{bmatrix} \\
&\approx \prod_{k=0}^{N-1} [(\mathrm{e}^{-\varepsilon c_k} + \mathrm{e}^{\varepsilon c_k})\cosh(\varepsilon b1_k)\cosh(\varepsilon b2_k) \\
&\quad + (-\mathrm{e}^{-\varepsilon c_k} + \mathrm{e}^{\varepsilon c_k})\sinh(\varepsilon b1_k)\sinh(\varepsilon b2_k)] \\
&\approx \prod_{k=0}^{N-1} \cosh(\varepsilon c_k)\cosh(\varepsilon b1_k)\cosh(\varepsilon b2_k) + \sinh(\varepsilon c_k)\sinh(\varepsilon b1_k)\sinh(\varepsilon b2_k) \\
&\Pr(\Psi|S[x]_i \oplus m_i = 0) \\
&\approx \prod_{k=0}^{N-1} [\cosh(\varepsilon b1_k)\cosh(\varepsilon b2_k)\cosh(\varepsilon c_k) - \sinh(\varepsilon b1_k)\sinh(\varepsilon b2_k)\sinh(\varepsilon c_k)].
\end{aligned}$$

Next we can further extend the expressions to directly solve for probabilities. Using the following notation, let

$$\begin{aligned}
\delta 1(key) = \{k|S[key \oplus p]_i = 1, \text{or} S'[key \oplus p]_i \oplus M_1[key \oplus p]_i \\
\oplus M_2[key \oplus p]_i = 1, \quad \text{where } m_i = 0\}
\end{aligned}$$

and

$$\begin{aligned}\delta 0(key) = \{k \mid S[key \oplus p]_i = 0, \text{ or } S'[key \oplus p]_i \oplus M_1[key \oplus p]_i \\ \oplus M_2[key \oplus p]_i = 0 \quad \text{where } m_i = 0\}\end{aligned}$$

to support partitioning of power traces (based upon the users input plaintext and 8 key bits guess), we obtain the following statistics:

$$\begin{aligned}\Delta_{m_i=1} \approx & \prod_{k|k\in\delta 1} (\cosh(\varepsilon b1_k)\cosh(\varepsilon b2_k)\cosh(\varepsilon c_k) - \sinh(\varepsilon b1_k)\sinh(\varepsilon b2_k)\sinh(\varepsilon c_k))) \\ & \times \prod_{k|k\in\delta 0} (\cosh(\varepsilon b1_k)\cosh(\varepsilon b2_k)\cosh(\varepsilon c_k) + \sinh(\varepsilon b1_k)\sinh(\varepsilon b2_k)\sinh(\varepsilon c_k))) \\ \Delta_{m_i=0} \approx & \prod_{k|k\in\delta 0} (\cosh(\varepsilon b1_k)\cosh(\varepsilon b2_k)\cosh(\varepsilon c_k) - \sinh(\varepsilon b1_k)\sinh(\varepsilon b2_k)\sinh(\varepsilon c_k))) \\ & \times \prod_{k|k\in\delta 1} (\cosh(\varepsilon b1_k)\cosh(\varepsilon b2_k)\cosh(\varepsilon c_k) + \sinh(\varepsilon b1_k)\sinh(\varepsilon b2_k)\sinh(\varepsilon c_k)))\end{aligned}$$

for known m.

Alternatively one can represent the statistic Λ as given below, where the statistic is positive or negative to indicate that the m_i value is 1 or 0, respectively.

$$\begin{aligned}\Lambda &= \left|\Delta_{m_i=1} - \Delta_{m_i=0}\right| \\ &= \left| \begin{aligned} & \prod_{k|k\in\delta 1} (\cosh(\varepsilon b1_k)\cosh(\varepsilon b2_k)\cosh(\varepsilon c_k) - \sinh(\varepsilon b1_k)\sinh(\varepsilon b2_k)\sinh(\varepsilon c_k))) \\ & \times \prod_{k|k\in\delta 0} (\cosh(\varepsilon b1_k)\cosh(\varepsilon b2_k)\cosh(\varepsilon c_k) + \sinh(\varepsilon b1_k)\sinh(\varepsilon b2_k)\sinh(\varepsilon c_k))) \\ & - \prod_{k|k\in\delta 0} (\cosh(\varepsilon b1_k)\cosh(\varepsilon b2_k)\cosh(\varepsilon c_k) - \sinh(\varepsilon b1_k)\sinh(\varepsilon b2_k)\sinh(\varepsilon c_k))) \\ & \times \prod_{k|k\in\delta 1} (\cosh(\varepsilon b1_k)\cosh(\varepsilon b2_k)\cosh(\varepsilon c_k) + \sinh(\varepsilon b1_k)\sinh(\varepsilon b2_k)\sinh(\varepsilon c_k))) \end{aligned} \right|\end{aligned}$$

for unknown m.

Note that if the value of m changes for each trace, it may not be possible to formulate a statistic unless we have a sample of m in which case it will lead to an even higher order statistic.

References

Chudnovsky DV, Chudnovsky GV (1986) Sequences of numbers generated by addition in formal groups and new primality and factorization tests, Appl Math 7:385–434

Coron J-S and Kizhvatov I (2009) Analysis of the split mask countermeasure for embedded systems, ACM workshop on embedded systems security (WESS) 2009

Gebotys C (2004) Design of secure cryptography against the threat of power-attacks in DSP embedded processors, ACM Trans Embed Comput Syst 3(1):92–113

Gebotys C (2006a) A table masking countermeasure for low energy secure embedded systems, IEEE Trans VLSI 14(7):740–753

Gebotys C (2006b) A split mask countermeasure for low energy secure embedded systems, ACM Trans Embed Comput syst 5(3):577–612

Gebotys C, Zhang Y (2003) Security wrappers and power analysis for SoC technologies ACM/IEEE ISSS-CODES, CA, USA, pp 162–167

Gebotys C, Ho S, Tiu CC (2005) EM analysis of Rijndael an ECC on a wireless java-based PDA, CHES 2005, LNCS 3659. GmbH, pp 250–265

Hancke GP (2007) Modulating noisy carrier modulation for HF RFID. In: Proceedings of Eurasip, RFID 2007

Itoh K, Takenaka M, Torii N (2002) DPA countermeasure based on the masking method, LNCS 2288, pp 440–456

Khatibzadeh A, Gebotys C (2007) Enhanced current-balanced logic (ECBL): an area efficient solution to secure smart cards against differential power attack information technology, 2007. ITNG apos;07. In: Fourth international conference on volume, issue, 2–4 April 2007, pp 898–899. doi: 10.1109/ITNG.2007.89

Liao W et al (2002) Leakage power modeling and reduction with data retention, IEEE ICCAD: 714–719

Messerges T (2000) Using 2nd order power analysis to attack DPA resistant software. LNCS 1965, pp 238–251

Messerges T (2001) Securing the AES finalists against power analysis attacks, LNCS 1978, pp 150–164

Rosing M (1999) Implementing elliptic curve cryptography. Manning Publishing, Greenwich,CT

Shamir A (2000) Protecting smart cards from passive power analysis with detached power supplies. In: Koc CK, Paar C (eds) Cryptographic hardware and embedded systems – CHES 2000, LNCS 1965. Springer, Berlin, pp 71–77

Tiri K, Verbauwhede I (2002) A dynamic and differential cmos logic with signal independent power consumption to withstand differential power analysis on smart cards. In: Proceedings of 28th European solid-state circuits conference (ESSCIRC), pp 403–406

Tiri K, Verbauwhede I (2004) A logic level design methodology for a secure dpa resistant asic or fpga implementation. In: Proceedings of design, automation and test in Europe conference (DATE), pp 246–251

Chapter 10
Reliable Testable Secure Systems

The problems of single-event-induced errors in integrated circuits riveted the attention of the commercial IC industry in the late 1970s, when it was recognized that the random, correctable errors in DRAMs within ground-based computers were due to single α-particles (May and Woods 1978)

Ma and Dressendorfer (1989)

Although reliability has been extensively studied for decades in the space industry, it is now becoming evident that even ground-based embedded systems are facing similar reliability issues. This chapter will briefly discuss the single-event-upset (SEU) phenomena, also known as soft errors, and provide several examples of how reliability can be designed into secure embedded systems. The chapter will also discuss testability issues that relate to security and present some recent research in this area.

Reliable security is an extremely important area of engineering. Failure of a security application may have significant consequences, such as significant financial losses, personal injury in automobiles, losing control of a nuclear station, etc. Not only do security functions require rigorous testing before being put into the field but they also should be as reliable as possible. There is no room for errors in security. For example, a single error in AES causing one bit flip causes over 50% of the ciphertext bits to be in error. This is a result of the diffusion property of ciphers, which increases the effect of one bit over many bits in the output.

Ross Anderson reported the interesting case of a credit card, which was read by a misaligned card reader (Anderson 2001). The card reader should have detected errors in both the cryptographic computation (to detect forgery) as well as the simpler exclusive-or checksum computation (to detect errors). Instead and most unfortunate for the card owner, the cryptographic checksum successfully detected errors but the simpler checksum did not. This indicated incorrectly that the card was a forged card and the owner was "…arrested…and beaten up by the police." (Anderson 2001). The error apparently had been masked in the checksum causing much pain for the owner. This is likely not the first example of the disastrous impact of unreliablesecurity. Clearly designing for security includes designing for reactions to errors as well as resistance to attacks.

C.H. Gebotys, *Security in Embedded Devices*, Embedded Systems,
DOI 10.1007/978-1-4419-1530-6_10,

Reliability is an important concern for space, military, satellite, avionics, and other applications where systems are operating in environments with significant radiation. Specifically cosmic rays, caused by heavy ions and protons emanating from the sun and galaxies, as well as alpha particles create soft error problems for circuitry in space. The cosmic rays are a strong function of altitude, whereas alpha particles on the other hand result largely from impurities in packaging and metal interconnects of the chip (Kent and Williams 1993). Cosmic rays have several orders of magnitude higher ionization values than the alpha particles and thus have more severe effects on the integrated circuits (Johnston 2000). Nevertheless, techniques for reducing alpha particle radiation generally reduce cosmic radiation at sea level more than alpha particle radiation (Kent and Williams 1993). Low earth orbits largely suffer from proton radiation. High-energy protons indirectly cause SEUs. These particles pass through the circuit depositing energy through ionization and radiation along their path. They can free electron–hole pairs and deposit energy as well as lose energy directly in the circuit (Dodd and Massengill 2003). These are referred to as single event effects and in particular SEU. From 1970 to 1982 there were many reported SEUs in space satellites that affected not only memory chips but other circuits as well (Ohring 1998). In commercial applications this type of error is referred to as a soft error, or dynamic error. Typically these SEUs can flip a bit in a storage element or cause the logic state of a transistor to flip dynamically. They might even cause permanent damage to the device. For example, latch-up is reduced by processing techniques such as epitaxial layers, silicon on insulator; however, even latch-up immune devices might be prone to SEUs. For example, rad-hard processes can also reduce the long-term effect of single event effects but these processes cannot completely resist the short-term effects of SEUs.

SEUs hit the headlines in the late 1970s when it was discovered that alpha particles were causing dynamic errors in commercial memory chips. For example, the difference between a "0" and a "1" in a bit of memory is typically less than 1 pC of charge transfer. In space applications, one high-energy particle hitting the spacecraft can generate many low-energy particles, which can create many SEUs. As device scaling produces smaller feature sizes with lower power requirements, the impact of SEUs will increase (Helvajian 1997). In nonspace or terrestrial environments and apart from memory chips, it is assumed that once a chip passes the testing phase, reliability will not be an issue apart from the normal wear and tear (following the standard bathtub curve for failure rate vs. time). However, process geometries have continued to scale down to such an extent that chips in normal commercial applications are now starting to suffer from radiation effects. In these commercial applications radiation may originate from chip packaging. For example, alpha particles are emitted by uranium and thorium impurities in packaging materials (Baumann 2005). Other types of radiation are cosmic in nature. Although in the past this type of radiation only affected memory chips, now the device scaling is reaching a point where the charge injection is now sufficient to change the logical "0" to a "1" or "1" to a "0" of nonmemory chips. For example, the amount of charge representing a bit value in a 0.13-μm SRAM is approximately 1/16 of what is required in a 0.25-μm process. Hence, each new process technology makes the circuitry close to an order of magnitude more susceptible to SEUs (Helvajian 1997).

In DRAMs even less charge is required to represent a bit value, thus they have always been vulnerable to SEUs. Process technologies limiting the accumulation of electron–hole pairs, as well as improved packaging with fewer emissions of alpha particles, and other methods help to minimize a chip's susceptibility to SEUs. However, additional methods are required in order to ensure that chips are reliable.

Testing of advanced circuits for space includes subjecting them to bombardment of high-energy protons as well as heavy ions. Different intensities are used to study the SEU impact. For example, the intensity is measured with LETs, where the LET of a 5-MeV alpha particle is $1\,\text{MeV-cm}^2/\text{mg}$. There are standards for irradiation testing such as MIL-STD-883 Test Method 1019 (MILSTD 1996). But this radiation testing is extremely expensive. As technology scales down, the critical charge required to upset a transistor state decreases. However, it should be noted that there are also other process factors that come into play creating a more complex relationship (Johnston 2000). Although technology scaling reduces the chip sizes for a specific function, designers typically integrate more hardware onto a single die and die sizes become larger. The increase in die size increases the probability of a SEU hitting the chip. It is typical to measure the probability of being hit by a SEU, with failures in time metric (or FIT). In summary, if reliability is an important criterion of the embedded system, as such is the case for security in space or safety critical applications, then in addition to choosing appropriate processes that may be SEU-resistant, SEUs must also be mitigated through architectural and recovery techniques.

This chapter will review the impact of reliability or SEU resistance on security and discuss various options for designing reliable systems. Application-specific architectures for reliable security are the focus of this chapter since they offer the lowest overheads and therefore are important for embedded systems.

10.1 Reliability and Fault Tolerance

The most common solution for military or aerospace applications is to design embedded systems using radiation hardened (or rad-hard) processes. However, rad-hard technology is typically slower, less dense, and very expensive. For example, in 2007 a rad-hard 4M chip (AT68166) cost $1,710 each in volume of 100 (Gold 2007), contrasting to the same year when DRAM chips (512 Mb, 667-MHz DDR2) had contract prices below $2.00 (Nystedt 2007). There are also rad-hard ASICs, processors, and FPGAs on the market targeting space and military applications. In addition to these radiation-resistant process technologies other mitigation techniques are also needed to achieve higher reliability. This section will examine metrics as well as some techniques that have been used for radiation resistance.

Metrics for reliability include the failure in time (FIT), which is defined as the number of failures per 10^9 h. This measurement is quite common in space, safety critical, and military application areas. The FIT is typically estimated per component through accelerated testing by using, for instance, voltage and/or temperature, to age the component faster in order to acquire useful statistics. It is also important to note that as area increases, higher rates of SEUs are expected. For example, a rad-hard FPGA device from Actel indicates 1×10^{-7} and 1×10^{-6} SEUs per bit per day for their combinational and sequential modules, respectively (Actel 2005). Many FPGAs have been characterized in terms of SEU rates, which vary from 1 in 19,000 years to 1 in 1.1 days in Roosta (2004). Other metrics include mean time to failure and mean time between failures.

Rad-hard processors, such as the LEON3, claim to be both fault-tolerant and SEU-proof. The LEON3 is a synthesizable 32-bit soft core processor based on SPARC v8 architecture (Aeroflex 2009). Specifically it is considered to be SEU-proof since it uses SEU error detection and correction in its register files and cache memories. It can handle up to four errors per 32-bit word. Error correcting code is also used in all of its on-chip RAM blocks. This correction is totally transparent to the software. As an alternative to error correcting code, memory blocks may use a secondary memory, or mirror memory. Hence, an error detected in a memory block invokes a correction cycle, which reloads the memory from the mirror block. Error correcting code is done in parallel to avoid degradation of the processor speed. The fault-tolerant Leon3 uses 1.15 times more radiation tolerant (RTAX) cells (7,500 vs. 6,500) or 1.1 times more (equivalent ASIC) gates (22K vs. 20K) than a nonfault-tolerant version of Leon3 (Aeroflex 2009). Tests conducted to verify rad-hard claims include passing extremely expensive tests in a heavy-ion error-injection chamber, where the chip is irradiated by SEUs for 1 week (Gaisler 2005).

Rad-hard FPGAs use robust flip-flops to resist upset from particle radiation. Since FPGAs are cost effective at low volumes, they are quite useful in space applications. For example, there were over 50 rad-hard FPGA chips in the NASA Mars rover mission (Cataldo 2004). Most rad-hard FPGAs are not reprogrammable, such as Actel's antifuse technology (Actel 2009). In general, SRAM-based FPGAs have higher densities than their rad-hard counterparts, but are also more sensitive to SEUs. Hence, some vendors have used CRC circuitry (Altera 2008) to verify that the configuration data have not been corrupted by SEUs. For example, the configuration data are read and checked for integrity. Then, if any error is found, the FPGA is reconfigured. However, this does not support SEU-resistance for user data within the design. An alternative to rad-hard is to design using triple modular redundancy (TMR) and other fault-tolerant techniques. These will be briefly discussed next.

General design for reliability techniques includes hardware redundancy, such as TMR. In TMR three instances of the same hardware module are used, and an output majority voter selects which of the three hardware module outputs to propagate to the next hardware module. This arrangement is shown in Fig. 10.1, where three instances of module M are used along with a majority voter.

Another technique includes time redundancy where the function is recomputed at different times and latched into flip-flops. A majority voter can be used to select the correct output from the three computations as shown in Fig. 10.2. The function is latched into the flip-flop at three different times, and the majority voter then

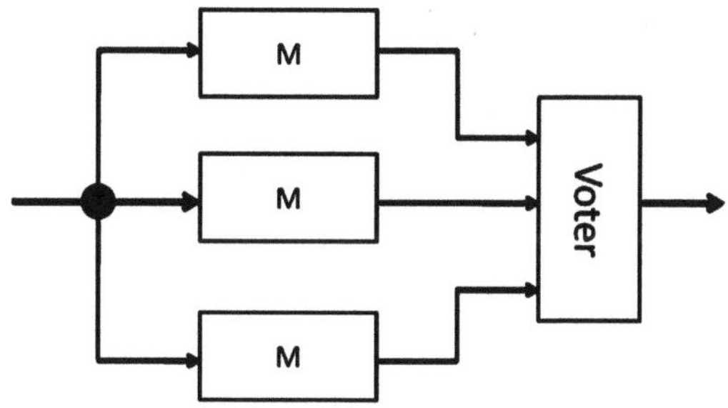

Fig. 10.1 Triple modular redundancy

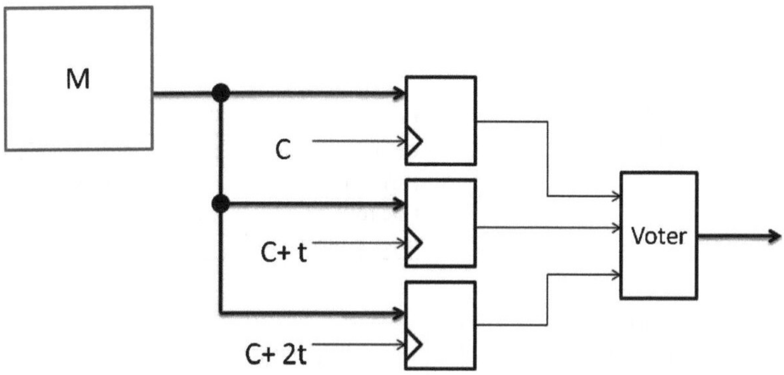

Fig. 10.2 Time redundancy example

determines the correct value. For example, a SEU might have hit the logic within M at a time when this error on the output was latched into one of the flip-flops. At a later time the correct value propagates to the other two flip-flops, since the effect of the SEU was only temporary. Time redundancy can also be used in software applications. When an error is detected, the processor stops executing and resets its state from an earlier point in time. Next, the processor recomputes the application. Typically this is referred to as rollback. In this case if an error is detected, the processor rolls back to the last point before the error hit and recomputes the function again.

Information redundancy can be used to support error detection and correction. Parity-based techniques can be used in applications where only one or an odd number of soft errors are expected to occur. Hamming codes are an example of data redundancy, which provide both error detection and error correction. When an error occurs, the Hamming codes are able to locate the erroneous bit within the codeword and correct it. Figure 10.3 introduces the notion of Hamming encoding and decoding. In Fig. 10.3 there are three parity bits, c_1, c_2, c_3, added to the 4-bits of data, $d_0, \ldots, d_3$. The parity bits are computed as shown at the top, based upon the exclusive-or of three bits of correct data. When the data correctness needs to be verified, each of the three parity bits is exclusive-ored with the exclusive-or of the three relevant data bits. If the values computed are all zero for all three computations, then there are no errors detected. Otherwise, if one or more parity computations are nonzero an error has occurred. In this case the value of $(c_1\ c_2\ c_3)$ identifies where

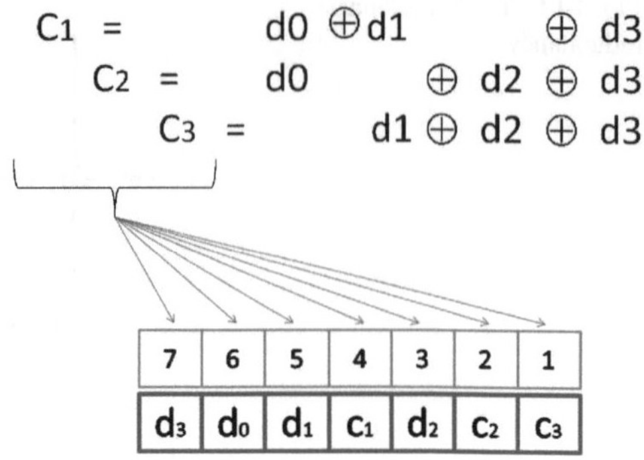

Fig. 10.3 Error correction for data $d_0, \ldots, d_3$ using parity bits c_1, c_2, c_3

Table 10.1 Techniques for detection or correction

Error detection	Error detection and correction
Duplicate modules	Triple modular redundancy
Parity codes	Hamming codes
One rollback	Rollback two times
Inverse function	

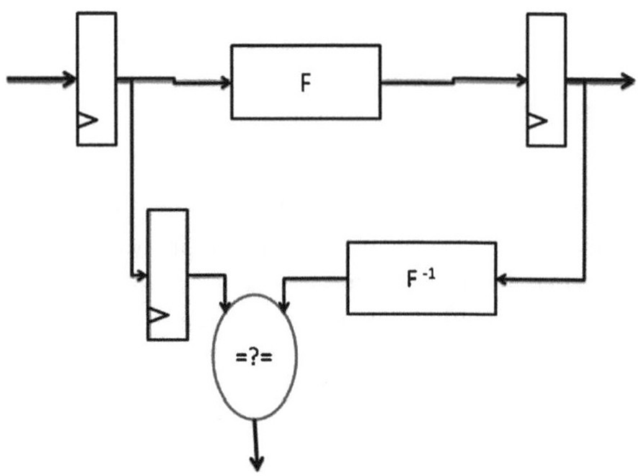

Fig. 10.4 Inverse function example for error detection

the error is located. For example, the 7-bit word is stored as shown at the bottom of Fig. 10.3 where the seventh position stores d_3 and the first position stores c_3, etc. Now consider an error in d_0. This error would create $(c_1\ c_2\ c_3) = (110) = 6$, which would identify the sixth position in error. Note that the sixth position stores d_0, thus correctly identifying the error.

In general, techniques for error detection have lower overheads than error correction approaches. For example, two redundant modules (duplication) vs. three redundant modules (TMR) as shown in Table 10.1 are used for detection vs. correction. Alternatively if an inverse function is available, then there is an alternative for error detection shown in the last row of column 1 in Table 10.1. Figure 10.4 illustrates the inverse function approach to error detection. For example, if a function, f, has an inverse function, f^{-1}, then this can be used to check for SEUs. An example of this is to run a decryption function on the ciphertext in order to check that the generated plaintext matches the original plaintext thus verifying that the cipher did not suffer from SEUs. The computation $f(x) = y$ can be checked for the presence of a SEU by further sending the result through the inverse function. Thus, computing $f^{-1}(y)$ should be equivalent to $f^{-1}(y) = f^{-1}(f(x)) = x$ if there are no SEUs present. Notice, however, that if f^{-1} suffers a SEU, the encryption would erroneously assumed to be corrupt. Since soft errors are dynamic, one method for correcting them is to reset the circuit (or power down and then power it up again). Thus, error detection is a viable option if the chip or module reset rate (proportional to expected SEU hit rate) is suitable for the embedded system. Typical analyses are performed to determine whether the error detection/correction times or down times for the device are acceptable or not. Take for example a low earth orbit satellite, which is expected to suffer 1 SEU per hour over an area equivalent to three chips. The FPGA chip required 40 ms for error detection and correction, hence the device was reliably operating for approximately 99.9989% of the time (Carmichael et al. 1999). Of course, as is the case for FPGAs, there may be no choice other than to reconfigure the whole or part of the chip. Other technologies may provide better alternatives such as error correcting codes where the area overheads and slight increase in FIT (due to area increase) might be traded off with lower down times.

Standard TMR and Hamming error detection and correction may incur unacceptable overheads in terms of area or performance for many embedded systems. Hence, an alternative approach would be to optimize the reliability techniques specifically for the application at hand. For example, it might make sense to implement Hamming codes in one part of the design, whereas TMR could be implemented in another part in order to optimize the overheads. Thus, application-specific techniques optimized for SEU-resistance are of interest in embedded systems. The next section will focus on SRAM-based FPGAs, which are highly susceptible to SEUs. Section 10.2 will discuss the impact of SEUs in FPGAs. Sections 10.3 and 10.4 will illustrate design for SEU-resistance for FPGA implementations of the AES cipher and the SHA-2 hash functions, respectively.

10.2 SEU-Resistant FPGA Design

Most often the focus of SEU-resistance is on the more susceptible registers and memory where SEUs result in bit flips. There is some ongoing work to determine what impact SEUs might have on combinational parts of the circuitry but for now the majority of SEU focus is on the storage elements. In SRAM-based FPGAs, storage elements are used to function as combinational logic and interconnect (routing) as well as storage. Whereas ASIC design can totally focus on storage elements, FPGAs must examine all components of their design for SEU-resistance. For example, consider the FPGA interconnections. There are numerous metal wires interconnected with pass transistors. The pass transistors are in the open or closed position according to a storage element that stores a "0" or "1," respectively. Thus, if a SEU hits one of these storage elements, controlling a pass transistor, a short circuit or open circuit may result. In an ASIC implementation the SEU at most might cause a temporary setting of a wire to flip but it likely would not propagate into a storage element and hence is only temporary. Thus, in the ASIC case, there would likely be no errors caused by this SEU. Whereas SEUs in FPGAs can create dynamic shorts and open circuits as well as modify the functionality of combinational logic. For example, consider a lookup table (LUT), representing some combinational logic (whose truth table is stored in the LUT). If a SEU hit this storage element it would flip one of the bits in the lookup (or truth) table, thus modifying the function of the logic. In an ASIC, a SEU that hit this logic would again likely have only a temporary affect (though it may become an issue in future technologies). For example, in a V1000 Xilinx FPGA (Carmichael et al. 1999), it has been estimated that half of the die area represents storage elements. Of these storage elements, approximately 85% of this area is devoted to storage elements controlling interconnects. However, in a typical design, only 10% of the interconnect controls is typically used, thus the probability of hitting an unused interconnect control is 76%. Nevertheless, both storage elements, interconnects, and combinational logic must be considered when designing for reliability or SEU-resistance with FPGAs. Correction typically involves reconfiguration of the FPGA to address both SEUs in configuration data as well as user data. In a previous study (Carmichael et al. 1999), a partial reconfiguration of a FPGA reduced the correction time from 20 ms (full reconfiguration) to 3 μs, thus reducing overheads associated with removing SEUs. Hence, tradeoffs in detailed SEU detection vs. reconfiguration times can also be made to optimize SEU-resistance in FPGAs. This may be quite important for satellite applications where total chip reconfiguration may dissipate too much energy.

It is also interesting to note that although FPGAs are highly susceptible to SEUs, they can provide low cost at low volumes and reconfigurability to space systems. Antifuse and flash devices provide higher reliability, but still must have their registers and memory elements protected against SEUs. Hence SEU-resistance in FPGAs is an important problem of study. Consider designing an embedded system for a satellite. Satellites have stringent constraints on mass, power dissipation, weight, and cost. Additionally low-energy dissipation is crucial since the satellite is solar powered. For some applications, SEU-resistance through TMR or rollback may demand unsatisfactory area overheads or energy dissipation overheads, respectively. Section 10.3 will discuss how SEU-resistance, in particular SEU detection, is added to the AES cipher under these constraints of low area and low-energy dissipation. Section 10.4 will discuss how both error detection as well as error correction can be designed into SHA-2, also providing lower overheads than a TMR approach.

10.3 SEU-Resistant AES Design

Although the AES cipher was discussed in Chap. 6, the FPGA implementation is presented here along with the detailed SEU-resistant design. This design was taken from Ghaznavi and Gebotys (2009). It will be assumed that the major objective is to modify the cipher design so that a SEU or single bit flip is detected in the FPGA. When the SEU is detected, it is further assumed that a reconfiguration of the FPGA will occur or the approximate location of the SEU will be identified followed by a partial reconfiguration. Thus, the design for reliability addressed in this section will only consider SEU detection.

Details on AES were given in Chap. 6; however, the AES round is shown in Fig. 10.5. In particular, the SubBytes or S-box stage is further defined as a multiplicative inverse followed by an affine transformation. This SubByte function will be discussed first with respect to SEU-resistance.

Typically SEU-detection for memories such as the S-box is achieved using duplication (or TMR if one is interested in error correction as well as detection) or through parity (or Hamming codes if error correction is also desired). All of these approaches add significant overheads in duplicating, or supporting extra hardware for an inverse function or through adding extra bits per word in memory. Nevertheless, consider the inverse function approach. Although no inverse function is used for the S-box, the multiplicative inverse, $()^{-1}$, or *Inverse* in Fig. 10.5, has the property that for any 8-bit input, a, taking the inverse twice should result in the original input or $((a)^{-1})^{-1} = a$. This can be further optimized in the design by taking advantage of dual-ported memory blocks within most SRAM-based FPGAs. Figure 10.6 illustrates the resulting SEU-resistant pipelined architecture for the SubBytes function. Apart from the inverse function, the remaining logic uses parity checks for SEU-resistance of the affine function. The operation of this architecture

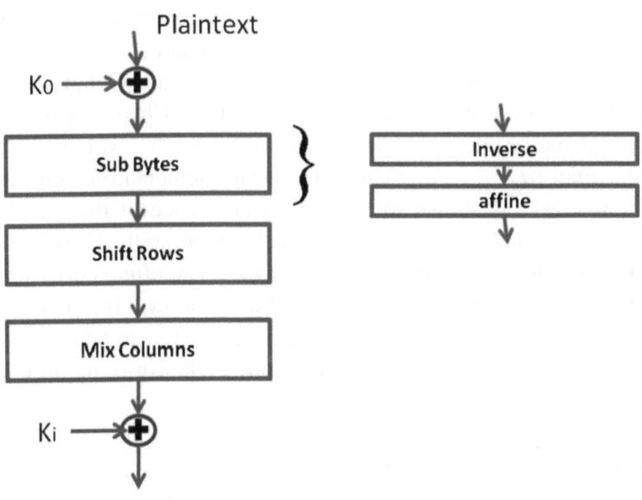

Fig. 10.5 AES S-box decomposition

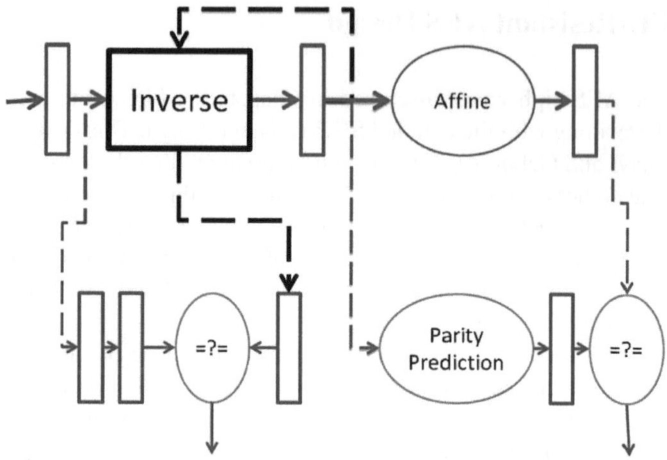

Fig. 10.6 SEU-detection in S-box where *Inverse* uses dual-ported memory

is as follows: Let the data at time t be represented by x_t. After one clock cycle the architecture computes $(x_t)^{-1}$. During the second clock cycle the architecture computes: $(x_{t+1})^{-1}$ and $((x_t)^{-1})^{-1})$. During the third clock cycle the architecture computes $(x_{t+2})^{-1}$, $((x_{t+1})^{-1})^{-1})$, $((x_t)^{-1})^{-1}) = ? = x_t$, which verifies that the operation performed at time t was correctly performed, etc.

This approach provides a more efficient implementation than using duplication or error detection codes. It uses fewer memory cells, provides 100% single-bit SEU coverage, and achieves a low failure in time (FIT). Other approaches used combinational representations of AES (Zhang and Parhi 2006) or required the AES decryption module (Karri et al. 2001). It should be noted that many modes of operation including GCM and CCM do not require the use of the decryption module; hence, this later technique incurs a large overhead in these cases. Other techniques that used the block memory approach used more memory (Karri et al. 2001) or provided only partial single bit error coverage (Bertoni et al. 2003; Breveglieri and Maistri 2007). Further details can be found in Ghaznavi and Gebotys (2008).

The affine function was previously defined in (6.4) in Chap. 6. This equation performs four exclusive-or operations for the 8-bit input affine value $(a_7^{-1}, \ldots, a_0^{-1})$, where we take *affine* $(a_7^{-1}, \ldots, a_0^{-1})$ in (6.4). The implementation of the affine transformation using LUTs is shown in Fig. 10.7 where each of the four LUTs is an exclusive-or. Any single SEU in the LUTs results in either three errors or one error at the 8-bit result, thus detectable by parity checks. The affine parity predictor implements (10.1) by using three additional LUTs, not shown. In (10.1), $P_{\text{out},a}$ is the affine output parity bit of an 8-bit element a, and a_i^{-1} refers to bit i of a corresponding inverse a^{-1}. Equation (10.1) is derived from exclusive-oring all entries in the output vector of (6.4)

$$(P_{\text{out},a} = A_0 \oplus A_1 \oplus A_2 \oplus A_3 \oplus A_4 \oplus A_5 \oplus A_6 \oplus A_7).$$
$$P_{\text{out},a} = \overline{a_0}^{-1} \oplus \overline{a_1}^{-1} \oplus a_2^{-1} \oplus a_3^{-1} \oplus a_4^{-1} \oplus \overline{a_5}^{-1} \oplus \overline{a_6}^{-1} \oplus a_7^{-1}. \quad (10.1)$$

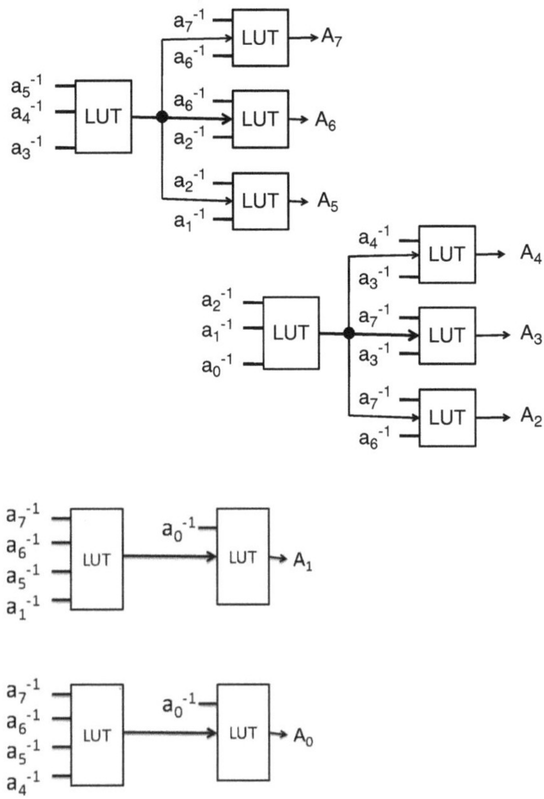

Fig. 10.7 Affine transformation implemented with LUTs

There are 16 of these 1-bit parity values computed in (10.1) for the 128-bit cipher. One parity bit representing the parity of each 8-bits affine value is produced.

The Shift Rows is strictly a routing of the signals; hence, parity schemes are not needed since reliability is strictly determined by the configuration data. The Mix Column module uses a parity scheme to detect the presence of an SEU. This transformation, previously presented in Sect. 6.2.1.3, on a state column $[a\ b\ c\ d]'$ (where $[\,]'$ is transpose of the row vector) is transformed for LUT implementation and shown in Fig. 10.8 where the bit positions i and j belong to {0, 2, 5, 6, 7} and {1, 3, 4}, respectively, in a and b. This difference in indices is due to the exclusive-or equations previously presented in algorithm 6.1 and 6.2. The equations were transformed to the form below in (10.2) and (10.3) for i and j belonging to {0, 2, 5, 6, 7} and {1, 3, 4}, respectively. For example, in Sect. 6.2.1.3 we derived the 0th bit as $a_0' = (a_7) \oplus (b_0 \oplus b_7) \oplus c_0 \oplus d_0$. Now using our transformed formula optimized for FPGA LUT implementation we get from the first row of (10.2) $a_0' = (a_0 \oplus b_0 \oplus c_0 \oplus d_0) \oplus b_7 \oplus a_7 \oplus a_0 = b_0 \oplus c_0 \oplus d_0 \oplus b_7 \oplus a_7$, which is equivalent to our previously derived equation.

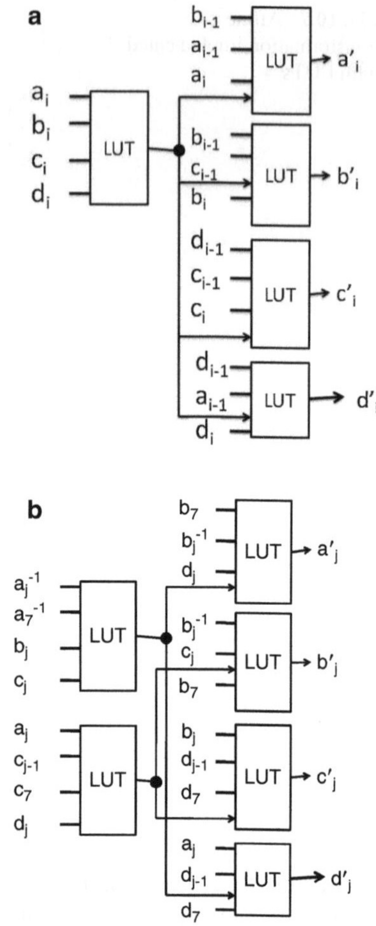

Fig. 10.8 Mix columns for $i = \{0, 2, 5, 6, 7\}$ in (**a**) and $j = \{1, 3, 4\}$ in (**b**)

$$\begin{bmatrix} a'_i \\ b'_i \\ c'_i \\ d'_i \end{bmatrix} = \begin{bmatrix} a_i \oplus b_i \oplus c_i \oplus d_i \\ a_i \oplus b_i \oplus c_i \oplus d_i \\ a_i \oplus b_i \oplus c_i \oplus d_i \\ a_i \oplus b_i \oplus c_i \oplus d_i \end{bmatrix} \oplus \begin{bmatrix} b_{i-1} \oplus a_{i-1} \oplus a_i \\ b_{i-1} \oplus c_{i-1} \oplus b_i \\ d_{i-1} \oplus c_{i-1} \oplus c_i \\ d_{i-1} \oplus a_{i-1} \oplus d_i \end{bmatrix}. \tag{10.2}$$

$$\begin{bmatrix} a'_i \\ b'_i \\ c'_i \\ d'_i \end{bmatrix} = \begin{bmatrix} a_{j-1} \oplus a_7 \oplus b_j \oplus c_j \\ a_j \oplus c_{j-1} \oplus c_7 \oplus d_j \\ a_j \oplus c_{j-1} \oplus c_7 \oplus d_j \\ a_{j-1} \oplus a_7 \oplus b_j \oplus c_j \end{bmatrix} \oplus \begin{bmatrix} b_{j-1} \oplus b_7 \oplus d_j \\ b_{j-1} \oplus b_7 \oplus c_j \\ b_j \oplus d_{j-1} \oplus d_7 \\ a_j \oplus d_{j-1} \oplus d_7 \end{bmatrix}. \tag{10.3}$$

In Fig. 10.8, there is LUT sharing between different 8-bit elements of a state column. For instance, there is no LUT sharing between a_i and a_j but there is a LUT shared between a_i and b_i in Fig. 10.8. Therefore, a single error in a shared LUT is detected by different bits of the 16-bit error flag register. The parity prediction from Satoh et al. (2001) is expanded and rearranged as it is shown in (10.4) with respect to the 4-input LUT structure of an FPGA to reduce hardware. $P_{\text{out},x}$ and $P_{\text{in},x}$ correspond to the output and input parities of an 8-bit element x in (10.4). The error detection hardware overhead is five 4-input LUTs per column of the state.

$$\begin{bmatrix} P_{\text{out},a} \\ P_{\text{out},b} \\ P_{\text{out},c} \\ P_{\text{out},d} \end{bmatrix} = \begin{bmatrix} P_{\text{in},a} \oplus P_{\text{in},b} \oplus P_{\text{in},c} \oplus P_{\text{in},d} \\ P_{\text{in},a} \oplus P_{\text{in},b} \oplus P_{\text{in},c} \oplus P_{\text{in},d} \\ P_{\text{in},a} \oplus P_{\text{in},b} \oplus P_{\text{in},a} \oplus P_{\text{in},d} \\ P_{\text{in},a} \oplus P_{\text{in},b} \oplus P_{\text{in},c} \oplus P_{\text{in},d} \end{bmatrix} \oplus \begin{bmatrix} P_{\text{in},b} \oplus a_7 \oplus b_7 \\ P_{\text{in},c} \oplus b_7 \oplus c_7 \\ P_{\text{in},d} \oplus c_7 \oplus d_7 \\ P_{\text{in},a} \oplus d_7 \oplus d_7 \end{bmatrix}. \quad (10.4)$$

The AddRoundKey uses one 4-input LUT altogether for each output bit. Therefore, there is no LUT sharing in the AddRoundKey implementation. Equation (10.5) is used for the parity prediction of the AddRoundkey transformation. In this equation, $P_{\text{out},b}$ represents the output parity generated for b, which is an 8-bit element of the state, and k_{bi} is the corresponding expanded key bit for bit b_i. The error detection of AddRoundKey uses three LUTs for each 8-bit element of the state.

$$P_{\text{out},b} = k_{b7} \oplus k_{b6} \oplus k_{b5} \oplus k_{b4} \oplus k_{b3} \oplus k_{b2} \oplus k_{b1} \oplus k_{b0} \oplus P_{\text{in},b}. \quad (10.5)$$

It is interesting to note that there has been some work on error detection for AES, but this work did not consider SEUs present in FPGA devices. Hence, the work focused on error detection and correction for registers and memory in AES using Hamming codes or was limited to error detection alone using parity bits. Additionally, multiple faults were of interest for detecting fault attacks, which are discussed briefly in Chap. 8. The presented reliable design of the AES forward cipher, which can detect SEUs (and single event transients), had a 29.5% higher FIT, 8% higher dynamic power dissipation, and 2.5% decrease in frequency compared with the AES design without SEU detection. The FIT rate was estimated by adding the FIT values for individual components together. The FIT generally increases with area; however, the area increase was 68% (Ghaznavi and Gebotys 2009).

It is important to emphasize again that if the embedded system of interest is using FPGAs and radiation or SEUs are a concern, then SEU-resistance of the combinational logic is crucial since it is implemented with LUTs that are prone to SEUs.

10.4 SEU-Resistant SHA-2 Design

This section will discuss the incorporation of SEU detection and correction for the SHA-2 hash function, previously introduced in Sect. 7.3.1. Unlike Sect. 10.3 that provides an example of SEU-resistance, SEU detection and correction is discussed in this section. The architecture for SHA-2 is based on the presented architecture in Sect. 7.4, taken from Juliato et al. (2009). The incorporation of SEU-resistance is presented in this section.

Hamming codes can be used to protect the register contents in SHA-2 from SEUs. These Hamming codes will detect a single bit flip within any cycle of SHA-2.

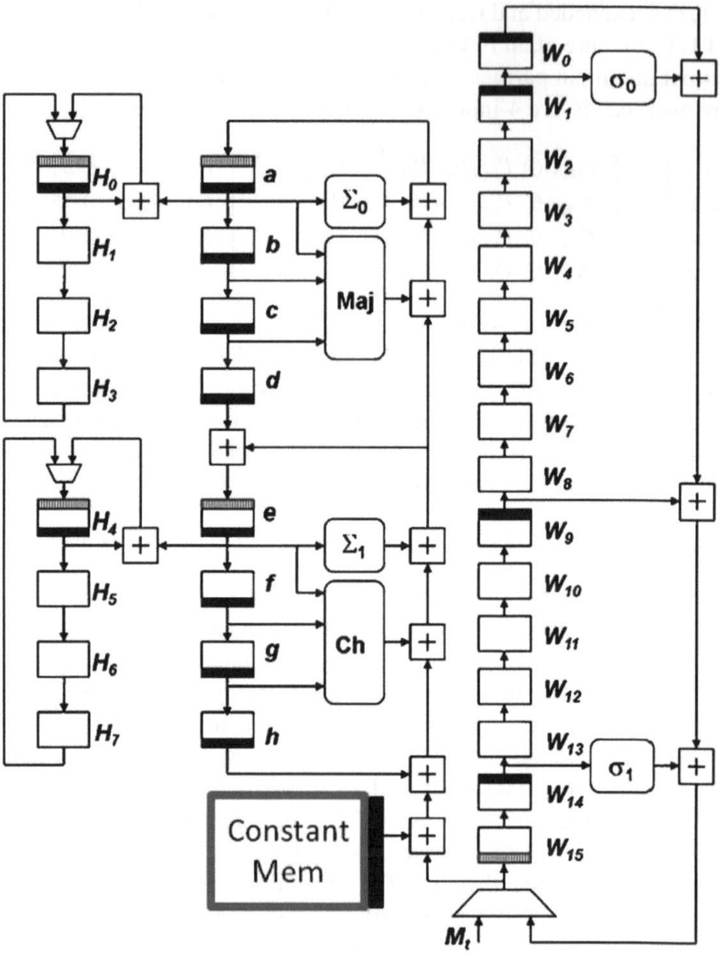

Fig. 10.9 Hamming (de)coders incorporated in the SHA-2 function

In general, the register contents are encoded before their values are written into the register. When the register is read, the Hamming codes are decoded in order to detect and correct any single error. However, performing Hamming encoding and decoding for every register leads to significant overheads. The use of Hamming decoders and encoders can be optimized as illustrated in the SHA-2 design in Fig. 10.9, where registers are indicated by rectangles. Encoders are indicated in Fig. 10.9 by the gray bar at one end of the rectangle representing the register input. Decoders are indicated by the black bar at the output end of the rectangles. Hamming encoders are only needed when writing new data (produced by an operation) into a register. Hamming decoders are only needed when data from a register are to be used by an operation. Thus, hamming circuitry is not needed when data are transferred from one register to another register. Only the registers directly involved in operations are designed with the encoding and decoding Hamming code circuitry. This can lead to significant savings by minimizing the number of registers with encoding and decoding circuitry, without sacrificing reliability. For example, in the third column in Fig. 10.9 the output of register labeled W_9 is sent to an adder block. It has a Hamming decoder at its output since the data output will be used directly in a computation. Register W_8, however, takes the data from W_9 and outputs the data into register W_7. There are no computations performed on the data directly output from W_7; hence, there is no Hamming weight decoding. Finally the output from W_0 is used in a computational block; hence, Hamming weight decoding is performed before the data are input to the computation. Register W_{15} is the only register in this column to use a Hamming encoder since it is the only register that receives newly computed data from a computational block (multiplexor, adder block, etc.). The constant memory includes storage for Hamming bits as well, and a Hamming decoder at the output of the constant memory is used. Reliability is maintained by assuming that the time between SEUs hitting the device is greater than 64 clock cycles (between intermediate hash computations using $H_0, \ldots, H_7$), which is the maximum time taken between two Hamming decodings of the same register. This is quite a reasonable assumption. If SEUs were to hit at a higher rate, the risk of having two errors in the same register would increase and the Hamming coder could not correct this case.

Unlike the limited past research in SEU-resistance in SHA functions (Ahmad and Das 2007), the design shown in Fig. 10.9 corrects any SEUs found in the registers or constant memory. This approach uses up to 32% less area and consumes 43% less power than the traditional TMR approach (using Altera Cyclone II EP2C35F672C6 technology and Quartus II power estimation tools). Furthermore, the memory and registers are estimated to be 435 times and 175 times more resistant to SEUs than that in TMR. More details can be found in Juliato et al. (2009).

10.5 Reliable Key Recovery for Satellites

> *Hacking a commercial satellite that's been up there more than 10 years is very easy for some people, if you have the right equipment*
>
> Jim Geovedi, Bellusa Asia Pacific from Kent (2006).

Multinational companies, military, governments, and citizens heavily rely on commercial satellites for communication. Thus, threats to satellites in general could pose risks to a country's communications infrastructure (USGAO 2002). There are many reports of satellites being compromised. One can also refer to a report (USGAO 2002) presenting various threats to satellite systems, which concludes that the security of commercial satellites should be more fully addressed. For many reasons incorporating security into satellites makes sense. For example, confidentiality of communications and authentication are typically needed. However, embedding security into satellites is not a straightforward process due to reliability issues and challenging constraints of satellite systems. Satellites have extremely stringent constraints including high reliability, low mass (low weight, low area), low energy, low power, etc. Depending upon which orbit a satellite is launched into, it has varying amounts of radiation it must withstand. The satellite must have a low mass and weight to minimize launch costs and to help maintain the satellite's orbit. Furthermore, the satellites are solar powered and heat dissipation is a problem; hence, low energy and power dissipation are crucial. Most important of course, field servicing is out of the question. Once launched, there is no possibility of repairing/replacing components; hence, the design of the satellite must be failsafe, fault tolerant, and remotely reprogrammable or resettable in the case of (power or other) failures. Alternatively one could replace a damaged satellite by launching another satellite. However, considering that a satellite launch costs \$11.5 million to \$100 million (Flanigan 2007), it is typically not a viable option for many. In general satellites traditionally communicate with two types of ground stations: communications stations and control stations. The latter will be referred to as "ground stations" for short. Satellite links can be divided into three groups:

1. Telemetry, tracking and control links (TT&C)
2. Data links
3. Cross-links

The first two links involve communications between the satellite and the ground station. The cross-links involve communication between satellites in orbit. The TT&C link is important since it provides mission information and control of the satellite. Hence, it requires both confidentiality and authentication. In commercial applications the data link may carry important information. One may not think that cross-links require much security; however, given the large number of multicountry satellites in orbit [likely well over 4,000 satellites according to Butterworth and Palmer (2005)], there is a concern for security on these satellite network links as well. However, these links involve different types of data with different requirements. For example, the command link involves short data messages where error-free reliable transmission is crucial but throughput or encryption speed is not as important. In contrast, the data link may involve transmission of large chunks of image data with extremely high throughput requirements (e.g., supporting data transmission over the S-band or X-band that operates at 2–3 GHz or 709 GHz) where some errors can be tolerated. In general, good cryptographic strength lasting for a long period of time is required. Components on the satellite should be long lived, adaptable, and of low cost.

There are many challenging and interesting problems in the area of satellite design; one of particular interest that will be discussed in this section is cipher key reliability. Consider the following scenario. One is communicating over a confidential channel with the satellite. The satellite passes through a radiation storm and the cipher key gets corrupt (perhaps just one bit is flipped due to a SEU). Enciphered communications with the satellite are now corrupt. What do you do? Do you turn off the encryption? Noting the quote (Kent 2006) at the beginning of this section, one may be worried about an attacker seeing the command to turn off encryption and hence using it whenever they like during the satellite's secret mission to remove confidentiality. Do you send the satellite a universal command to reset keys?

Same problem here, an attacker could launch a denial of service attack by continually sending a reset type command. Do you support more than one key at a time? Hence, if one key is corrupt how do you confidentially tell the satellite to switch keys. Of course, it is best to avoid getting the cipher key corrupt in the first place using techniques such as TMR, error detection/correction, etc. An alternative approach is described in Sect. 10.5.1 which can be used to establish new keys when these other approaches still fail.

State of the art design for reliability follows the direction of TMR (Blain et al. 1999) in addition to readback and reconfiguration of FPGAs. Other approaches (Samudrala et al. 2004) suggest applying reliability to only the most critical components of a design. Some research (Ingemarsson and Wong 1981; Roy-Chowdhury et al. 2005; Arslan and Alagoz 2006) proposes the use of satellites for key distribution and authentication in communication systems. However, the security of the satellite is not addressed. In Vladimirova et al. (2005) a fault-tolerant AES based on parity-based fault detection scheme and Hamming codes is suggested to mitigate SEUs. In Papoutsis et al. (2007) a proposal is presented to generate cryptographic keys from features associated directly with the actual satellite. However, such a technique is only valid if we assume that an attacker never gains control over the satellite. Otherwise, an intruder could learn the satellite characteristics and then have enough information to easily derive future cryptographic keys. Apart from the research outlined above, there is limited research addressing the recovery of cryptographic capabilities of satellites that have suffered faults due to SEUs or from attacks leading to compromised keys. The next section proposes an approach based on trusted modules, which can recover a satellite damaged by SEUs or even when an attacker has broken into it.

10.5.1 Satellite Recovery from Faults and Attacks

Consider the following security requirements of a satellite containing FPGA device configured for security functions.

1. The satellite must keep the integrity of cryptographic (resistant to SEUs and attack)
2. It must be able to authenticate and reinitialize the computational platform
3. It must be able to renegotiate authenticated keys with the ground station

The three functions are required in case the satellite suffers from corruption due to SEUs or attack or power failure or some other invasive event. Clearly a public key cryptographic approach can support authentication as well as key establishment in a secure way. However, public key systems are quite expensive in terms of area (as shown in Chaps. 5 and 6) as well as energy dissipation. Hence, they are currently not suitable for satellite applications. There may be different security approaches depending upon which key is corrupt. For example, key transport protocols covered in Chap. 4 may be used to transport new keys in place of keys with expired lifetimes, such as the Needham–Schroeder Protocol 4.5. The next section will discuss a solution when there is no cipher key available due to SEU corruption. The Hash approach is described in Sect. 10.5.1.1 and the trusted module approach is described in Sect. 10.5.1.2.

10.5.1.1 Hash Approach

Assume that an attacker is able to eavesdrop on the ground station to satellite communication and send control signals to the satellite. Also assume that a satellite stores hardwired secret information, k_0, shared only with the ground station. Hence, it is assumed that the agency that built the satellite will disclose this information only to the agency that will control it. Notice that k_0 must not be stored in the FPGA since it could get corrupted by SEUs. Further, k_0 must never be used as a key since once this value is discovered or leaked by any means, the satellite will never be able to reestablish a secured channel with the ground station. As an example, k_0 could be a serial number, an ID number, or any other data that only the satellite and the ground station know, and which is readable by the FPGA. It would be better still if k_0 was randomly created during the satellite construction or hardwired in the satellite circuitry and be readable by the FPGA. This permanent storage of information would then be immune to SEUs for the whole lifetime of the satellite. The recovery can be performed as shown in Protocol 10.1. One assumes that the ground station has detected communication problems, possibly due to SEU corrupt encryption key or the ground station has decided that it is time to update the key. The Protocol 10.1 assumes that communication has been lost due to a corrupt session key or the ground station would like to set a new session key. It is assumed that the satellite has a secret k_0, which can never be corrupt. The protocol resets the shared symmetric encryption key to k.

Protocol 10.1. Satellite key recovery

(1) *GroundStation*: $H(nonce, k_0) = k$, $E_k(GS\_nonce) = Check_S$
(2) *GroundStation* → *Satellite*: $nonce$, $Check_S$

(2.1) *Satellite*: $H(nonce, k_0) = k$
(2.2) *Satellite*: $D_k(Check_S) = GS\_nonce$
(2.3) *Satellite*: $E_k\ (GS\_nonce\text{-}1) = Check_G$

(3) *Satellite* → *GroundStation*: $Check_G$

(3.1) *GroundStation*: $D_k(Check_G) = ? = GS\_nonce\text{-}1$

The ground station creates a nonce (recall that a nonce is defined as a random number used no more than once) and with this nonce it creates a new session key, k. The session key is created by hashing the concatenation of the nonce with the secret shared value k_0. The ground station then creates another separate unique nonce, *GS_nonce*, and encrypts it with the new session key, k, and sends this along with the original nonce, in the clear, to the satellite. The satellite concatenates the shared secret key, k_0, with the nonce and hashes it to obtain the new session key. The satellite then decrypts the GS_nonce with the session key. The decremented *GS_nonce* is then encrypted by the satellite and sent back to the ground station. The ground station decrypts this and verifies that the value is equivalent to *GS_nonce*-1. This verifies that the satellite now has the same shared session key as the ground station, similar to the Needham–Schroeder Protocol 4.5 in Chap. 4. Notice that a separate nonce must be used since the original nonce was sent in the clear; hence, an attacker could decrement it and attack the session key knowing both the plaintext and ciphertext. This approach is relatively simple, only requiring the satellite to compute one hash, one encryption, and one decryption. An alternative protocol is described in Juliato and Gebotys (2008).

10.5.1.2 Trusted Modules Approach

Previously it was assumed that the satellite FPGA could be remotely reconfigured and the configuration file (which may be corrupt) was stored in a programmable read-only memory (PROM). This permanently stored configuration data could also include some mandatory functions for the recovery process, such as basic arithmetic operations and a hash function. This assumption is no longer valid if we assume that the attacker can remotely break into the satellite. The attacker can now modify the FPGA configuration, modify the program memory, and perhaps control the entire satellite. For example, there is no guarantee anymore that the satellite has a hash function implemented in the FPGA to create a new key. Since an attacker can modify the satellite's FPGA system, it becomes impossible to recover the control over the satellite without the help of trusted modules.

Trusted modules were introduced in Chap. 3 as trusted platform modules. In a satellite scenario it is assumed that the trusted modules are additionally implemented with fault tolerance and are not remotely accessible by attackers. It is also mandatory to have a secure information path between the radio-frequency circuitry and the trusted modules both in the satellite. The proposed approach (Juliato and Gebotys 2008) uses three trusted modules, including a trusted random number generator available only to the trusted modules.

The *trusted hash and configuration module*, shown in Fig. 10.10, is responsible for checking the integrity of some system security components including program memory, the current FPGA configuration, and the FPGA configuration file. For example, using hash tree schemes presented in Chap. 7, the integrity of the untrusted storage elements can be verified. The trusted module also contains some elementary circuitry to reinitialize the program memory and to configure the FPGA from the minimal configuration file. Given that the efficient hashing techniques were previously discussed in Chap. 7, the remainder of this section will focus on the details of the trusted reset and key recovery modules.

The *trusted key recovery module* is responsible for recovering cryptographic keys. This module stores l keys (k_1,k_2 ..., k_l) in a secrets table. The index of this

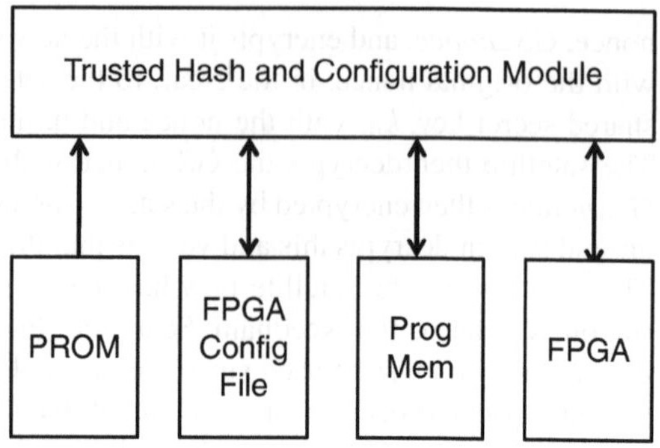

Fig. 10.10 Trusted module example

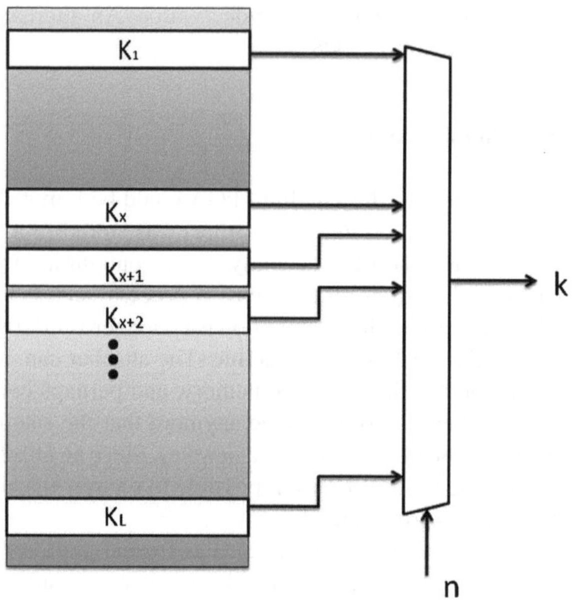

Fig. 10.11 Secrets table

table is a b-bit value, n, denoted as a one-time key recovery secret. The key recovery secret is sent to the satellite in the clear (i.e., no encryption is needed) by the ground station following the recovery protocol. If n points to a position holding a key, such a key is output. The bottom arrow in Fig. 10.11 represents the input n. The strength of this module relies on two fundamental features: (1) the l keys are randomly distributed throughout the 2^b positions of the table, and (2) b is made big enough to become infeasible for an attacker using brute force to find a position containing a key. It is feasible to implement the proposed scheme by using selection circuitry, as shown in Fig. 10.11. With that, only l storage elements are required for the keys, even maintaining a huge address space. This way, up to l key recoveries can be performed by the ground station. Therefore, the length of l will dictate the satellite's protection against key losses. However, increasing the length of l will require more storage in the satellite. For example, one could use 128-bit keys, $l = 64, 128$, or 256, and $b = 64, 128$, or 256 bits. Another important feature of the trusted key recovery module is that, once a given key is read, it is removed from the secrets table and never used again. This also means that a given one-time key recovery secret is used only once. If reused, attackers could perform replay attacks and easily break the scheme.

The *trusted reset module*, when necessary, issues a general reset signal to bring the satellite to a reliable state from which the rest of the system can be recovered. This module can also be thought as a table, but instead of outputting a key when it receives a valid one-time reset secret s, it issues a general reset. The reset signal causes the FPGA to be reconfigured (with minimum capabilities) and the program memory to be restored. Furthermore, once a reset secret is used, it is destroyed and can never be accessed again. This prevents an attacker from resetting the satellite through a replay attack using an old reset secret s. When these three trusted modules are implemented in a satellite, it becomes possible to detect when one of its crucial components is corrupt, bring the satellite to a reliable state, and restore its cryptographic capabilities.

From time to time, during the normal operation of the satellite, the trusted hash module checks the integrity of the program memory, the FPGA's current configuration, and the FPGA's configuration file. The computed hashes are then sent to the ground station. Since the trusted hash module is out of reach of attackers, it can detect any data corruption on those components due to intentional and unintentional causes, for example, SEUs and attacks. Given that the ground station knows the hashes of the FPGA configuration and program memory, a simple comparison determines whether the satellite has some crucial component corrupted. If that is the case, the ground station broadcasts a message to the communicating parties stating that the faulty satellite is not reliable and imposes that all parties stop communicating with it. It is up to the ground station to decide what to do next with the satellite. In some cases only a cryptographic key recovery may be necessary, whereas in more severe cases, a reset followed by a key recovery may be performed. The latter case is more complex and is exemplified below. In order to proceed with the recovery as the entity controlling the satellite, the ground station uses the one-time secrets n for key recovery and s for the reset. A challenge–response Protocol 10.2 is used to establish a key recovery procedure. The ground station first requests the initiation of the trusted reset protocol by sending the request, *init_req*. Then the satellite generates a random number r and sends it to the ground station. After that, the ground station performs an exclusive-or of s and r, and it sends the value to the satellite.

Protocol 10.2. Challenge–response reset

(1) *GroundStation* → *Satellite: init_req*
(2) *Satellite* → *GroundStation: r*

  (2.1) *GroundStation*: $(s \oplus r) = c$

(3) *GroundStation* → *Satellite*: c

  (3.1) *Satellite*: $c \oplus r = s$

Finally, the satellite recovers s in (3.1) of the protocol. For the key recovery procedure, the protocol is exactly the same, but n is used instead of s. When the trusted reset module receives s, it indexes its secrets table. If s corresponds to a valid address, a general reset signal is issued. Next, the contents of the accessed address in the secrets table are destroyed to avoid any possible replay attack. After the FPGA and the program memory are reconfigured, the satellite will send a status message to the ground station, which contains information on the success of the reset operation and also the new hashes of the FPGA configuration file and program memory. Finally the satellite needs to establish a session key with the ground station. Using Protocol 10.2 but with a different random number the satellite can again compute n to index its secrets table and obtain a session key.

Additionally, the ground station can use the new secured channel to send new program and FPGA configuration files to the satellite. This may be an important step, since attacks or failures may have occurred due to bugs or security holes in the previous system's embedded hardware or software. Alternatively, it may be sufficient to recover a cryptographic key as opposed to running a satellite reset operation. Also instead of the challenge–response protocol, it may be possible to send s and n directly to the satellite in the clear; however, these support lower levels of security.

Details of the actual implementation of this scheme can be found in Juliato and Gebotys (2008). For example, the implementation of the trusted modules consists of four components: control unit, address counter, secrets and keys table, and compare secret unit. The secrets table stores l secrets, each of them b-bits wide, which are protected against errors by the use of Hamming codes. When the memory is read, an internal Hamming decoder corrects a potential bit flip caused by SEUs and sends the secret to the secret compare unit. The implementation of the trusted reset module considers secrets of 64, 128, and 256 bits, which require the addition of 7, 8, and 9 parity bits, respectively. In the case of the trusted key recovery module, the secrets and keys are stored together in the same storage element. This saves some parity bits in its Hamming encoded form, thus requiring only 8, 8, and 9 parity bits, respectively, to encode the 64, 128, and 256-bit secrets along with the 128-bit keys. Further, if an unrecoverable error is found in the stored secret or key, e.g., two bit flips, the control unit is informed. Consequently, the address counter is incremented and the ground station is instructed to use the next one-time secret. One-time secrets are indexed sequentially through a pointer generated by the address counter unit. This unit consists of a fault-tolerant counter, allowing for l resets/key recoveries. It can correct one bit flip and detect two bit flips.

Once the stored secret is read from the secrets table, it is sent to the compare secret unit. Then, the compare secret unit performs a bit-wise comparison between the stored secret and the one-time secret under test. From the comparison results, the control unit determines whether or not to issue a reset signal (which is really a key, in the case of the key recovery module). If the one-time secret was successfully "guessed," the new key is reestablished in the satellite, and the stored secret is zeroed in the secrets table (along with its corresponding key, in the case of the key recovery module). Finally, the control unit increments the address counter in preparation for the next recovery process. In case of an error in the address counter, its value can be reset by the control unit. The control unit executes a series of counter increments, until it detects that the secret coming from the secrets table is nonzero. At this point, it knows that the address pointer has been recovered to its correct position. Likewise, the control unit follows this procedure to recover the address counter after FPGA reconfigurations. An actual implementation of the secrets table in a satellite should use ROM memory, so that its contents are not lost in an eventual FPGA reconfiguration.

Several implementations (using Altera CycloneII EP2C35F672C6) of the trusted modules using various levels of security and number of recoveries were presented in Juliato and Gebotys (2008). For instance, a trusted reset module working with 256-bit one-time secrets and allowing for 256 satellite resets utilizes 1,654 LEs and 68,096 memory bits. It issues a reset signal in only 73.83 ns when operating at 54.18 MHz, while its dynamic and total power consumption is 20.56 and 177.19 mW, respectively. In contrast, a trusted key recovery module allowing for 256 key recoveries and using 256-bit one-time secrets occupies 2,372 LEs and 100,864 memory bits. This module can recover a key in 78.85 ns when operating at 50.73 MHz, with dynamic and total power consumption of 40.65 and 266.02 mW, respectively.

The time t spent in the communications between the ground station and the satellite is quite significant and is an important parameter in determining satellite key size as discussed previously in Sect. 10.3.4. The communication time t (in seconds) is determined by $t = d/c$, where d is the distance (in km) between Earth and the satellite, and c is the speed of light in vacuum (299,792.458 km/s). For example, a ground station that knows the one-time secrets would spend 2.4 ms on the challenge response protocol (three steps) while recovering a low-earth orbit (UCS 2008) satellite. If the satellite was orbiting Mars (UCS 2008), it would spend 2,350 s (about 39 min). A brute force attack against these modules using 256-bit secrets would take approximately 10^{66} years, when applied to LEO satellites. Since the time spent in a brute force attack is completely independent of the computing power of the attackers, the proposed scheme can be considered very secure.

In summary, the problem of bringing satellites to a safe state after key losses, major faults, or attacks is an interesting problem. This section has provided an example of addressing this problem through reliable security techniques, which allow restoration of important cryptographic capabilities. This research is important for supporting security within satellites.

10.6 Testable and FPGA Security

Although testability is normally a separate field of study from reliability, this section will briefly touch upon testability. Design for test strategies modify the system architecture to make internal nodes of a circuit observable and controllable. For example, the JTAG standard uses scan chains to control/observe internal registered values, which are not directly connected to I/O pins of the chip. The internal registered values are serially shifted in/out of a test pin from the chip. This helps to verify that there are no faults in the registers or no faults in the logic whose outputs are captured in a register. This JTAG standard incorporated into most chips allows testing at any time including when the chip is in the field. However from a security point of view, having a mechanism on chip, such as JTAG, which allows users to serially shift data including the keys out of a chip (through a JTAG), is not a good idea at all, and certainly it makes for an easy attack. Nevertheless, during chip testing stages, controllability and observability are crucial to ensure that the chip is functioning correctly. The use of built-in self-test where linear feedback shift registers are used to generate and compress data on-chip at speed may seem to be useful since the LFSRs could be shared for stream cipher implementations, etc. However, this sharing would be an extremely bad idea since it would not be difficult for attackers to gain further information on the particular LFSR polynomial, random seed, etc., thus recreating the key stream. Hence, testability in many ways works against security. Nevertheless, there are ways of embedding security into this type of device so that it remains testable and secure. This can be achieved by encrypting the data so that it leaves the chip as cipher text and enters into the chip as plaintext. The encryptor can be placed at the periphery of the chip. An interesting example of this is the security of the bitstream, which will be discussed next.

FPGAs typically load their configuration bits through the JTAG interface or other more parallel interfaces. As discussed in Chap. 1, FPGAs need security to keep designs represented by bitstreams confidential, such as IP cores, etc. Furthermore, reliable systems must ensure that attacks on the bitstream have not taken place, making integrity an important issue. An excellent reference on security in FPGAs can be found in Drimer (2007). One avenue for bitstream confidentiality is to share a key between the FPGA and the designer, thus supporting off-chip encrypted bitstreams and on-chip unencrypted bitstream configuration data. Many FPGA vendors already support encrypted bitstreams, for example, using a secure configuration module. Additionally some vendors support CRC checks; however, these checks do not verify that the bitstream has not been modified and thus do not provide integrity. Researchers (Drimer 2007) as well as FPGA vendors have proposed bitstream authentication at power up or reset. However, in these cases, the authentication did not prevent older bitstreams from passing authentication. Protocols that support the confidentiality as well as integrity required by FPGA industry are discussed in Badrignans et al. (2008) and Sect. 7.3. This approach additionally precludes the replay of older configurations, thus preventing downgrade attacks. Bitstreams are encrypted and signed in this approach, and remote monitoring of the FPGA is possible including version authentication. The hardware support required for this includes a secure bitstream manager. TPM implementation on FPGAs (Schellekens et al. 2008; Eisenbarth et al. 2007) has also been suggested.

10.7 Summary

This chapter has discussed reliable security for embedded systems. Some architectural approaches to SEU-resistance as well as security design for resetting devices after they have been hit by SEUs have been covered. Although standard approaches to reliability, such as TMR, lead to large overheads, optimized approaches to reliability can be achieved through detailed analysis of inverse functions, or redundancy, within the security function.

In general for error-prone environments where SEUs may be present, cipher modes such as counter mode, CTR, or output feedback mode, OFB, are recommended since one error bit in the ciphertext creates just one bit error in the plaintext. However, higher level issues with respect to resetting keys and recovery protocols are an important part of reliable secure design of embedded systems. This chapter discussed one such recovery protocol for a satellite system faced with a corrupt encryption key; however, consideration for other embedded systems such as safety critical systems is crucial. The next chapter will discuss higher level security issues for embedded systems, including standards.

References

Actel (2005) Radiation-hardened FPGAs. http://www.actel.com/documents/RadHard_DS.pdf

Actel (2009) RTAX-S/SL RadTolerant FPGAs. http://www.actel.com/documents/RTAXS_DS.pdf

Aeroflex Gaisler (2009) Leon3 Processor. http://www.gaisler.com/cms/index.php?option=com_content\&task=view\&id=194\&Itemid=139

Ahmad I, Das AS (2007) Analysis and detection of errors in implementation of SHA-512 algorithms on FPGAs. Comput J 50(6):728–738

Altera (2008) Error detection and recovery using CRC in Altera FPGA devices. Application Note 357, Vers. 1.4, http://www.altera.com/literature/an/an357.pdf

Anderson R (2001) Security engineering a guide to building dependable distributed systems. John Wiley, New York

Arslan M, Alagoz F (2006) Security issues and performance study of key management techniques over satellite links. In: 11th intenational workshop on computer-aided modeling, analysis and design of communication links and networks, 122–128

Badrignans B, Elbaz R, Torres L (2008) Secure FPGA configuration technique preventing system downgrade. In: Proc. of the 18th IEEE international conference on field programmable logic and applications (FPL)

Baumann R (2005) Soft errors in advanced computer systems. IEEE Des test comput 22(3): 258–266

Bertoni G, Breveglieri L, Koren I, Maistri P, Piuri V (2003) Error analysis and detection procedures for a hardware implementation of the advanced encryption standard, IEEE Trans Comput 52(4):492–505

Blain P, Carmichael C, Fuller E, Caffrey M (1999) SEU mitigation techniques for Virtex FPGAs in space applications. In: MAPLD Proceedings, September 1999

Breveglieri L and Maistri P (2007) An operation-centered approach to fault detection in symmetric cryptography ciphers, IEEE Transactions on Computers, 56(5):635–649

Butterworth P and Palmer D (2005) Ask an Astrophysicist. http://imagine.gsfc.nasa.gov/docs/ask_astro/answers/980202e.html

Carmichael C et al (1999) SEU Mitigation techniques for Virtex FPGAs in space applications. MAPLD 1999, http://china.xilinx.com/esp/mil_aero/collateral/presentations/SEU_mitigation_technique.pdf

Cataldo A (2004) In Hot market, only reliable Rad-hard need apply. EE Times 05/24/04 articleID=20900592

Dodd P, Massengill L (2003) Basic mechanisms and modeling of singe-event upset in digital microelectronics. IEEE Trans Nucl sci 50:583–602

Drimer S (2007) Volatile FPGA design security – a survey, http://www.cl.cam.ac.uk/~sd410

Eisenbarth T, Güneysu T, Paar C, Sadeghi A, Schellekens D, Wolf M (2007) Reconfigurable trusted computing in hardware. In: Proc. of the ACM workshop on scalable trusted computing, 15–20, 2007

Flanigan J (2007) U.S. working to develop and launch cheaper satellites. New York Times, http://www.nytimes.com/2007/11/15/business/smallbusiness/15edge.html

Gaisler J (2005) LEON3-FT-RTAX SEU Test results. Issue 1, Gaisler research http://gaisler.com/doc/leon3ft-rtax.pdf

Ghaznavi S, Gebotys C (2008) A SEU-resistant, FPGA-based implementation of the substitution transformation in AES for security on satellites. In: Proc of Int'l workshop on Signal processing for space communications, SPSC 2008:1–5

Ghaznavi S, Gebotys C (2009) Error Detectino of AES implemented on an SRAM FPGA. CACR Tech Rept, University of Waterloo

Gold M (2007) Atmel offers Rad-Hard 16-Mbit SRAM for space apps. http://www.eeproductcenter.com/memory/brief/showArticle.jhtml?articleID=196802389

Helvajian (1997) Microengineering technology for space systems. Aerospace press.

Ingemarsson I, Wong C (1981) Encryption and authentication in on-board processing satellite communication systems. IEEE Trans Commun 29(11):1684–1687

Johnston AH (2000) Scaling and technology issues for soft error rates. 4th Annual Research Conference on Reliability, Stanford University, http://nepp.nasa.gov/docuploads/40d7d6c9-d5aa-40fc-829dc2f6a71b02e9/scal-00.pdf

Juliato M, Gebotys C (2008) An approach for recovering satellites and their cryptographic capabilities in the presence of SEUs and attacks. In: Proc of NASA/ESA conference on adaptive hardware and systems, AHS, Issue 22–25, doi 10.1109/AHS.2008.57, pp 101–108

Juliato M, Gebotys C, Elbaz R (2009) Efficient fault tolerant SHA-2 hash functions for space applications. Proc of IEEE Aerosp Conf. doi 10.1109/AERO.2009.4839503, pp 1–16

Karri R, Wu K, Mishra P, Kim Y (2001) fault-based side-channel crypto-analysis tolerant Rijndael symmetric block cipher architecture, DFT'01, IEEE Int'l Symp on defect and fault tolerance in VLSI Sys. 427–435

Kent J (2006) Security fears raised at conference, http://news.bbc.co.uk/2/hi/technology/5399050.stm

Kent, Williams (1993) Encyclopedia of microcomputers, CRC Press

Ma TP, Dressendorfer PV (1989) Ionizing radiation effects in MOS devices and circuits. Chapter 9. Transient-ionization and single-event phenomena by Kerns SE with contributions by Shafer BD.

MILSTD (1996) Department of defense test method standard microcircuits, http://scipp.ucsc.edu/groups/glast/electronics/mil-std-883.pdf

Nystedt D (2007) DRAM price crashes through $2. http://www.infoworld.com/t/hardware/dram-price-crashes-through-2--002

Ohring M (1998) Reliability and failure of electronic materials and devices. Academic Press, USA

Papoutsis E, Howells G, Hopkins A, McDonald- Maier K (2007) Key generation for secure inter-satellite communication. In: Second NASA/ESA conference on adaptive hardware and systems, AHS 2007, IEEE Computer Society, 671–681

Roosta R (2004) A comparison of radiation-hard and radiation-tolerant FPGAs for space applications, NASA JPL, JPL D-31228

Roy-Chowdhury A, Baras J, Hadjitheodosiou M, Papademetriou S (2005) Security issues in hybrid networks with a satellite component. IEEE Wireless Commun 12(6):50–61

Samudrala P, Ramos J, Katkoori S (2004) Selective triple modular redundancy (STMR) based single-event upset (SEU) tolerant synthesis for FPGAs. IEEE Trans Nucl Sci 51:2957–2969

Satoh A, Morioka S, Takano K, Munetoh S (2001) A compact rijndael hardware architecture with s-box optimization. ASIACRYPT'01: Proceedings of the 7th international conference on the theory and application of cryptology and information security. Springer, London, UK, 239–254

Schellekens D, Tuyls T, Preneel B (2008) Embedded trusted computing with authenticated non-volatile memory, In Proc. of TRUST 2008, LNCS 4968, Springer, New York

UCS (2008) U. of Concerned Scientists satellite database, http://www.ucsusa.org/globalsecurity/spaceweapons

USGAO (2002) Critical infrastructure protection: commercial satellite security should be more fully addressed. Technical Report GAO-02–781, United States General Accounting Office. http://www.gao.gov/new.items/d02781.pdf

Vladimirova T, Banu R, Sweeting M (2005) On-board security services in small satellites. In MAPLD Proceedings, 2005

Zhang X, Parhi K (2006) On the optimum constructions of composite field for the AES algorithm, IEEE Trans Circuits Syst II Express Briefs 53(10):1153–1157

Chapter 11
Summary, Standards, and Ongoing Efforts

... the latest in quantum crypto gear...

Spielberg (2009)

This book started with examining security issues in a number of embedded systems. It continued to further introduce the concept of keys, trusted platform modules, network on chips security, using keys (symmetric and public key) protocols, ciphers, integrity, authentication, and integrity trees. Attacks were introduced focusing on the side channel analysis. Analysis of real embedded systems including attack on a PDA, a DSP processor core and eavesdropping on contactless cards were presented. The purpose and design of countermeasures was discussed. Finally the design of reliable and testable security was examined with a satellite embedded system case study. There is, of course, much more to security in embedded systems. Since security is an ever-changing field with on-going discovery of new attacks, it is always recommended that designers stay up to date. Before implementing any security standards, a look at recent publications and attacks should be performed. There are many references throughout the book so that readers can obtain further details. This book would have been too long to cover details of all protocols/attacks/etc. Hence in addition to references, Table 11.1 provides a list of worldwide standards which is also a very useful reference for readers. To conclude this book, we will discuss some general issues and again emphasize that security should be an important criteria for embedded designers. In fact, high security designs must consider security at all levels of design, from the lower level (process) up to the higher level (applications).

Embedding security into a system requires consideration at all levels of the design from the circuit up to the higher level algorithm and protocols. Determining how much security should be put into an embedded system is also a complex process. There are many hidden costs that must be considered which include not only silicon cost (for cryptocore design) but also key management costs (e.g. cost to update keys throughout lifetime of product), etc. High-level issues include consideration of the impact of system characteristics on security as well as the interaction of high-level algorithms with the security algorithms. Even when an embedded system has been designed, it should be evaluated to verify security. For high security products this includes subjecting the product to security analysis including side channel leakage analysis, etc.

C.H. Gebotys, *Security in Embedded Devices*, Embedded Systems,
DOI 10.1007/978-1-4419-1530-6_11,

Table 11.1 Standards

Standards	Location
FIPS, NIST	http://www.itl.nist.gov/fipspubs, http://www.nist.gov
ANSI	http://www.ansi.org
X9	http://www.x9.org
P1363	http://grouper.ieee.org/groups/1363
ISO/IEC/SC 27	http://www.iso.ch, www.iec.ch, http://www.din.de/ni/sc27
SECG/SEC	http://www.secg.org, www.secg.org/secg_docs.htm
NESSIE	http://www.cryptonessie.org
IPA/CRYPTREC	http://www.ipa.go.jp/security/enc/cryptrec/index-e.html

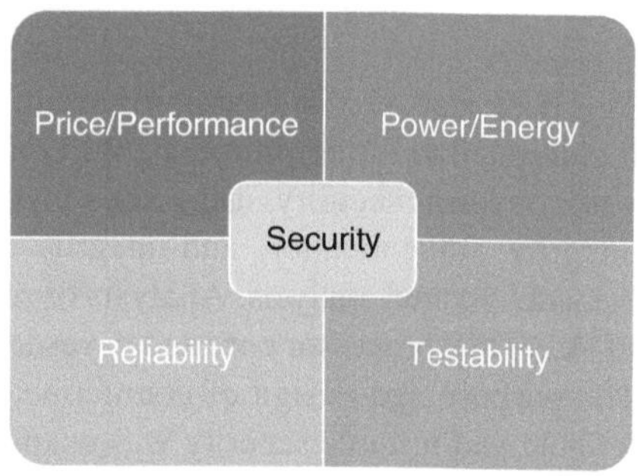

Fig. 11.1 Impact of security

Figure 11.1 illustrates the impact of security on various design constraints. For example, security has an important influence on the power/energy, performance, and price constraints of an embedded system. The incorporation of security into an embedded design does not come for free. Furthermore there are inherent tradeoffs when designing a secure embedded system. For example, reliable embedded systems require security so that they do not become a focus of attack. On the other hand reliability implemented through dual or triple modular redundancy may be detrimental to security. In particular, the redundancy may increase the side channel leakage. Another example which illustrates this tradeoff is the relationship between testability and security. All systems need to be testable and therefore need security to protect the observability and controllability of state or data from being subject to attack. High levels of testability provide controlability and observability of all registers within the chip. Systems may even be designed to capture these register values during a specific clock cycle and serially shift them off through the JTAG interface while the chip continues to function. This increased testability reduces the security of the chip since attackers may be able to trigger this capture and possibly acquire information such as key or confidential data. Even with JTAG encryption primitives attackers may be able to launch attacks by triggering this capture utilizing the zero data in registers after power up, etc. The tradeoff between testability and security is an important design decision which has important implications for attacks.

When high security is required, often designers will integrate the security functions with memory onto the same silicon as the embedded processor (as done with current chip cards, see Fig. 8.37). This integration thwarts any chip to chip bus probing attacks and also eliminates the stronger interchip side channel signals. The use of these cryptocores integrated with processors and on-chip memory can also improve performance. Nevertheless design of the cryptocore architecture should be performed with side channel leakage considerations (for example 128-bit register switching may actually help a side channel analysis attack). Additionally as security standards change these cores may become obsolete (if they are not fully reprogrammable). On the other hand the use of existing embedded processors alone for implementing security functions may be more cost effective, but they may not provide sufficient levels of performance. Secure memory is also an important design aspect in order to store keys, tags, and other important cryptographic primitives requiring confidentiality. In summary there are many other types of design decisions which have to be optimized for the specific embedded system so that the constraints on the objectives in Fig. 11.1 are achieved.

There are many other important issues in security. For example, often the order of operations can have a significant impact on implementation security. We have already noted this within the context of security (authentication and encryption order). However, consider an embedded system that performs compression, error detection/correction, and security functions. Now the designer has to consider the order of applying security, compression, and error detecting and correcting codes. Recall that encryption aims to transform the data into a pseudo-random form that is essentially decorrelated from the original plaintext. Hence compression after encryption may not produce good compression ratios. In general, it is always recommended to compress and then encrypt. The only down side to this is that many compression algorithms produce predictable data (code) that may be used as plaintext to attack the encryption. Furthermore, even the length of the data after compression can reveal information about the data that may be used in an attack (Hollenbeck 2004). Nevertheless, the benefits from the compression (e.g., good compression ratios) are usually greater than the cost of increasing the security strength (through larger key size) to offset this consideration. This is especially true in low-power wireless embedded devices where the energy to transmit is orders of magnitude higher than the energy of an operation or processor instruction, thus good compression ratios are extremely important.

In some cases it may be necessary to perform compression after encryption. For example, consider a compression engine that is located at the transmitter chip within the system. The transmitter chip may be quite a distance away from the TPM chip that performs the encryption. Given the stringent timing constraints of an embedded system, it may not be possible to send the data to the transmitter chip for compression, then send the data back into the TPM for encryption, and finally send the data back to the transmitter. Thus, designers may be forced to encrypt before compression. Furthermore, in embedded systems where security is more important than compression, then encryption followed by compression is preferred. It has been shown that encryption followed by compression is as secure as the encryption algorithm (Jakimoski and Subbalakshmi 2007), and furthermore, the theoretical compression gain is unchanged (Mark et al. 2004).

Clearly the last step should be to incorporate error detection and correction techniques after all compression, encryption, etc. have been incorporated. Consider the implications of performing error detection/correction bits before encryption. Recall that one bit error in ciphertext can cause over 50% of errors in plaintext during decryption, thus it is clearly better to leave error detection and correction codes to the end.

Security is a fascinating area of study. Not only is it continually changing, as new stronger cryptographic functions and protocols are developed due to advanced research as well as attacks, but it is also a field which continually is finding new applications in many embedded systems. Design for security in these embedded systems is indeed a challenging area due to the multitude of constraints (see Fig. 11.1) which must be satisfied for each embedded device. However an important emerging focus is 'design for privacy'. Today embedded systems are being designed with objectives such as cost minimization and security defined solely for the embedded system without the user in mind. For example RFID tags are extremely low cost commercial products which have been designed to function within inventory warehouses, or stores, etc. Consider the store scenario, where the user has purchased the product which contains the RFID tag and leaves the store. The RFID tag will still function when activated by a tag reader. This functioning tag has privacy implications since it can be used to track the individual/etc. There are many other examples of systems with privacy issues (VANET, chipcards with weak or no authentication of card readers, lack of security in satellites, etc). Largely the solution to these privacy issues is cost. But given the major constraint of commercial products is cost, the embedded/security industry may not yet be willing to adopt a more costly 'design for privacy' approach for products where privacy issues are important. Hence finding lower cost solutions for privacy (which will then be more easily adopted by industry) will likely come from researchers, unless significant pressure is placed on industry earlier. The future of security will likely include exciting advances in quantum cryptography and many secure applications using nanotechnology.

References

Spielberg S (2009) Transformers revenge of the fallen. Paramount Pictures

Jakimoski G, Subbalakshmi (2007) Security of compressing encrypted sources. ASILOMAR conference, Pacific Grove, CA

Hollenbeck S (2004) Transport Layer security protocol compression methods. http://www.ietf.org/rfc/rfc3749.txt. Network Working Group, request for comments

Mark J, Wagner D, Ramchandran K (2004) On compressing encrypted data without the encryption key. TCC 2004: theory of cryptography conference. LNCS:491–504

Index

[illegible]
[illegible] the EU authorised representative:
Springer Nature Customer Service Center GmbH
Europaplatz 3, 69115 Heidelberg, Germany
[illegible]

MIX
Papier aus verantwortungsvollen Quellen
Paper from responsible sources
FSC® C105338

If you have any concerns about our products,
you can contact us on
ProductSafety@springernature.com

In case Publisher is established outside the EU,
the EU authorized representative is:
Springer Nature Customer Service Center GmbH
Europaplatz 3, 69115 Heidelberg, Germany

Printed by Libri Plureos GmbH
in Hamburg, Germany